THE GREAT SCIENTISTS

THE GREAT SCIENTISTS

Jack Meadows

Oxford University Press
New York 1987

Editor Bill MacKeith
Designer/Art Editor Chris Munday
Picture Researchers Sharon Hutton, Linda Proud, Alison Renney

Design Consultant John Ridgeway
Project Editor Lawrence Clarke

Contributing Editor
Professor A.J. Meadows, Department of Library and Information Studies, Loughborough University (chapters 4, 6, 7, 12)

Contributors
Dr W.H. Brock, Director, Victorian Studies Centre, University of Leicester (5, 8, 9, 11)
Dr A.G. Keller, Department of History, University of Leicester (1, 2, 3, 10)

AN EQUINOX BOOK

Planned and produced by:
Equinox (Oxford) Ltd
Littlegate House
St Ebbe's Street
Oxford, OX1 1SQ

Published in the United States of America by Oxford University Press, Inc., 200 Madison Avenue, New York, NY 10016

Library of Congress Cataloging-in-Publication Data

Meadows, A.J. (Arthur Jack)
The great scientists.

1. Scientists–Biography. 2. Science–History. 3. Science–Social aspects–History. I. Title.
Q141.M38 1987 509.2'2 [B]
87-18567

ISBN 0-19-520620-7

Printing (last digit):
9 8 7 6 5 4 3 2 1
Printed in Spain

Introductory pictures (pages 1-8)
1 Louis Pasteur (➧ page 169)
2-3 Early X-ray "seaside postcard" from Germany (➧ page 193)
Right Illuminations at the 1885 International Invention Exhibition, Crystal Palace, London (➧ page 144)
7 World's first atomic blast, 16 July 1945 (➧ page 243)
8 Antoine and Marie-Anne Pierrette Lavoisier (➧ page 96)

Contents

Introduction

"History is more or less bunk," said Henry Ford. Whether you agree or not, it is certainly true that writing about the history of science does sometimes require justification. Most obviously, science has a major impact on our lives today; yet many people find science so impenetrable that they give up trying to understand it. The history of science can help here by providing some of the background, so that how scientists work becomes clearer. Science also has a unique historical property – its worldwide acceptability. Fundamental disagreement has always been typical of, say, politics or religion. In science, disagreements occur, but they are usually resolved fairly quickly: scientists in all countries can normally agree on what constitutes an important piece of scientific research.

Science as a cumulative subject

Science is historically unique in another sense. Aristotle was the foremost systematizer of the knowledge of the ancient world and the views he expressed on biology, physics and cosmology were dominant for some 2,000 years. However, the modern scientific way of looking at the world appeared in one place – western Europe – at one particular period – the 16th and 17th centuries. Looking backward is therefore essential if we are to understand what distinguishes science from other activities. In addition, old science contributes to modern science directly. Science is a cumulative subject – that is, what scientists have done in the past forms the essential basis for what scientists do today. Indeed, modern scientists may find it necessary to refer directly in their work to research done long ago: this is commonplace in observational subjects, such as astronomy and botany. This contrasts with, say, art, where contemporary artists may feel they owe little to their distant predecessors.

History also becomes important when science undergoes one of its occasional upheavals. For example, Albert Einstein changed our way of looking at the Universe during the early years of the present century. In the process, he overthrew some of the basic assumptions made by Isaac Newton more than two centuries before. A significant part of this revolution consisted of reviewing how the discussion initiated by Newton had developed down the years.

Science through the lives of its great practitioners

So there are many reasons for writing on the history of science. But something else requires justification – the approach taken in this book. Why do we concentrate here on the lives of just twelve scientists? Part of the reason is that the way science develops depends particularly on its most eminent practitioners. The world around us has many aspects that might be studied – some selection is always necessary in science. The influence of the great scientists often decided which aspects are selected and how they are studied. Another reason is that many people, including scientists, have a highly simplified view of the contributions made by scientists of the past. Historians of science can provide a useful service by demonstrating that the lives and work of scientists are usually less straightforward and less unambiguous than tradition supposes.

A more specific question must be – why have we chosen this particular group of individual scientists? Some clearly choose themselves. No list of great scientists could exclude such names as Galileo, Newton, Darwin or Einstein. But others who appear here – Alexander von Humboldt, for example – are less obvious candidates. A key factor in our selection of these has been their importance in opening up new areas of science. So Antoine Lavoisier is included not because he was necessarily the greatest chemist of all time – that might well be disputed – but because his work was of prime importance in establishing chemistry as a modern discipline. Our twelve scientists also provide a chronological spread and international selection, and illustrate a wide range of disciplines.

It was through the new chemistry that Lavoisier was able to explain the task which oxygen performs in the vasco-respiratory system. This problem had exercised William Harvey who, as the founder of experimental physiology, also finds a place in this volume. As for Humboldt, he both pioneered and stimulated work across a broad front of scientific endeavor – in biogeography (which he founded) and ecology, geology and cartography, and meteorology; he attempted to present a unified view of the complexities of nature and he was also instrumental in the development of international scientific collaboration.

The work of Michael Faraday in the first half of the 19th century made possible the age of electricity which dawned in the 1870s and helped to create the modern discipline of physics. Darwin and Freud revolutionized people's view of themselves and their place in nature, and thus completed the process of secularizing nature which Copernicus, Galileo and Newton had begun in the 16th and 17th centuries. The careers of Louis Pasteur and Marie Curie illustrate how scientific problems may arise from industry and medicine. Pasteur's life, in particular, shows how the practical applications of a scientist's work can benefit national and international economies. Curie is our only woman scientist and she symbolizes the way women were able to break into the essentially male world of scientific research by the end of the 19th century. Her work formed the leading edge for the creation of the nuclear physics which, together with relativity and quantum mechanics, became the foundation of 20th-century physical science.

The broader cultural context

The greater prominence of the biographical approach in recent writing in history of science is a reflection of the tendency among its practitioners to move away from the inward-looking "science-centered" tradition of writing about the scientific past, toward an attempt to set scientific advances (and setbacks) in the broader cultural context of the society of the day, including its religious and philosophical ideas, art and politics, and the institutions and societies that are their expression. The biographical approach enables the historian to show how scientists invent, articulate and transform ideas, and gain consent for them, through observation, experiment, calculation, argument and controversy, while at the same time showing them as ordinary mortals embedded in their culture. Such a perspective involves consideration of changing ideas about the role of science in a culture, questions concerning, for example, the demarcation between science and "pseudoscience", and the reasons for the continuing under-representation of women among scientists.

These are among the themes treated in the four separate boxed "features" that accompany the main text in each of the twelve chapters of the book. Every one of the boxed subjects arises from the career of the scientist under discussion and is categorized under one of four headings: Technology (symbolized by the Sextant), including instrumentation, applications of science, and scientific exploration; Institutions (symbolized by the Neoclassical Façade), including institutions of learning, and the organization of the scientific community; Politics, Conflict, Religion and Society (symbolized by the Balance); and the Popular Face of Science (symbolized by the Crowd), in the arts, literature and the dissemination of scientific knowledge to the public. These features have made it possible to focus on particular aspects of the life and work of the scientists concerned, and to view through a wider frame their contemporaries and their impact upon society.

Finally, the work, life and times of the leading scientists discussed in the book are summarized in the chronological tables. These provide a readily accessible reference element which illustrates, at a glance, the relationship between the individual and his or her scientific and social milieu.

Aristotle 384-322 BC

Beginnings of natural science...Atoms and the four prime elements...Hippocrates...The student of Plato... Aristotle on Assos and Lesbos...Writings...The Lyceum...Aristotle's cosmology and teleology... PERSPECTIVE...Centers of learning...Wheels, screws and cogs...Tutor to Alexander the Great...Aristotle through the Middle Ages

"Through wonder, philosophy begins," wrote Aristotle, "first about ordinary things, then about Sun and Moon and the origins of the Universe." But why did this wonder lead to a rational study of nature in ancient Greece, rather than in Babylon and Egypt, where precise observations of the stars had been made much earlier than in Greece? These older civilizations were more advanced technically than the Greeks, who were also introduced to sea-going trade across the Mediterranean, by the Phoenicians. Perhaps this clash of cultures gave a shock to the simple society described in the early Greek epics. Certainly, the Greeks soon outpaced their Phoenician teachers. Their enthusiasm for foreign parts may perhaps have made the Greeks sceptical about their own traditions, for example their old creation myths, like those of most other peoples, in which sexual mating was the metaphor for the emergence of nature, as "Night gave birth to Day and bright Sky"; or as the Sky-father impregnated the Earth-mother.

▲▼ By the 4th century BC, the Greeks had moved away from representing the forces of nature as traditional deities like Artemis (below), the Mistress of the Wild Animals. Aristotle (above) was the foremost natural philosopher in the ancient world. For two thousand years the arguments and theories of the exact sciences, as well as of natural history, were a long commentary on his work.

Aristotle lays the foundations

At the fountainhead of modern scientific thought stands Aristotle, who wished to find out all that had been, or could be, known about the natural world. Before him, the sages of ancient Greece had begun to assume that nature behaves in a constant, uniform manner, and had tried to find one universal substance that underlay all changes.

Aristotle believed that detailed and objective descriptions of each particular thing would have to precede any overall theories. He had hoped to cover all aspects of nature, but where he could not attain his aims, as in his cosmology, he had to fall back on certain basic principles, apparently drawn from common sense but actually taken from his assumptions about physics. In subjects like zoology, where he could get close to his material, he virtually created a new science. He made reproduction the main issue in biology, and stressed that the structure of an organ must be related to its function in the whole animal.

Subsequently, under the Roman empire, scholars began to take his work as their chief guide to nature. Galileo challenged much of his mechanics, and saw that Aristotle's cosmology would have to be jettisoned. Nevertheless, Galileo's new physics has similar aims to Aristotle's: he simply showed that Aristotle had jumped to the wrong conclusions. In the 17th century, Aristotle became a byword for the old errors that would be overthrown by a new science employing instruments and experimental techniques unknown to him. Yet, in many ways, the new science fulfilled Artistotle's ideal of an impersonal objective science, self-consistent and all-embracing.

Greek Science

From creation myths to natural science

Almost all early cultures portrayed the natural world about them as the activity of beings who thought and felt, loved and hated, behaved violently or kindly, much as they did: themselves writ large. The old creation myths were an example of this. About 600 BC these images of gods and spirits began to be translated by the Greeks into analogies drawn not from humans, but from craft processes. Nature then lost its sacred character, and became an array of objects with no will of their own. Only living things, it seemed now, had the power to move themselves, and have senses, so the inanimate was for the first time clearly separated from what is alive, and feels.

A similar view of nature as "furniture" may be found, however, in the Bible, where a clear distinction is also made between creator and creatures. From the available evidence, it does look as if, at times, the ancient Greek thinkers were also moving toward a monotheistic view of an abstract, impersonal deity beyond the sky, the Mind which is the source of reason in nature. The earliest of their "philosophers", as they were called, tried particularly to explain the more dramatic and terrible manifestations of nature, and to show that the gods, or god, were not the lustful and vindictive beings of legend. They wanted to show that thunder did not express the anger of an all-too-human Zeus; nor did the sea rage because the god Poseidon was offended. Such natural phenomena as earthquakes, floods and eclipses could all be reduced to simple causes similar to those which could be observed operating all the time around us.

Most often the early Greek sages used comparisons with the things people make, and so, presumably, understand and control. For example, in cookery and pottery and glassmaking, loose or fluid substances are, through the application of heat, given solid and precisely bounded forms with properties different from the raw materials. In buildings, for example, where many basic components go together to make a complex, stable structure, or in the art of sculpture, through which a shapeless block of stone receives a form that mimics the human body, they could find fertile means of expressing analogy. So, while heat or crude mechanical action seemed sometimes to be at work, the early Greek philosophers could also explain processes in terms of ordinary natural events, such as evaporation and condensation, or the eddies in streams.

The first such naturalist thinkers lived in the Greek towns on what is now the west coast of Turkey. It was then called Ionia, the first region to which the Greeks had moved out of their crowded homeland (the legend of the Trojan War is perhaps an echo of this migration). As the Greeks began to colonize farther north, and westward to Sicily and Italy, their new ideas went vith them. According to Aristotle, who is the source of much of what we know of his predecessors, the founder of natural philosophy was Thales (c. 640-546 BC) of Miletus, halfway down the Ionian coast. He is supposed to have held that all things come from water, and that the Earth floats on water – its slight motion thus accounting for earthquakes. He also declared that the world is "ensouled" and full of gods; so perhaps in his case naturalism had brought him to a kind of pantheism. More obviously important for the development of science, from the time of Thales on, a succession of Ionian philosophers sought to discover what is the basic stuff that remains the same through all the changes in nature: when water freezes to ice, or evaporates to steam; when a log burns to ash, giving off smoke; when a plant grows, blooms, then dies and decays.

Historical Background

478-404 BC Athenian empire in the Aegean

447-432 BC Parthenon buil

3,000 BC Mathematics, astronomy, writing already developed; from Sumeria/Babylonia and Egypt knowledge spread to Greeks

7th century BC Ionian Greek philosophers began to translate creation myths and personified forces of nature (e.g. Zeus, Poseidon) into analogies drawn from craft processes

KEY
- Biology and botany
- Medicine
- Mechanics
- Matter
- Mathematics, astronomy, geometr

c. 640-546 BC Thales, philosopher of Miletus; proposed that all things are formed of one prime material, water

c. 450 BC Leucippus (?of Miletus) conceived idea of indivisible "atoms"

c. 460-370 BC Democritus, Ionian philosopher; developed theory of atoms, least units of all matter, differing only in size and shape

c. 490-430 BC Empedocles, Sicilian poet and philosopher; proposed four elements (earth, water, air, fire), saw cosmos as field of conflict beween forces of Love and Strife

c. 610-545 BC Anaximander of Miletus, philosopher; held that life originated in the sea, that Universe is unbounded, that Sun, Moon and stars are hoops of fire circling Earth

c. 500-428 BC Anaxagoras, Ionian philosopher; explained that Sun is a gigantic red-hot stone larger than Earth; Moon Earth-like reflecting Sun's light

600 BC | 550 BC | 500 BC | 450

c. 460-377 BC Hippocrates of Cos, physician; wrote on scientific medicine; over 60 "Hippocratic writings" survive

c. 470-399 BC Socrates, Athenian philosopher; turned philosophy to moral rather than to natural issues

c. 427-347 BC Plato, Athen philosopher; organized Academy; taught knowled through theory of Ideas, prima of geometrical reasonir studied cosmology, physiolo

c. 570-500 BC Pythagoras of Samos, philosopher; established theory of numbers, taught (at Croton, S Italy) that number and geometry are keys to understanding Universe

■ **334 BC** Alexander the Great, succeeding his father Philip, invaded the Persian empire

390 BC Rome sacked by Gauls

-**371 BC** Sparta dominant in ...ece and Asia Minor

■ **146 BC** Romans completed conquest of Greece

■ **31 BC -AD 14** Reign of Octavian, first Roman emperor

AD 313 Conversion of Constantine to Christianity ■

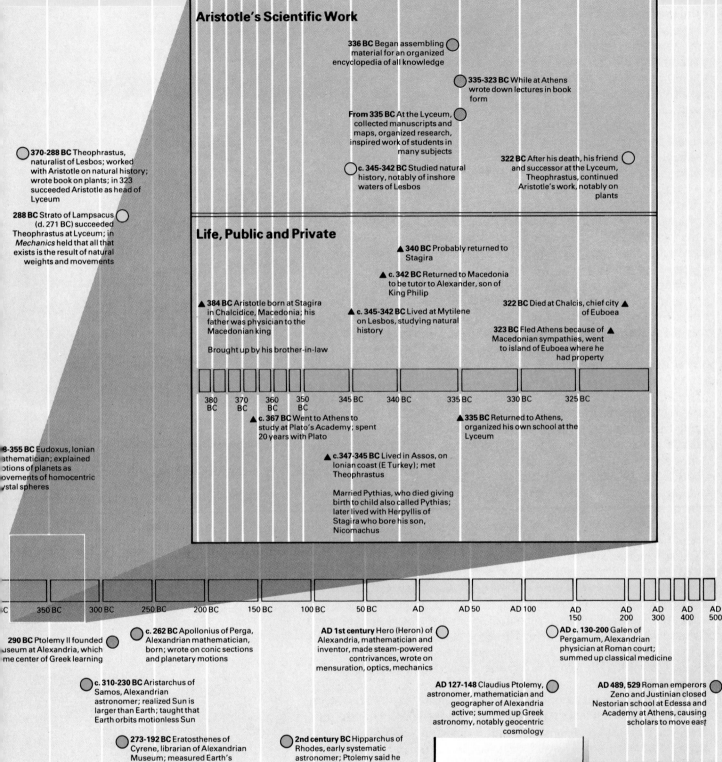

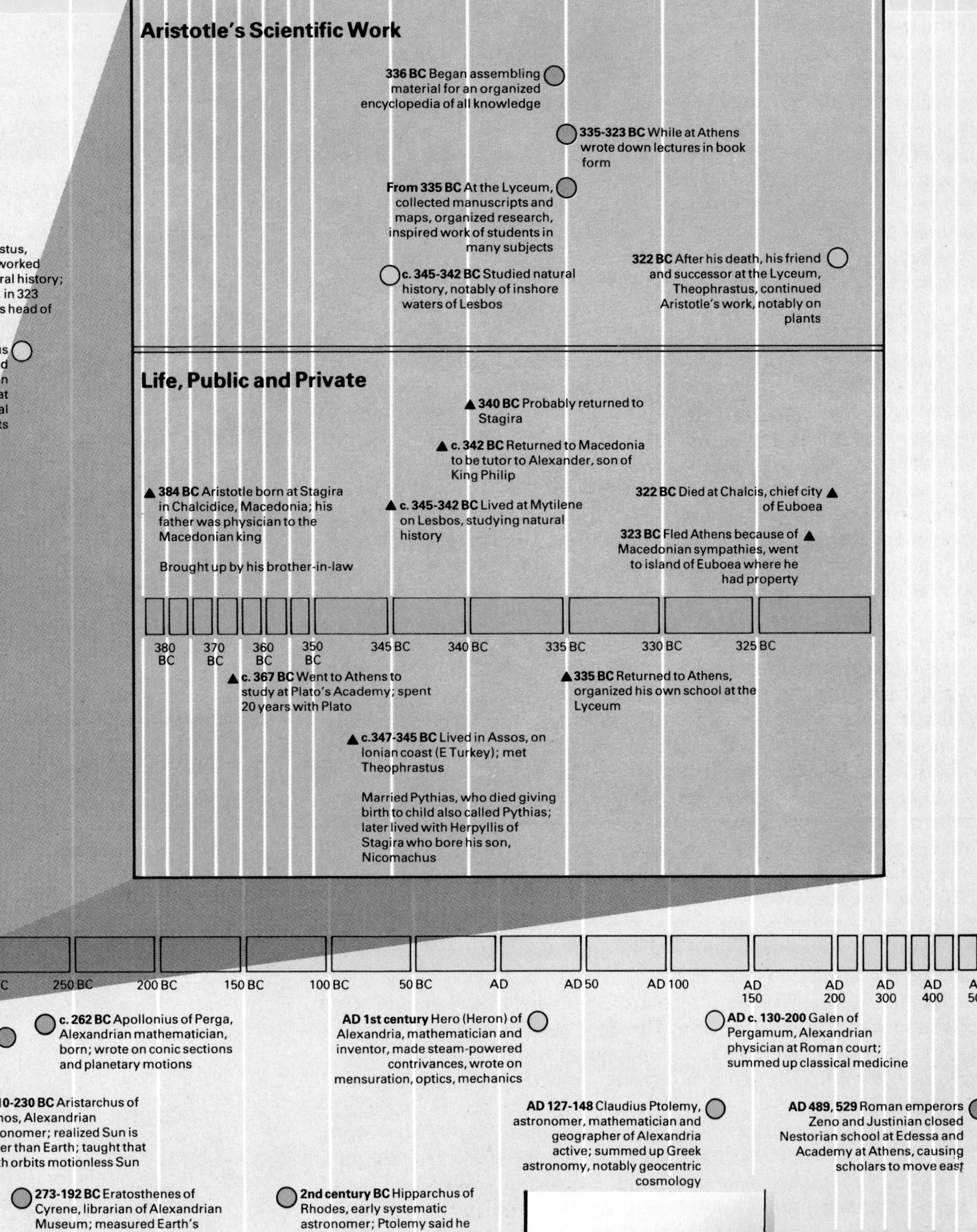

Aristotle's Scientific Work
336 BC Began assembling material for an organized encyclopedia of all knowledge
335-323 BC While at Athens wrote down lectures in book form
From 335 BC At the Lyceum, collected manuscripts and maps, organized research, inspired work of students in many subjects
c. 345-342 BC Studied natural history, notably of inshore waters of Lesbos
322 BC After his death, his friend and successor at the Lyceum, Theophrastus, continued Aristotle's work, notably on plants
Life, Public and Private
340 BC Probably returned to Stagira
c. 342 BC Returned to Macedonia to be tutor to Alexander, son of King Philip
384 BC Aristotle born at Stagira in Chalcidice, Macedonia; his father was physician to the Macedonian king
Brought up by his brother-in-law
c. 345-342 BC Lived at Mytilene on Lesbos, studying natural history
322 BC Died at Chalcis, chief city of Euboea
323 BC Fled Athens because of Macedonian sympathies, went to island of Euboea where he had property
380 BC
370 BC
360 BC
350 BC
345 BC
340 BC
335 BC
330 BC
325 BC
c. 367 BC Went to Athens to study at Plato's Academy; spent 20 years with Plato
335 BC Returned to Athens, organized his own school at the Lyceum
c.347-345 BC Lived in Assos, on Ionian coast (E Turkey); met Theophrastus
Married Pythias, who died giving birth to child also called Pythias; later lived with Herpyllis of Stagira who bore his son, Nicomachus

370-288 BC Theophrastus, naturalist of Lesbos; worked with Aristotle on natural history; wrote book on plants; in 323 succeeded Aristotle as head of Lyceum

288 BC Strato of Lampsacus (d. 271 BC) succeeded Theophrastus at Lyceum; in *Mechanics* held that all that exists is the result of natural weights and movements

...**8-355 BC** Eudoxus, Ionian ...athematician; explained ...otions of planets as ...ovements of homocentric ...ystal spheres

...C | 350 BC | 300 BC | 250 BC | 200 BC | 150 BC | 100 BC | 50 BC | AD | AD 50 | AD 100 | AD 150 | AD 200 | AD 300 | AD 400 | AD 500

290 BC Ptolemy II founded ...useum at Alexandria, which ...me center of Greek learning

c. 262 BC Apollonius of Perga, Alexandrian mathematician, born; wrote on conic sections and planetary motions

c. 310-230 BC Aristarchus of Samos, Alexandrian astronomer; realized Sun is larger than Earth; taught that Earth orbits motionless Sun

273-192 BC Eratosthenes of Cyrene, librarian of Alexandrian Museum; measured Earth's circumference, mapped known world

c. 300 BC Euclid, Alexandrian mathematician, active; wrote *Elements*, still the basis of much geometry today

c. 287-212 BC Archimedes of Syracuse, mathematician and physicist; wrote on mathematics, hydrostatics, mechanics; invented war machines; made planetarium

AD 1st century Hero (Heron) of Alexandria, mathematician and inventor, made steam-powered contrivances, wrote on mensuration, optics, mechanics

2nd century BC Hipparchus of Rhodes, early systematic astronomer; Ptolemy said he discovered precession of equinoxes, made first known full star catalog, worked on trigonometry

AD c. 130-200 Galen of Pergamum, Alexandrian physician at Roman court; summed up classical medicine

AD 127-148 Claudius Ptolemy, astronomer, mathematician and geographer of Alexandria active; summed up Greek astronomy, notably geocentric cosmology

AD 489, 529 Roman emperors Zeno and Justinian closed Nestorian school at Edessa and Academy at Athens, causing scholars to move east

8th-9th centuries Aristotle and other Greek scientists translated into Syriac

9th-10th centuries Aristotle and other Greek scientists translated into Arabic

11th-12th centuries Arab scholars in the tradition of Aristotle included the Persians Avicenna (Ibn Sina, 980-1037) and, in Spain, Averroës (Ibn Rushd, 1125-1198)

12th and 13th centuries Most surviving works of Aristotle translated into Latin

1543 Aristotelian-Ptolemaic Earth-centered cosmology challenged by Nicolaus Copernicus; Galileo Galilei used observations by telescope to attack geocentric theory

1859 Aristotle's "teleology" laid to rest by Charles Darwin in *Origin of Species*

Atoms and elements

About a century after Thales' death, the Greek philosopher Leucippus, probably also from Miletus, conceived the idea of "atoms", which are indivisible units (the very word means "uncut") which differ only in size and shape. Infinitesimally small atoms roam through infinite empty space, the void; as atoms bump into, or hook up with, one another, all conceivable combinations are produced. All the objects in the world, apparently so diverse, and all phenomena of color, texture, solidity or fluidity, indeed all change can be explained by the packing of atoms, or their kaleidoscopic shifts.

These ideas were developed in the late 5th and early 4th centuries BC by Leucippus' pupil Democritus, to whom Aristotle attributes the details of atomic theory. As Aristotle points out, words are built up from changes in the order of a few letters, so that an entire book with all its narrative and nuances of meaning can be reduced to a couple of dozen symbols. Great variety may be composed from very few simple objects.

A contemporary of Democritus, a Sicilian Greek called Empedocles, portrayed the Universe as a perpetual struggle between Love and Strife, first one prevailing, then the other. An harmonious world emerges in the Age of Love. For this, four prime elements or "roots", not one, would be needed – two of them passive, two active. They were earth, water, air and fire. Later, these four elements were taken up by Plato and themselves given a geometrical character.

No less important than attempts to interpret the whole world, and show how everything in it had arisen, were the first attempts to use rational principles to explain the diseases that afflict the human race. Although these first naturalistic physicians were influenced by the philosophers, and used their terms heat, cold, airs and waters as the fundamental causes of good and bad health, unlike the philosophers, they also investigated. The Greek physicians described all the symptoms of illness so that they would be able to diagnose, and perhaps even predict the course of a disease.

Hippocrates and a rational medicine

Among these physicians, Hippocrates of Cos (c. 460-377 BC) was reputed to be the greatest doctor of his time. More than 60 books were later ascribed to him and he was seen as the founder of scientific medicine. Although modern scholars feel he must have written some of them, they cannot agree on which. So it might be better to speak of the school of Hippocrates – of Hippocratic books, rather than Hippocrates' books. The story of careful observation and experiment begins with them.

If the gods do not send us diseases, or punish us for our sins – what then are the causes of our maladies? By the later 5th century BC, many people began to be worried that the new philosophy and medicine were undermining moral and social order. Turning way from the study of nature to human behavior, Socrates the Athenian (c. 470-399 BC) was concerned to construct a rational system of ethics, which would reject all assumptions that had not been examined. Unfortunately this procedure of questioning all conventions and revealing how many strongly noted beliefs were just prejudices soon led to accusations that he was subverting all that Athenians had always, and rightly, taken for granted. Aristophanes, the writer of comedies, lampooned him in his play *The Clouds* as a man trying to replace the gods with natural forces and overthrow traditional order.

◄► The tombstone of the 2nd-century AD Greek doctor, Jason (left), shows him treating a boy with a distended stomach. At his feet lies a gigantic cupping glass used for bleeding patients. Hippocratic physicians had an excellent knowledge of bones and joints because accidents in gymnastics and athletics provided them with opportunities for study and surgical manipulation. The 2nd-century BC portable surgical instruments (right), include scalpels, probes and forceps.

◄► Although the Greeks were the first civilization to separate medicine from myth and magic, rational therapy was often practiced alongside older magico-religious medical rituals. Asclepios (left), the mythical father of Greek medicine, was associated with "the cult of temple sleep". In the Asclepios Temple of the mid-4th century BC on the island of Cos (right), the sick would bring offerings, sleep in its precincts and dream that Asclepios, his symbolic snake, and his helpers, performed miraculous cures. Hippocratic medicine probably originated in such temples. Unlike philosophers, Hippocratics investigated; by describing the symptoms of illness in order to diagnose, they were able to demonstrate power and expertise by predicting the course of a disease.

◄ Hippocrates portrayed by a Byzantine artist in the 14th century. Greek myths frequently depict outbreaks of plague as punishments from the gods; in the 5th century BC they still called epilepsy the "sacred malady". To Hippocrates, all diseases were equally natural. Commenting on a belief among the Scythians, who lived along the Black Sea, that cases of transvestism among the nobles were of divine origin, he declares that "every natural event has a natural cause". Even if most of the natural causes which the Hippocratic books gave for various ailments were eventually proved to be wrong, their main idea was the key to future science. "To help, or at least to do no harm", the essence of the so-called Hippocratic Oath, gave medicine its ethics.

At the end of the play a lynch mob storms Socrates' "Thinkery" – and Aristophanes seems to want us to approve. Not surprisingly, in 399 BC, Socrates was condemned to death for his "impiety".

Socrates' friends defended his memory against these accusations. Plato (c. 427-347 BC), in particular, saw himself as Socrates' heir. He replaced the older literature of the philosophic poem, or treatise, by dialogues, in which the main speaker (usually Socrates) exposes the woolliness of contemporary thought, and presents a new doctrine. Whereas Socrates definitely thought that earlier philosophers had spent too much time on the study of nature, Plato became more and more interested in cosmology and in human physiology. He also insisted on the primacy of geometrical reasoning for the proper understanding of the sciences.

The Greek Academies

Early schools

Unlike the great civilizations of the Middle East, where priestly castes had institutionalized the preservation and transmission of learning, the Greeks at first taught and studied in the most informal way. The physicians were probably the first to organize themselves into schools, with a master teaching pupils; they may have originally formed a kind of guild, or even hereditary caste. The followers of one early philosopher, Pythagoras (c. 570-500 BC), formed a group that lived together and was committed to a special code of behavior. Another group, the sophists, who claimed they would teach wisdom as the physicians taught medicine, brought to a fine art this attempt to reform behavior through instruction and argument.

Plato's Academy

In the 5th and 4th centuries BC, the Athenians Socrates and Plato criticized the sophists' mode of teaching, but, for all that, their own concerns were derived from them. Around Plato a circle of students formed. They, too debated questions of social conduct, but gradually developed an interest in the questions about nature, which earlier natural philosophers had posed, but failed to answer satisfactorily. Plato's company would meet at the Academy, a public gymnasium outside Athens, where young men exercised and paraded as soldiers. When they rested, they could argue: the gymnasium included a grove where philosophers could stroll in the shade as they talked. The dramatist Aristophanes described the maps and globes of Socrates' "Thinkery" and Plato also doubtless owned and used such apparatus.

Plato owned land in and around the city, including a garden near the Academy, which could also be used. So, unlike the sophists, he could easily refuse to take money for his instruction.

Aristotle's Lyceum

Aristotle kept up that custom, although, as an alien from Stagira, he could not possess property in Athens (naturalization was all but unknown). When he returned to Athens in 335 BC, he set up a new "academy" to rival that of Plato's disciples. There was another gymnasium, with grove attached, called the Lyceum and dedicated to the god Apollo Lyceius (Apollo of the wolves). There Aristotle began to lecture. His teaching was more a matter of lectures, and less of debate, than Plato's. Plato had tried public lecturing, but was not successful. Aristotle made a practice of speaking at regular hours in the evenings in the Lyceum, to all who would come and listen. His lectures often required diagrams which he could pin on the wall, or specimens, and he certainly used such devices as a celestial globe. All the Lyceums in the world take their name from these gatherings. To his pupils and friends, Aristotle talked in the Peripatos, the colonnade that went all round the building, and so his style of philosophy is often called "peripatetic". His students also helped in his research, for example in compiling the constitutions of over 150 Greek states. Friends and students all over the Greek world also supplied information on local animals and plants, and curious natural features.

Theophrastus and Strato

After Plato died, his students chose a successor: fresh heads of Plato's school, the Academy, were elected thereafter for about three centuries. Similarly, after Aristotle's death, Theophrastus was chosen to lead the peripatetics in his place. A more popular figure than Aristotle, he sometimes lectured to large numbers, and political circumstances enabled him to acquire a garden where the school could also meet. His will and that of his successor, Strato, suggest that the school had taken over some responsibility for the upkeep of premises in the Lyceum, and of its shrines.

▲ The Olympic Games still remind us that athletics formed the basis of Greek males' education. Plato's Academy and Aristotle's Lyceum were elaborations of gymnastic clubs.

▼ Plato's and Aristotle's schools led to other institutions, such as the great museum and library at Alexandria in Egypt, which was founded by the first Macedonian king, Ptolemy, in 290 BC.

▲ **Much of Greek philosophy took its inspiration from open-air discussion – Aristotelian philosophers were often known as "the peripatetic school". Later writers often wrote discussions – in the style of Plato's dialogues – between eminent past philosophers. In this mosaic from Pompeii, Plato (pointing to a globe) is implausibly surrounded by (clockwise) Theophrastus, Socrates, Epicurus, Pythagoras, Aristotle and Zeno. Zeno's "paradoxes" concerning space and time include the race between the hare and the tortoise.**

Theophrastus had a set of maps illustrating exploration painted on the wall of one of the buildings. Members of the school sometimes ate together, and had their communal cups and cushions. But books and instruments belonged to the head of the school, and its future was jeopardized when Theophrastus left his own and Aristotle's library not to Strato, but to another member, who took it off to his home in Ionia.

The Museum at Alexandria

Intellectual leadership gradually shifted from Athens to the capitals of Alexander's successors. In Alexandria, the Ptolemies founded a Museum. Originally this was to be a shrine of the goddesses known as the Muses, inspirers of art and learning, found at the Lyceum, and many other places.

But the Alexandrian Museum dwarfed all that had gone before. The ruling Ptolemies endowed it with a communal dining room for the philosophers. A great library was provided, with apparatus, and probably a botanical garden. Research was pursued in geometry, astronomy, geography, medicine, botany (as the source of medicinal drugs) and also into the development of more powerful siege engines; new techniques were devised for firing missiles at enemy fortresses. After the fall of the Ptolemies, the Museum continued. Some of the Roman emperors also endowed chairs in philosophy at Alexandria, or at Athens which had become a center for study again. But original ideas were not expected, for the professors simply expounded what Aristotle or other founders of schools had written long ago.

Technology in the Greek World

Craftsmen and slaves

There is little truth in the view that the ancient Greeks failed to develop technology because their society was founded on slavery, that Greek scholars and philosophers despised trade and industry because they were the business of slaves, while technicians, being presumably slaves, had neither motive nor training to apply "scientific" methods to their techniques. Aristotle and his school took great interest in the crafts of their day, and were keen to talk with craftsmen about their work; that was in fact the chief source of analogies to explain nature. One member of the Peripatetics – possibly Strato, who succeeded Theophrastus in 288 BC as the school's head – wrote a little book on "mechanical problems", in which he asked how a number of common tools functioned, and tried to reduce them all to the law of the lever. This book was later known simply as "Mechanics" – which is how the modern science of mechanics gets its name. But he does not suggest how his theory could be employed to improve existing technology, and there were but few instances where the ideas of theoreticians could be put to practical use. Most of these were economically marginal: for example, siege engines, cranes, or automatic puppets (which were believed to demonstrate physical principles, while entertaining the viewer). All the same, many techniques did progress in the ancient Greek world.

Watermills, screws and gears

In the course of the next two centuries, someone, probably in the Egypt of the Ptolemies, invented the endless chain of pots to raise water – still a popular technique in that area. This was transformed into a giant wheel which lifted water in boxes around its circumference as it turned. The use of watermills to harness power from water is rarely mentioned in ancient literature; it was once assumed that the invention was ignored, because slaves were cheaper. Actually slaves did not grind all the grain: more than one young man paid for his studies by grinding corn in a bakery, as it was night work and could be combined with leisure by day. Humans free or unfree had anyway been largely replaced by donkeys before water power took over. Watermills, both vertical and of the horizontal gearless variety, were quite common in the later Roman empire. The gear wheel, essential for the vertical watermill, was a crucial bequest of Greek civilization to all later industrial revolutions.

Screws, too, may be an invention of that era. The earliest known form was "Archimedes' screw", used to raise water, named for the Greek mathematician and physicist. Small precision metal gears had also appeared in the later years of the Hellenistic kingdoms, as in what seems to be a mechanized calendar fished up off Anticythera. This machine related solar and lunar dates, so as to calculate dates, perhaps for astrological purposes. Recently, a similar but simpler, late Roman device has been found. These are more intricate and mechanically sophisticated than anything described in the ancient literature, and show that the Greeks could devise complex instruments which brought together science, mathematics and technology.

◄ The 1st century BC waterwheel, used for grinding corn, had a vertical axis, the power being delivered by the water to scoops or vanes immersed in it. The Greeks also devised waterwheels in which the axle was horizontal and the wheel vertical – as in this Syrian example, which dates form the Roman era. Such wheels were "undershot", the water passing under the wheel.

▲ In addition to his contribution to hydrostatics, Archimedes was famous for his mathematical discoveries and for his studies in mechanics and the properties of the lever. These he applied to the design of military catapults and grappling irons. Here he is seen devising machines to hold the Roman army at bay during the siege of his native Syracuse in 212 BC.

◄ The four magnificent bronze horses which stand outside St Mark's church in Venice are probably Greek originals dating from the 4th century BC. They illustrate the Greeks' advanced bronze-casting technology. Such high-level technology is also implicit in their artefacts and temple remains.

▼ In 1900 a fragment of an astronomical clock was recovered from a ship wrecked of the island of Anticythera in the first century BC. This, and other fragments, confirm the Greeks employed toothed wheels to transmit motion. By contrast the Babylonians made astronomical calculations on clay tablets.

Aristotle, the student of philosophy

The northern part of mainland Greece formed in ancient times the kingdom of Macedonia. Along its coast lay several colonies, where Greeks from farther south had settled. At one of these – Stagira – Aristotle was born in 384 BC. His father, Nicomachus, was personal physician to King Amyntas of Macedonia, but Aristotle was brought up by his brother-in-law. Probably when no more than 17, Aristotle went to Athens, chief of Greek cities, where he soon began to study with Plato. Apparently he had a reputation as something of a dandy, but he was also known as "the reader", eager for knowledge. Socrates had been ready to argue about any subject anywhere, but Plato was more organized, meeting regularly with his friends and students at an exercise-ground, or gymnasium, called the Academy. Aristotle spent some 20 years with Plato and wrote philosophic dialogues in Plato's style, but all have disappeared.

Plato died in 347 BC, and about that time Aristotle left Athens. Perhaps the political atmosphere was a problem for anyone linked to the Macedonian court. For that kingdom was expanding, and had clashed with Athens. But perhaps Aristotle simply felt that, without Plato, the Academy had no more to offer him. For some three years he lived at Assos, a small port on the Ionian coast, not far from the site of Troy. It had become part of the territory of Hermias, a self-made man who had established a principality of his own in the area, more or less accepted by the Persian empire to which the whole region officially belonged. He was interested in Plato's philosophy, had ex-students of Plato as his advisors, and befriended Aristotle, and may indeed have invited him in the first place. Hermias is supposed to have been a eunuch, and had adopted a niece, Pythias, whom Aristotle was to marry. The couple had one daughter, who was given her mother's name. Possibly the mother died in childbirth. Certainly she died long before Aristotle, for, according to his will, she had left instructions that she was to be re-interred with Aristotle when he in turn should die. That was not usual, and suggests a powerful affection between them. Still, that did not prevent Aristotle sharing the full male Greek assumptions about the mental and physical inferiority of women. Some time after Pythias' death, Herpyllis, who came from his native town of Stagira, entered his household. She bore Aristotle a son, called Nicomachus for his grandfather. The arrangements made for her in the will suggest more the devoted housekeeper than a wife, and they were never formally married.

◄ Philosophers of Greece inherited the knowledge of the astrologer-priests in Babylonia from c. 3000 BC onward. To draw omens from celestial phenomena, their observations had to be recorded, as in this clay astrolabe of the 7th century BC. The Babylonians named constellations and planets and invented the 12 signs of the zodiac.

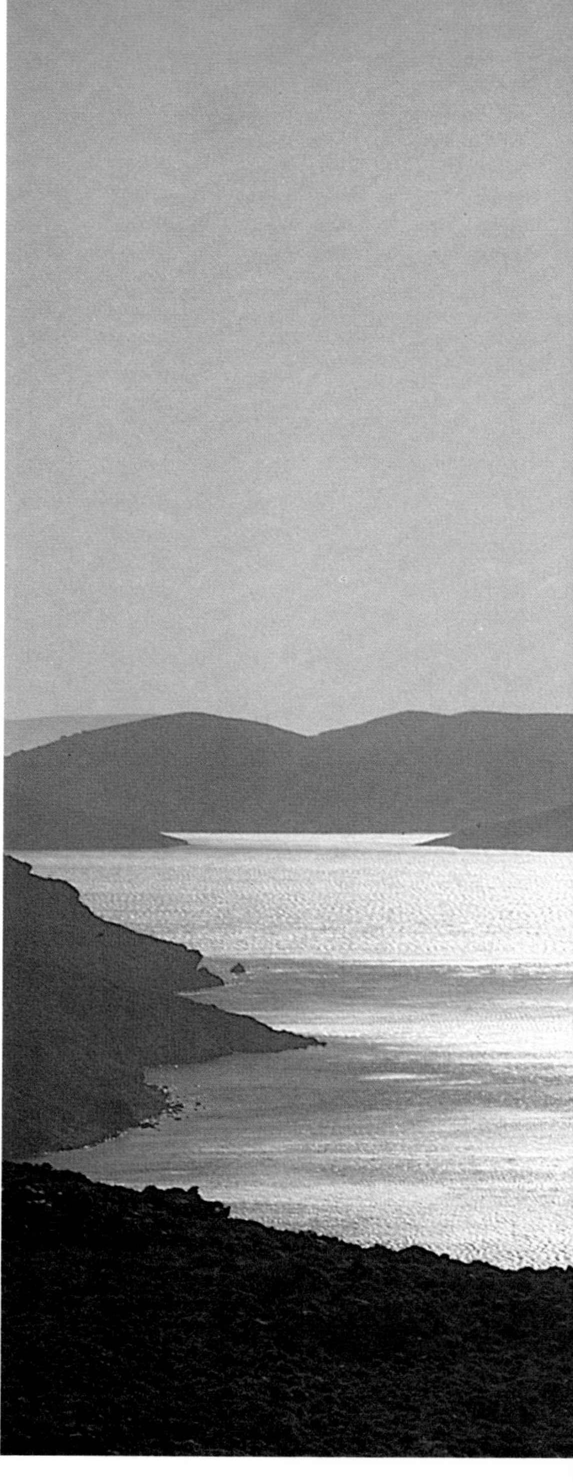

At Mytilene on Lesbos, Aristotle talked with fishermen about their catch and dissected creatures from the inshore waters

Aristotle the naturalist

While at Assos, Aristotle met Theophrastus, who was to become his right-hand man and closest intellectual companion. This new friendship may account for Aristotle's move across the straits from Assos to the island of Lesbos, where Theophrastus had been born (he died about 287 BC). While living there, at the town of Mytilene, Aristotle became fascinated with the life of the inshore waters, from which many of the islanders gained their living. Aristotle had a much greater interest in biological questions than any earlier thinker, apart from the medical writers. He saw that human biology could never be understood except through the investigation of simpler creatures. No one else before the 19th century developed such a broad knowledge of marine biology. Beside what Aristotle observed for himself – and he dissected several marine animals – he learnt much by talking with fishermen, whom he would ask about the character and behavior of their catch, squids, octopus and dogfish as well as fish.

Although Aristotle spent his time on Lesbos well, he was there for only two or three years before King Philip invited him back to Macedonia, to be tutor to his son Alexander. Not long after Aristotle's return to his native land, Hermias was seized by the Persians, who feared that he had been enlarging his miniature state in order to make it a bridgehead on the Asian mainland for King Philip. Macedonia had become the dominant power in Greece over the previous decade, and Philip was already thinking of an invasion of Persian territory. Hermias was dragged off to the Persian capital, tortured to confess his dealings with Philip, and crucified. His last message from the cross to his old friends was to assure them that he had "done nothing that would disgrace philosophy". To Aristotle he was a martyr who exemplified the special virtues of Greece: he wrote a poem in honor of Hermias, and dedicated a monument to him.

Once Alexander had succeeded to his father's troubled throne, he had little time for science. Aristotle, now away from the court, began to assemble material for a set of books on all existing knowledge – an organized encyclopedia that would contain detailed information about the Universe, the natural environment, and that little universe which is human life, private and public. Every entity would be defined, in such a way as to show its particular characteristics; facts should always prevail over what might be supposed to follow from general principles – this was central to Aristotle's thinking.

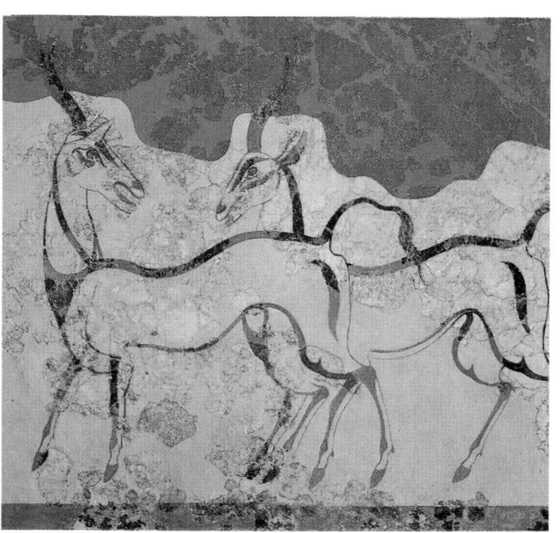

◀ An antelope fresco at Santorini (16th century BC). Although Aristotle had a greater interest in biology than earlier thinkers, such Bronze Age depictions of animals and plants as this show brilliance and vigor, and considerable powers of observation. However, Aristotle went beyond mere observation and searched for the function of each particular organ. By this means, he believed, the philosopher would be led to admire the necessary way in which each plant and animal had grown into the form which suited it best, and that logically each stage of development would lead inevitably to the subsequent stage.

▲▶ The exuberant variety of nature was observed both in this octopus flask from Minoan Crete and in the Nile riverscape mosaics which were very popular in Pompeii in Hellenistic times, some 15 centuries later.

► *"Aristotle's lantern". Parts of the body which may seem revolting in isolation work together to form an harmonious whole. For this reason Aristotle was interested in lowly marine creatures no less than in humans. Sea urchins, which lack jaws, possess five retractable calcified plates which allow them to "chew" tough seaweeds and even shells. The device was first described by Aristotle as "lantern teeth".*

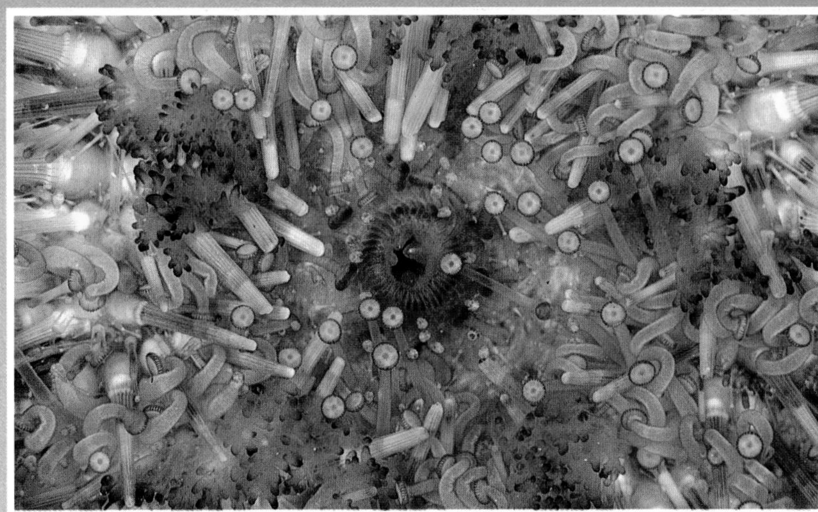

▲ *The sea was the basis of Greek civilization. It was the Greeks' source of living, trade and colonization, while geographically its hundreds of islands encouraged political fragmentation and competition. The changing images of light, cloud and water were also the inspiration for speculations concerning the nature of matter, its transmutations, permanence and change. The sense that everything has a purpose "for the sake of which each thing, and each part of each thing, exists" – a purpose which the philosopher can find out if he looks hard enough – was the great joy of science for Aristotle.*

Explaining his attitude to fact and theory, Aristotle, puzzled at how bees reproduce, wrote "the facts of the matter have not been sufficiently estabished. If at some future time they should be established, more trust should be put in the evidence of sense perception than in theories, save if these are shown to correspond to observation."

In 335 BC Aristotle went back to Athens, whose resistance to Macedonian domination had been crushed. Alexander is said to have listened to Aristotle's pleas on Athens' behalf, and to have made him donations to pursue his researches. Aristotle certainly enjoyed the support of Antipater, whom Alexander had left as his viceroy in Greece when he set forth with his army into Asia. Now Aristotle had both the opportunity to bring his ideas before the intellectual cream of Greece, and the motive to put his notes into order. Those books of Aristotle that survive are essentially his lecture notes, compiled over 20 years, and do not make easy reading. It was during these years that Aristotle established his school of philosophy, known sometimes as the Lyceum or the Peripatetics (◀ page 14).

On the death of Alexander in 323 BC, a fresh revolution threw the Macedonians temporarily out of Athens. Aristotle left in a hurry, while accusations that he talked and behaved contrary to established religion flew about his head. For fear, as he claimed, that the Athenians might be tempted to commit a second crime against philosophy, and put him to death as they had done to Socrates, he moved across the straits of Euripos to the island of Euboea. He had inherited property there, and there in the following year (322 BC) he died.

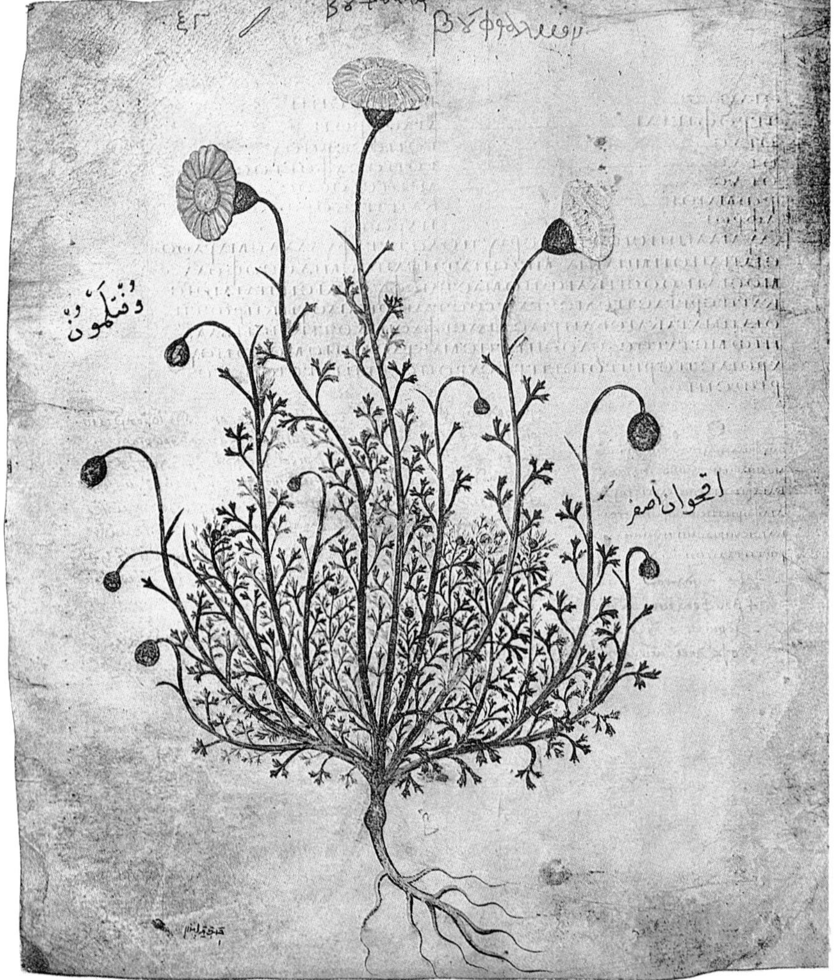

◀ Theophrastus's excellent description of plants, their germination and parts, helped establish today's botanical language. Most of his work descended to the Middle Ages through the "Materia Medica" of Dioscorides, composed in the 1st century AD. Many of Dioscorides' manuscripts, as in this copy, were accompanied by botanical paintings.

◀ Although Aristotle established a formal proof that the Earth was round, medieval manuscripts and early books frequently portrayed it as flat. In a philosophical treatise published in Switzerland in 1508 (left), Thales' concept of the Earth as a flat disk floating on water is portrayed. To this, however, the artist has anachronistically added the two other Aristotelian elements of air and fire, and made a synthesis of Greek philosophy and Christianity by showing God creating Eve from the sleeping Adam. Aristotle believed that because elements held qualities in common which could be exchanged (earth, dry and cold; water, wet and cold) elements could be transmuted into each other. This provided the philosophical justification for the science of alchemy.

The structure of the heavens

In the opening passage of one of his books, *Meteorology*, Aristotle explains that he will then proceed to "the region that borders most nearly on the stars", to write about comets, on wind and weather, and on what we would now call geology, before ending with animals and plants. However, the animals took up so much of his efforts that the plants, together with mineralogy, had to be left to his successor, Theophrastus. In an earlier work, Aristotle concluded that the idea of a single prime matter was less satisfactory than Empedocles' system of four elements (◀ page 12). These correspond in a way to the basic states of matter in modern science: solid earth, fluid water, gaseous air, and fire as the source of heat and energy. But these four elements could not be identified by their resistance to analysis, for they might well be transformed one into another.

In reality, then, the elements were mixed up, but Aristotle thought it possible to establish the order that they would adopt naturally, if they had not been disturbed. Undisturbed they would then settle in layers. Earth falls through water, and water through air. Air bubbles up through water, and flames of fire rise through air. Air and fire do not find it easy to rise through earth, but there is much air, and fire, trapped in the earth, which from time to time breaks through the surface with a violence proportionately greater than bubbles of air in water. That was his explanation for earthquakes and volcanic eruptions. If he had been asked the question that recurred to Newton – why does an apple drop straight off a tree? – he would have said that the earthy and watery parts of the apple were moving to their natural place in the world, to join the rest of the earth and water. For the stratification of the elements had earth at the center, water lying on its surface, air above that, and fire at the top. Similarly, at death, animals and plants fall down and return to the earth and water of which they are mostly made. Thus all movement here on Earth sooner or later comes to an end.

Was the sky made from the same substances? It does not fall in on the earthbound observer. Presumably then it does not contain anything earthy. Aristotle felt there were good reasons to doubt if even our "superior" elements (air and fire) stretched so far above the Earth. Contemporary astronomy was already showing how small the Earth was: smaller than many of the stars, which were so distant that the diameter of the Earth itself was virtually nothing in comparison.

▼ A printed edition of Aristotle's "Meteorology" in 1512 shows the comets and meteors which Aristotle explained as fiery phenomena occurring beneath the heavenly sphere and within the Earth's atmosphere. Besides dealing with wind and weather, the book also explained the growth of minerals within the Earth. A dry exhalation arising from the bowels of the Earth produced "fossils", while a moister exhalation produced metals.

▶ On a bronze coin of Samos of the Roman imperial period, Pythagoras, the 6th-century BC philosopher, is portrayed pointing to a globe. Unlike Aristotle, he interpreted the world through numbers, the systematic study of which he is said to have originated. Thus he noticed the relationship between musical pitch and the lengths and thicknesses of strings. He ascribed such ratios to the distances of the planets which made a "harmony of the spheres". The theorem relating the lengths of right-angled triangles was named for him. He also believed in the transmigration of souls.

Anaxagoras was expelled from Athens for stating the Sun was a rocky mass the size of the Peloponnese

It had long been known that the motions of all the celestial bodies continue with a wonderful regularity. Anaxagoras, in the mid-5th century BC, had been expelled from Athens for asserting that the Sun is a body of stone the size of the Peloponnese (the peninsula of southernmost Greece), but after him all had agreed that the Sun and Moon are plainly spheres. By the same arguments we first learn at school, astronomers well before Aristotle were demonstrating that the Earth too is spherical. They pointed to the ship's mast, or the mountain top that appears first above the horizon, the appearance of new stars in the sky as you travel south, the eclipses of the Moon, which they realized were due to the movement of the Earth's circular shadow across the Moon. Aristotle added that water, air and fire must form spherical shells round the central sphere of earth. Finally, the night sky might seem hemispherical, but, since the journeys of traders and soldiers had disclosed more than half a sphere, that too must be a full sphere, carrying the fixed stars on it. A body carried on the surface of a rotating sphere will describe a perfect circle round its center, which could account for the daily revolution of the sky, with Sun and Moon. Why not then for their annual and monthly motions? On the analogy of the turning sphere which carries the stars round each day, it came to be believed that other spheres carried these seven bodies: Moon, Mercury, Venus, the Sun, Mars, Jupiter and Saturn. Each sphere rotated on its own axis in the appropriate period. As the Greeks became aware of the records which the Babylonians and Egyptians had maintained before them, they were confirmed in the belief that these seven "wanderers" or planets had always traveled the same route, in exactly the same periods of time. The year had always lasted 365 days, so many hours and so many minutes; astronomers could even try to calculate the precise number of minutes. On Earth all is change and mortality, only celestial motions are perpetual, outlasting the generations of mankind. That is why people had always revered the Sun, Moon and stars and believed them, understandably, to be the homes of high gods, while departed heroes, so legends said, had been turned into stars.

Since the circle is a closed line, and perfectly symmetrical, it provided a symbol of whatever is unending, and absolutely harmonious. Only a body carried on a rotating sphere will describe a perfectly circular motion. Unfortunately, by Aristotle's time astronomers had already realized that the planets' motions are actually slightly irregular; they sometimes appear to move at varying speeds and they do not even always move in the same direction. Aristotle was not too worried by this. He considered that the main features of his model were so satisfactory that, at worst, a little modification, by the insertion of extra spheres, would solve the problem of these variations. Often dubious about mathematics, whose importance as the key to knowledge he thought Plato had exaggerated, he held that in the real world the accuracy needed for mathematical demonstration was so rarely attainable that a rational view of the cosmos would have to do without it. The spheres of heaven must, in any case, have different properties from terrestrial elements. They must be made of some superior matter, rigid yet transparent, but shining brilliantly where it collected together to form the visible stars and planets. To this substance he gave the name "ether" (which had originally meant something like "bright flame", but had come to mean the heat and light of the sky). Being imperishable, this ether was not subject to the laws that affect earth, water, air and fire.

The tutor and the king

While Alexander was in his teens, Aristotle was invited to the court of his father (Philip II of Macedonia) to be his tutor. This meeting of the greatest thinker and the greatest man of action of ancient Greece has caught the imagination of mythmakers. But what could Aristotle have taught this ambitious young prince? He might have encouraged Alexander to see himself as the nemesis of the Persian empire, the bearer of Greek culture to the limits of the known world. A lost treatise of Aristotle, entitled "Colonies, or Alexander", may have recorded their discussions. But Aristotle's political theories, as they have come down to us, would have been of little use in the administration of this vast empire. Indeed, Alexander had to free himself of the chauvinist prejudices which his old teacher could never shake off, and assimilate himself to Persian ways; he endeavored to promote assimilation among his men by encouraging marriages between his officers and Persian ladies.

Alexander's campaigns revealed to the Greeks lands of which they had but faint knowledge before. Greek cities were founded in the conquered lands, often named for their conqueror. One, Alexandria the Farthest, stood on the banks of the Syr-Darya, in the present Kazakh SSR. Callisthenes, Aristotle's nephew and ex-student who accompanied the great expedition as an official historian, was imprisoned on suspicion of conspiracy, and died in a cage to which he had been confined. Legend claims that Aristotle then turned against Alexander. The waters of the Underworld river Styx were supposed to corrode everything they touched except a horse's hoof. Aristotle therefore had a cup made of this latter and sent some Styx water to Alexander, who in his ignorance drank it, and perished horribly. On the strength of this fanciful tale, the Roman emperor Caracalla (AD 188-217), who idolized Alexander, persecuted any philosophers who followed Aristotle's ideas.

Aristotle and Alexander

► In 334 BC the armies of Alexander the Great crossed into Asia. In a few years Alexander (depicted here in a mosaic of c. 150 BC at Pompeii) overthrew the Persian empire, the largest the world had seen, larger even than China.

▼ The discovery of Greek coins in non-Greek areas reveals the extent of the Greek economy – the first civilization to use coins: top, a silver shekel from Tyre c. 360 BC; bottom, a silver tetradrachm from Macedonia showing Alexander as Hercules c. 330 BC.

▼ In an Islamic text on hygiene, written over 1,500 years after his time, Aristotle is pictured showing disapproval of his pupil Alexander, who is about to take a cup of wine. His disapproval was well placed – Alexander's drink problem precipitated his death at the age of 32.

Animals and plants from the East

Later, more plausible stories claimed that Alexander had, in fact, ordered that specimens of new kinds of animals and plants should be collected for Aristotle. There is no trace of this in Aristotle's works, most of which were already written before the great march was over. Aristotle knew no more of Persia and India than earlier scholars among the Greeks. From Ctesias, a Greek doctor who had practiced at the Persian court nearly a century before, he took a fabulous account of tigers, beasts so terrible they could fire darts from their tails. Yet Aristotle was dubious about the correct reports concerning the ease with which men could float in the Dead Sea. His friend Theophrastus had access to information on the plants of India, especially those of commercial value. The contrast between Aristotle's treatment of Indian animals, and Theophrastus's treatment of Indian plants, shows what a difference the expedition had made. Theophrastus describes the cultivation of cotton and citron, which was already moving westward out of India into the Middle East.

The legacy of Alexander

After Alexander's death, his generals soon began to quarrel over the succession. Unlike other generals, such as Seleucus, satrap of Babylon, Ptolemy, Alexander's general in Egypt, was relatively modest. He was happy to secure one province – so long as it was rich enough. So he ensured that he was absolute master of Egypt.

The Egypt of the Ptolemies became the main center of Greek learning; in their capital, Alexandria, the Museum provided a home for several leading scientists. Even after the Romans put an end to that kingdom, too, when its last queen, Cleopatra, committed suicide, Alexandria remained a cultural capital.

It was at Alexandria in the 2nd century AD that Claudius Ptolemy – not related to the erstwhile ruling dynasty of Ptolemies – summed up the achievements of Greek astronomers since Aristotle. He described the Earth-centered "Ptolemaic system" of astronomy which was generally accepted until it was finally supplanted by the Copernican system 14-15 centuries later (➧ page 38).

◀ The teaching of Socrates, whom the Delphic Oracle declared to be the world's wisest man, were recorded and elaborated in the "Dialogues" of his pupil, Plato. "There is only one good, namely knowledge, and only one evil, namely ignorance". His "Socratic" method consisted of getting pupils to think aloud by asking them leading questions. Socrates was also renowned for his ugliness – captured here in a Roman bronze from Pompeii.

▼ This magnificent 3rd-century BC gold wheat spray probably belonged to a priestess of Demeter, the Greek god of agricultural fertility whom the Romans called Ceres. Aristotle saw it more as the result of natural rainfall.

To everything its purpose

For Aristotle, one grand principle united sky above and earth beneath; the idea that everything has its function within a greater whole; and that, over time, the end of any change or movement explains the beginning and all intermediate stages. Because the stars are so marvellous and remote, we may esteem knowledge about them more highly. But we can much more easily find out about objects on Earth, like the "perishable plants and animals which grow beside us". Here, too, there are "immeasurable pleasures to be had by those who can learn the causes of them". These causes might include the "necessity" that makes the apple drop. But that is not enough. For Aristotle, to explain the nature of anything truly, you must explain "for the sake of what" that event has occurred, or that object is as it is. Such questions make the humble creatures fascinating. Thus, all stages in the growth of an acorn, from seedling to sapling, can be explained by their underlying destiny to become an oak. This final form of the oak should also explain the leaves, and why they are the shape they are, the bark, the roots and all other parts. The tails of lobsters "would be of no use to crabs as they live close to shore; in such crabs as live out to sea, the feet are less suited to moving about, and they walk little. Very tiny crabs have their hindmost feet flattened like fins or oar blades, so as to help in swimming. Prawns have tails, but not claws, because of the large number of their feet, on which has been expended the material for the growth of claws. But they do have numerous feet, to suit their mode of progression, which is by swimming".

The same principle could also help correlate differences between organs, by showing how they cooperate to maintain the organism as a whole. Thus, among birds, the raptors have hooked beaks and sharp, curved talons shaped "to obtain mastery over their prey, that being suited better for deeds of violence than any other form".

▲ ***The purpose of an acorn is to become an oak tree, of a caterpillar to become a butterfly. Aristotle saw each organism as a self-contained system which had to grow the way it did to fulfill its preordained purpose, or "teleos". In his account of the formation of the sea, he made a distinction between those, like Plato, who based their account on stories relating to the gods or to myths, and those concerned with "human wisdom"; he contrasted the metaphors of poetry with the precise analogies of science. "Spiny lobsters have tails because they swim about, and so a tail is of use to them, serving them for propulsion like an oar." Similarly, the eagle (above, on a Sicilian coin of c. 412 BC) possesses curved talons to capture its prey. Twenty-two centuries later Darwin rejected such teleology and viewed such contrivances and organisms themselves as adaptations developed by chance through natural selection.***

If all parts collaborate to maintain the whole, no part can ever change, for all would have to change at once. This state of equilibrium must have lasted for all eternity, for, to Aristotle, "time is infinite and the Universe perpetual" – it did not come into existence at some particular point in time.

But could the apparent adaptation of organs and organisms be historical accident? Perhaps there once were ill-adapted creatures, but they rapidly died out, just as highly deformed animals are occasionally born, but quickly perish? Aristotle asks: "Why should nature always act for the sake of something, because it is better it should be that way? Zeus does not send the rain in order to make the corn grow; it comes of necessity. What has been drawn up, must cool, turn to water, and come down; and it just so happens that that makes the corn grow. If a man's crop spoils on the threshing-floor, the rain did not fall for that purpose, to spoil the corn, that too just followed." Elsewhere Aristotle goes into some detail on this precipitation cycle, which he explains in terms not so different from those of a modern school textbook. This view contrasts with traditional belief that the crop was lost because a god wished to punish the farmer for some sin. In the previous century, the Athenian playwright Aristophanes had poked fun at Socrates for teaching this very lesson, that the clouds rain, or do not rain, for natural reasons.

Then, if some connections are accidental, perhaps our "front teeth grow sharp and serviceable for biting, the back teeth broad and useful for chewing, not for the sake of that, but by coincidence". And if any animals did not enjoy this happy accident, they died. But Aristotle argues that our concepts of chance only apply to occasional events. When a series of events occurs in a regular sequence with a specific final end, as with the growth of teeth, "all preceding stages are for the sake of that", and no event in the sequence can be accidental.

Aristotle in the Middle Ages

A lack of original scientific work

"He will teach you how long a gnat lives, how far sunlight penetrates water, what sort of a soul oysters have, he knows all about conceptions and birth and the formation of the embryo in the womb," wrote the Greek satirist Lucian about Aristotle in the 2nd century AD. Unfortunately, in these times of the Roman empire very few people wanted to know such things, and still fewer shared Aristotle's desire to find out for themselves. Lecturers were receiving stipends from Roman emperors, or sometimes from municipal funds, to lecture on Aristotelian philosophy, and this meant they could talk at length about its fundamental tenets, in particular its metaphysics and cosmology. But there was no enthusiasm for Aristotle's natural history, and anyone who adopted that philosophy necessarily adopted the explanations of nature that went with the package.

Medicine and astronomy continued to be studied, although, after these two sciences had been conveniently summarized, respectively, by the Greek physician Galen and the Alexandrian Claudius Ptolemy in the 2nd century AD, not much more was added. In so far as centers of learning, for example the Museum of Alexandria, survived, they did so as repositories of knowledge: nobody expected anything new from them.

If such centers were closed down by Christian emperors, or pagan philosophers forbidden to teach, it had no impact on original scientific work, for by then there was none. The Byzantine Greeks experienced a flurry of interest in their scientific heritage from time to time: fresh copies were made of the works of Aristotle, and fresh commentaries written on them.

▲ As a pagan philosopher whose determinism seemed to deny Christian free will, Aristotle was difficult to reconcile with Christian beliefs. One of the most important church leaders in the fusion of the Aristotelian world picture with Christian theology was the Dominican, Thomas Aquinas (1225-1274), portrayed here in his cell in a state of grace receiving divine words from the Holy Spirit. His commitment to the "double faith" doctrine that faith can be achieved through both revelation and reason ensured that science became part of the western intellectual tradition.

◄ Translations of Greek writings into Syriac and Persian were made in the 5th century AD. After the Arabs conquered Syria and Persia, Arabic translations began to appear, as in this 13th-century illustrated Arabic manuscript of Aristotle's biology.

► Ptolemy was renowned for his astronomy and for a geography of the ancient world. In the "Byzantine" frontispiece to a 15th-century edition of this "Geography", the artist has confused Ptolemy (using an astrolabe) with an Egyptian king of the same name.

▼ Aquinas was viewed as the culmination of a long tradition of classical learning. In this detail from "The Triumph of St Thomas Aquinas", a fresco in the Dominican church of S. Maria Novella in Florence painted in the 1360s by Andrea di Buonaiuto, four Greeks appear among the representatives of the liberal arts – left to right: Pythagoras, Euclid, Ptolemy and Aristotle; the tuneful blacksmith is the biblical character Tubalcain, son of Cain. Behind are female personifications of the arts they represent – mathematics, geometry, astronomy, music and logic.

Aristotle in the Arab world

In the Moslem world, these same texts were translated into Arabic and studied with renewed interest. The Arabs were mainly concerned with the medical, astronomical and mathematical heritage of ancient Greece. For them, Aristotle became the foundation of all natural knowledge. Averroës (Ibn Rushd, 1125-1198), who lived in both Spain and Morocco, taught that Aristotle "comprehended the whole truth – by which I mean that quantity which human nature, in so far as it is human, is capable of grasping." In Iran, Avicenna (Ibn Sina, 980-1037) brought Aristotle up to date, lecturing on all branches of Aristotle's learning. From Moslem Spain, Aristotelian tradition spread into western Europe. Over the 12th and 13th centuries, most of the surviving books were translated into Latin.

A pagan philosopher and Christianity

Christians found some of Aristotle's ideas dangerous. His assumption of an eternal Universe went against the Christian belief in a Creation at a specific time, while his description of the soul seemed to leave no place for personal immortality. Nevertheless, it did prove possible to absorb Aristotelian philosophy into Christian teaching, and Aristotle became orthodox. By the 16th century Aristotle was supreme, the rock on which all higher education was built; somewhere in Aristotle the answer to any question would be found. There might be difficulties over particular statements, but they merely presented scope for the ingenious commentator. No one thought Aristotle would ever be outmoded, or that almost everything he said would be tested, and much found wanting.

As Aristotle's ideas spread or were rediscovered, later civilizations strove to reconcile them with their theologies

▲ **Boethius, the 5th-century Roman philosopher and commentator on Aristotle, performs calculations using Arabic numerals. He competes for speed with Pythagoras, who uses an abacus. This page from Gregor Reisch's "Margarita Philosophica" of 1508 (➧ page 21) indicates the importance of the new invention of printing in making known the works of earlier generations.**

Aristotle's legacy

Teleology, the use of predetermined purpose to explain the Universe, survived until Charles Darwin (➧ page 149). The method was used to demonstrate the hand of a Creator. That does not seem to have been Aristotle's intention, for he saw each organism as a self-contained system which had to grow the way it did. In his account of the formation of the sea, he makes a distinction between those who based their account on stories relating to the gods, and those concerned with a "human wisdom"; he contrasts the metaphors of poetry with the precise analogies of science. In his way, he was a religious man; but he held that the divine in the Universe is pure intellect, so that the highest activity for human beings is intellectual activity – reasoning about the world. Finding the natural causes of things is the best thing a person can do. The objective impersonality of science – "something has been observed", rather than "I observed something", is Aristotle's way.

There is a certain loneliness in this procedure, even a certain aridity. Perhaps Aristotle sensed this himself, and began to feel that Plato had been right to express his profoundest thoughts in imagery, rather than Aristotle's endless series of definitions and abstract reasoning upon those definitions. In a letter to Antipater, he remarked, "the lonelier and more solitary I become, the more I have come to love myths".

Eventually, Aristotle was quite often proved wrong. Perhaps in later life his anxiety to explain everything did become a race against time, and his sense of isolation may have sprung from a realization that no one, except perhaps Theophrastus, could really share his enormous task. For all that, and despite his errors Aristotle had founded the method and the description of nature that is science, and it is this view which predominated for some 2,000 years after his death.

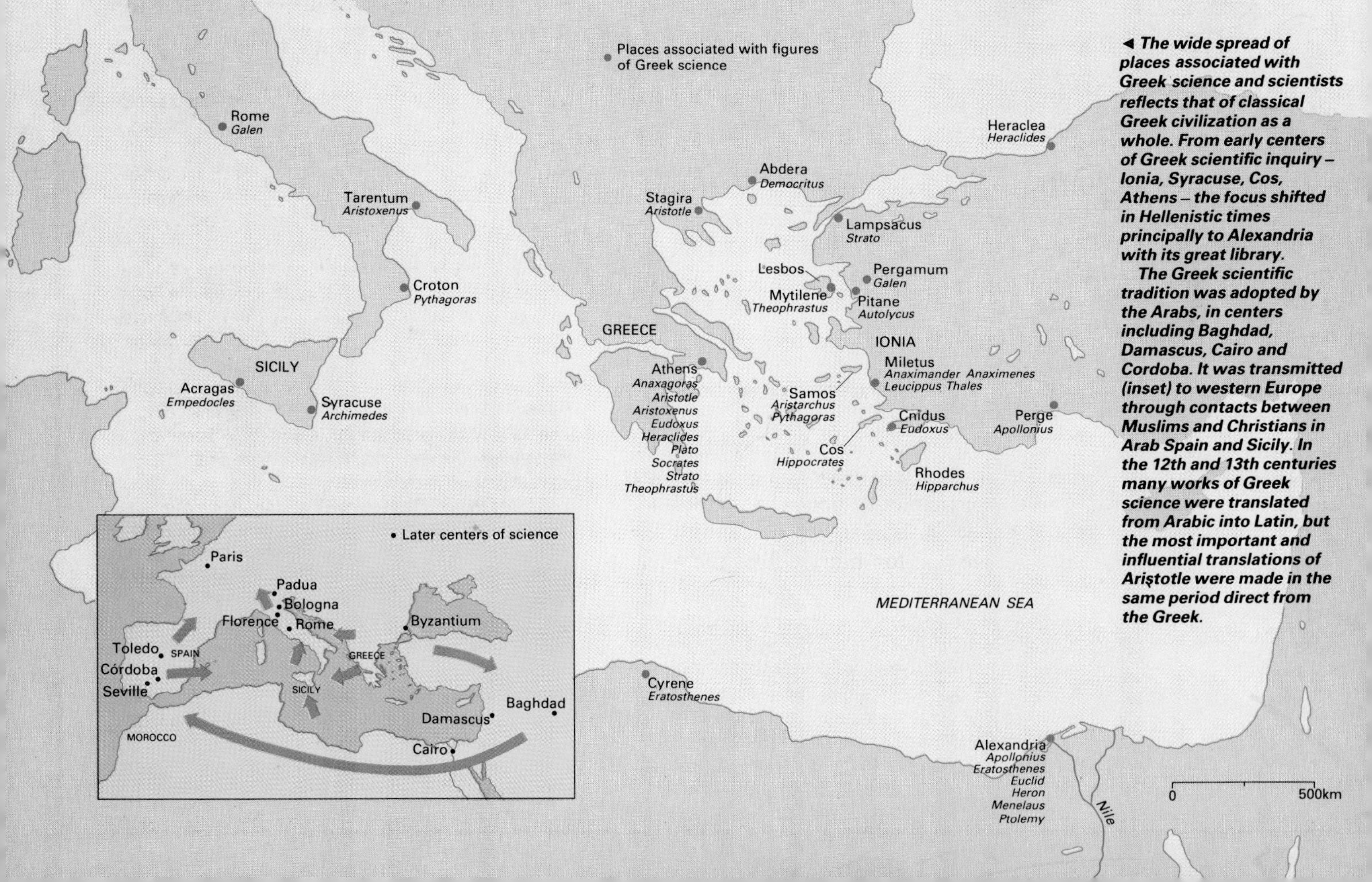

◄ **The wide spread of places associated with Greek science and scientists reflects that of classical Greek civilization as a whole. From early centers of Greek scientific inquiry – Ionia, Syracuse, Cos, Athens – the focus shifted in Hellenistic times principally to Alexandria with its great library.**

The Greek scientific tradition was adopted by the Arabs, in centers including Baghdad, Damascus, Cairo and Cordoba. It was transmitted (inset) to western Europe through contacts between Muslims and Christians in Arab Spain and Sicily. In the 12th and 13th centuries many works of Greek science were translated from Arabic into Latin, but the most important and influential translations of Aristotle were made in the same period direct from the Greek.

Galileo Galilei 1564-1642

2

A mathematical scientist...A Tuscan family...Galileo's law of fall...The pendulum...The basis of ballistics... Nicolaus Copernicus and Tycho Brahe...Galileo raises his telescope...and makes enemies...Trial by the Inquisition...An active old age...PERSPECTIVE... Academies of learning...Opposition to new ideas... Instruments on land and sea...Astrology – a science-art

"He hath first over thrown all former astronomy...and next all astrology." So reported the English ambassador at Venice, when he learnt of the discoveries published three days earlier by an obscure professor of mathematics at the nearby University of Padua, Galileo Galilei. Never before, and seldom since, has scientific news caused such a stir as Galileo's first observations with his telescope. Beside the seven planets known since the days of ancient Babylon, he had found four more, little ones which revolved around Jupiter. On the Moon he had seen mountains and plains like those on Earth. These were some of the revelations of a slim book called *The Sidereal Messenger*, which appeared in March 1610. To Galileo's contemporaries, and perhaps to us, Galileo is best known as the first man to raise a telescope to the sky, revealing something of the immensity of the Universe.

▲ *Galileo helped redefine our place in the cosmos.*

Prime mover of the Scientific Revolution

Yet Galileo is much more than that. If any one person can be said to have set the Scientific Revolution in motion and pulled modern science out of ancient natural philosophy, that man was Galileo. Of all the people of his time, he best realized that the old way of looking at the world would have to go; and he best knew how to begin constructing a new way. This he did by making physics mathematical. Events on Earth would help explain what could be seen in the sky, the sky could show us how things happened on Earth. Everywhere nature behaves in an orderly manner, which we can understand, provided our interpretation is couched in a mathematical language. For the proofs of geometry – so he thought – are absolutely certain, unlike other kinds of human reasoning. What is more, just as one theorem in geometry leads to the next, so one discovery will lead to another.

Some of Galileo's ideas are not wholly original, and can be traced back to the Middle Ages, even to ancient Greece. Although he often criticized Aristotle, Galileo realized that he had set out the basic questions we must answer, if we want to know how the world works. However, Aristotle's answers were inadequate, Galileo believed, because his physics was not mathematical. Galileo showed, too, how instruments designed according to the principles of optics, a mathematical science, could extend the powers of the human senses, making them stronger and more reliable. Above all, for him, unlike the ancient Greeks, geometry did not have to be restricted to the description of a static world. The movements of bodies can also be analyzed by means of lines, triangles, circles and numbers. So time can be treated mathematically in the same way as the three dimensions of space. Galileo bequeathed to later science the idea of "acceleration" as a mathematically defined concept. That enabled him to demonstrate a "law of falling bodies" which became the foundation of all later dynamics (i.e. studies of how objects move).

Mathematician, physicist, astronomer

By the time of Galileo, the institutionalized, 2,000 year-old tradition of Aristotelian science was already breaking down. A new way of thinking was taking shape: all real knowledge was to be expressed in mathematical terms, which, it was now believed, constituted the only objective and reliable language. Theories about nature had to be put to the test of carefully controlled experiments, whose results should take the form of measurements. The change of thought took almost two centuries to become established in western Europe; today this prolonged crisis is known as the Scientific Revolution.

Since physics explains the basic characters of things, it was the foundation or "cutting edge" of the new science. But only after Nicolaus Copernicus realized that the Earth is a planet of the Sun, not the center of the Universe, did it become possible to reason from Earth to sky and sky to Earth and so to construct a physics that would apply universally. That is why the greatest figures of the Scientific Revolution, Galileo and Newton, were both physicists and astronomers.

Galileo thrust Copernicus's theory upon the general public and showed how crucial it was for an understanding of our place in the cosmos. By recording his observations with the newly invented telescope, he launched an astronomy that studied the features of celestial bodies.

Galileo established mathematical laws describing the motion of falling bodies. Understanding the forces that cause bodies to fall, and that hold the entire system together, he had to leave to his successors, notably Newton. But Galileo's dynamics, however elementary, remain the foundation of classical mechanics.

Galileo Galilei – His Life, Work and Times

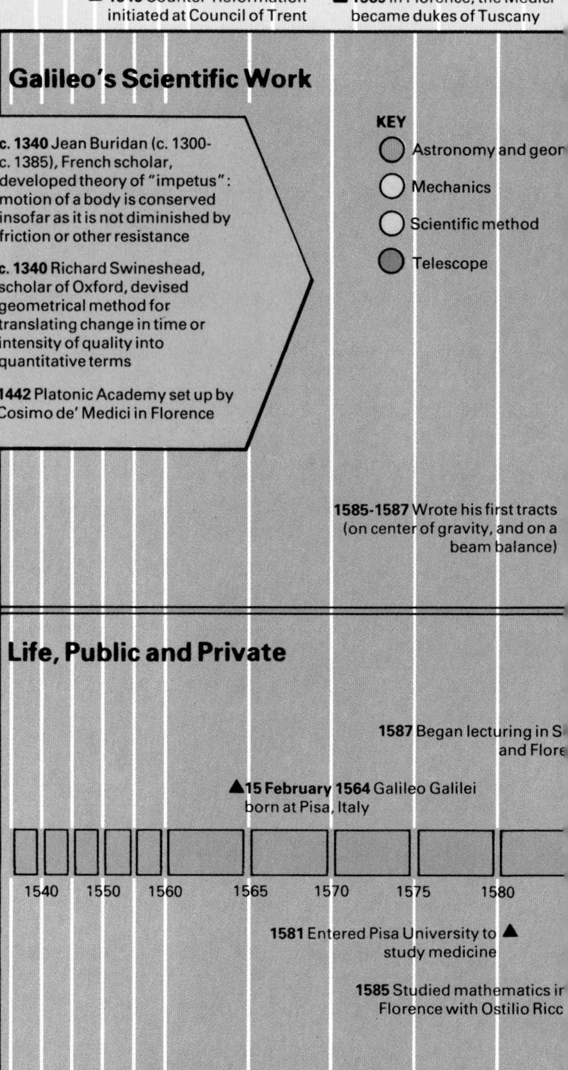

Last of the Renaissance, first of the New Age

Galileo can be seen as the culmination of the Italian Renaissance. He was born in Pisa on 15 February 1564 (just two months before Shakespeare). Both his parents came from long-established Tuscan families; his father, Vincenzio, was a native of Florence who eventually returned there, so it was at Florence that young Galileo was brought up from the age of ten. Vincenzio was a professional musician, interested in current debates on musical theory. Presumably Galileo did not have so much musical talent, although like his father, he enjoyed playing the lute, and always took an interest in acoustic questions. Vincenzio apparently intended Galileo, his eldest son, to be a doctor, and so Galileo went to medical school at the University of Pisa, in 1581. But he left after four years without taking his degree. On his return to Florence, he began to take a serious interest in mathematics, which he studied with Ostilio Ricci, who taught at the court of the Grand Duke of Tuscany.

Vincenzio seems to have been unhappy at the direction his son's interests were taking, but Galileo persisted and, indeed, was appointed to a junior post at his old university in 1589. There he took up the ideas of geometer Archimedes (◀ page 17), who had applied the methods of geometry to physical problems. Archimedes' principle states that bodies float or sink, according to the ratio of their density to that of the medium in which they are immersed. Galileo extended this, to consider the ratio of the speeds of falling bodies to the density of the media through which they fall. None of this was in itself so revolutionary, but it led him to contradict Aristotle's view that bodies "naturally" returned to their proper place. Galileo pointed out that the weight of the bodies was irrelevant to their speed of falling; only the density of the medium counted. If there were space without any medium – that is, completely empty space, a vacuum – all bodies would fall at the same speed. It is possible that he demonstrated his point by throwing different weights off the Leaning Tower of Pisa.

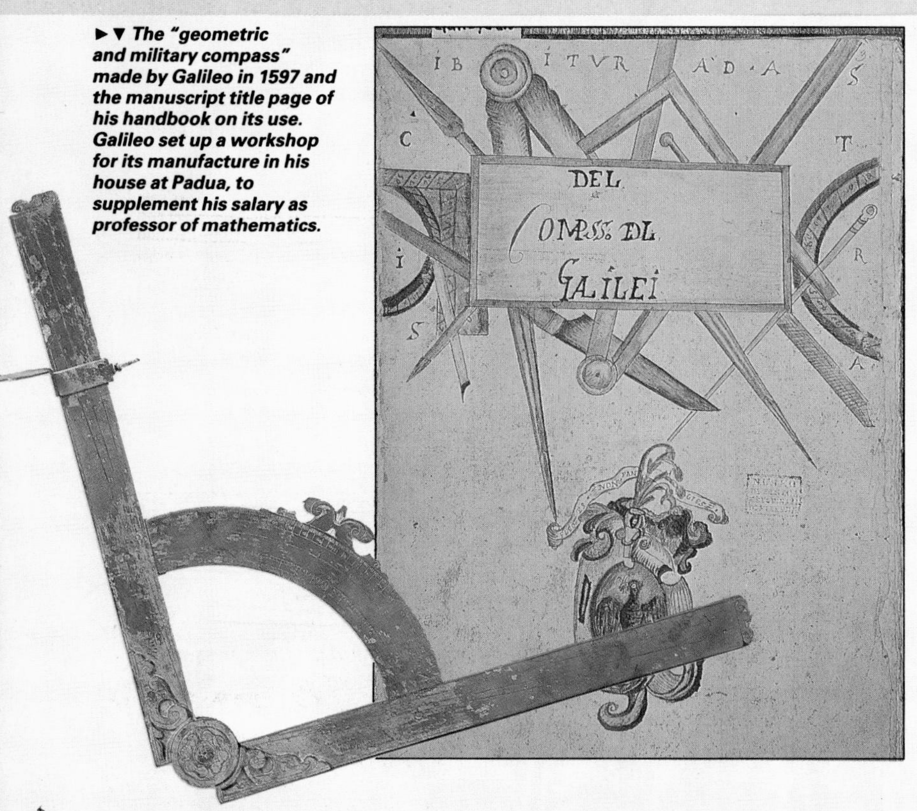

►▼ The "geometric and military compass" made by Galileo in 1597 and the manuscript title page of his handbook on its use. Galileo set up a workshop for its manufacture in his house at Padua, to supplement his salary as professor of mathematics.

Scientific Background

1577 Brahe showed that comets are interplanetary, not within Earth's atmosphere as previously supposed

1572 Tycho Brahe (1546-1601), Danish astronomer, showed that the new star in Cassiopeia was truly a star, and that the heavens were not immutable

1543 Nicolaus Copernicus (1473-1543), Polish astronomer, proposed heliocentric theory: that Sun is center of cosmos and Earth one of its planets

1536 Niccolò Tartaglia (1500-1557), Italian mathematician, launched "new science" of ballistics

1584 Giordano Bruno (1548-1600), Italian reformer, claimed that Universe is infinite; burned at stake by Inquisition

1552 Giovanni Battista Benedetti (1530-1590), Italian mathematician, showed that velocity of falling bodies is not related to their weights

1575 Academy of Mathematical Sciences set up in Madrid

1585 Simon Stevinus (c. 154 c. 1620), Dutch mathematicia introduced decimal fractio into common usa

1535 1545 1555 1565 1570 1575 1580

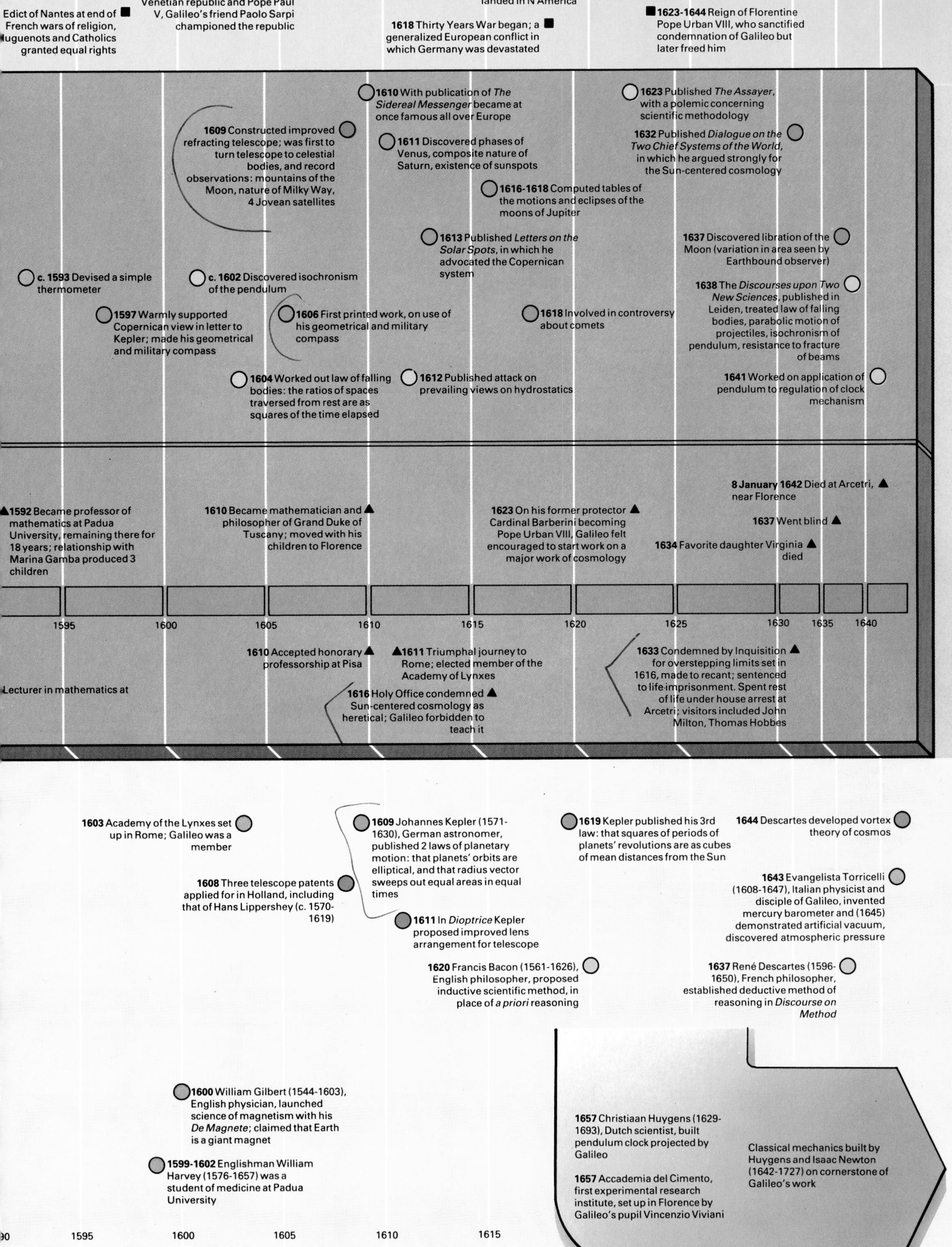

Edict of Nantes at end of French wars of religion, Huguenots and Catholics granted equal rights
1606 In contest between Venetian republic and Pope Paul V, Galileo's friend Paolo Sarpi championed the republic
1620 Pilgrim Fathers landed in N America
1618 Thirty Years War began; a generalized European conflict in which Germany was devastated
1623-1644 Reign of Florentine Pope Urban VIII, who sanctified condemnation of Galileo but later freed him
1610 With publication of The Sidereal Messenger became at once famous all over Europe
1623 Published The Assayer, with a polemic concerning scientific methodology
1609 Constructed improved refracting telescope; was first to turn telescope to celestial bodies, and record observations: mountains of the Moon, nature of Milky Way, 4 Jovean satellites
1611 Discovered phases of Venus, composite nature of Saturn, existence of sunspots
1632 Published Dialogue on the Two Chief Systems of the World, in which he argued strongly for the Sun-centered cosmology
1616-1618 Computed tables of the motions and eclipses of the moons of Jupiter
1613 Published Letters on the Solar Spots, in which he advocated the Copernican system
1637 Discovered libration of the Moon (variation in area seen by Earthbound observer)
c. 1593 Devised a simple thermometer
c. 1602 Discovered isochronism of the pendulum
1638 The Discourses upon Two New Sciences, published in Leiden, treated law of falling bodies, parabolic motion of projectiles, isochronism of pendulum, resistance to fracture of beams
1597 Warmly supported Copernican view in letter to Kepler; made his geometrical and military compass
1606 First printed work, on use of his geometrical and military compass
1618 Involved in controversy about comets
1604 Worked out law of falling bodies: the ratios of spaces traversed from rest are as squares of the time elapsed
1612 Published attack on prevailing views on hydrostatics
1641 Worked on application of pendulum to regulation of clock mechanism
8 January 1642 Died at Arcetri, near Florence
1592 Became professor of mathematics at Padua University, remaining there for 18 years; relationship with Marina Gamba produced 3 children
1610 Became mathematician and philosopher of Grand Duke of Tuscany; moved with his children to Florence
1623 On his former protector Cardinal Barberini becoming Pope Urban VIII, Galileo felt encouraged to start work on a major work of cosmology
1637 Went blind
1634 Favorite daughter Virginia died
1595
1600
1605
1610
1615
1620
1625
1630
1635
1640
1610 Accepted honorary professorship at Pisa
1611 Triumphal journey to Rome; elected member of the Academy of Lynxes
1633 Condemned by Inquisition for overstepping limits set in 1616, made to recant; sentenced to life imprisonment. Spent rest of life under house arrest at Arcetri; visitors included John Milton, Thomas Hobbes
Lecturer in mathematics at
1616 Holy Office condemned Sun-centered cosmology as heretical; Galileo forbidden to teach it
1603 Academy of the Lynxes set up in Rome; Galileo was a member
1609 Johannes Kepler (1571-1630), German astronomer, published 2 laws of planetary motion: that planets' orbits are elliptical, and that radius vector sweeps out equal areas in equal times
1619 Kepler published his 3rd law: that squares of periods of planets' revolutions are as cubes of mean distances from the Sun
1644 Descartes developed vortex theory of cosmos
1608 Three telescope patents applied for in Holland, including that of Hans Lippershey (c. 1570-1619)
1643 Evangelista Torricelli (1608-1647), Italian physicist and disciple of Galileo, invented mercury barometer and (1645) demonstrated artificial vacuum, discovered atmospheric pressure
1611 In Dioptrice Kepler proposed improved lens arrangement for telescope
1620 Francis Bacon (1561-1626), English philosopher, proposed inductive scientific method, in place of a priori reasoning
1637 René Descartes (1596-1650), French philosopher, established deductive method of reasoning in Discourse on Method
1600 William Gilbert (1544-1603), English physician, launched science of magnetism with his De Magnete; claimed that Earth is a giant magnet
1657 Christiaan Huygens (1629-1693), Dutch scientist, built pendulum clock projected by Galileo
Classical mechanics built by Huygens and Isaac Newton (1642-1727) on cornerstone of Galileo's work
1599-1602 Englishman William Harvey (1576-1657) was a student of medicine at Padua University
1657 Accademia del Cimento, first experimental research institute, set up in Florence by Galileo's pupil Vincenzio Viviani
1595
1600
1605
1610
1615

Galileo moves to Padua

In 1591 his father died, and Galileo found himself burdened with the duties of head of the family. His salary hardly matched this responsibility, being about one-thirtieth that of the professor of medicine. Besides, Galileo had made enemies in the faculty by writing a satirical poem poking fun at the academic gown, which university staff were expected to wear at all times.

In 1592 he obtained the better post he needed, teaching mathematics at the University of Padua, at three times his previous salary. He also added to his income by taking in students, and giving extra tuition. They mainly wanted mathematics for military engineering or navigation, and for them Galileo designed a new type of instrument, his "geometric and military compass".

Padua was then the premier university of Italy, and one of the best in Europe. There Galileo made friends with some of the leading minds of Italy. Although he never married, he entered into a relationship with a woman named Marina Gamba, from which three children were born. They parted when he left Padua in 1610, the children eventually joining him when their mother wed. The eldest, Virginia, the most perceptive of the three, was very close to him. In later years he leased a house to be near her nunnery in Arcetri, outside Florence.

► Galileo tests the "law of fall" by rolling a metal ball down an inclined groove. This 19th-century reconstruction includes Pisan landmarks, but Galileo had been in Padua for some time when the experiment was conducted in 1603-1604.

▼ The scientific Renaissance which flowered in the revolutionary theories of Copernicus, Galileo and others was built at least in part on the earlier rediscovery of writings by giants of Classical Greek science. Raphael's "School of Athens" (1509), which typifies Italian High Renaissance painting, includes Plato and Aristotle (center of painting), Socrates, Pythagoras, Euclid and Archimedes.

▲▼ *Tradition has it that Italy's greatest scientist disproved Aristotle by dropping different weights from Pisa's Leaning Tower. Galileo is said to have observed the pendulum swing of the chandelier in the cathedral next door.*

Acceleration, free-fall and inertia

At Padua, Galileo carried on his investigation of the simplest movements that we can observe. We should take this step first, he believed, before we try to understand the complexity of the world. He felt it would be possible to give an exact mathematical description of such movements, in terms that would be as certain and as precise as a theorem in geometry. Galileo believed that proceeding in this way, he could arrive at scientific laws of motion. He considered that all his predecessors had been vague and uncertain, because they would not use mathematical methods. But his program would depend on treating time as a geometrical dimension. He had already established that in empty space all bodies would fall at the same speed. But how do they fall, and how do they speed up as they fall?

First, he defined this speeding up, or acceleration, in a way that seemed to make obvious sense, as a motion that "from the point of rest, adds to itself equal moments of swiftness and equal times". From this he could deduce that "the spaces passed over in natural motion are in proportion to the squares of the times", as he wrote in 1604 to his friend the Venetian theologian and physicist Paolo Sarpi (1552-1623). This is essentially the "law of fall" that begins any modern textbook of dynamics.

To test his conclusions, Galileo devised a crude experiment in which a ball was allowed to roll down a groove in an inclined plane. The extremely short time intervals were measured by the human pulse and by the quantity of water that escaped from a large vessel. Such methods were not ideal but Galileo felt sure that he was being accurate enough to confirm his theoretical arguments, even though he was ignoring the effects of friction.

Galileo also found that a ball falling at the end of a piece of string presented the same sort of problem. From his theory it followed, he believed, that the ball would now oscillate in equal times, through very small arcs, depending only on the length of the cord. He made measurements which confirmed his hopes. In this way he had now discovered the law of how a pendulum works.

Galileo demonstrated that any projectile would move in a parabola

The foundation of ballistics

Once distinctions between the traditional four elements and their natural places are discarded, Galileo recognized, all motion is natural, and any force, however slight, can set a body in motion. Once moving, leaving friction aside, a body will continue until some other force brings it to a halt. Newton later saw this, correctly, as the basis for his own concept of what came to be called "inertia". Galileo also showed that if the vertical acceleration of a falling body is combined with horizontal motion, the resultant path of the body is parabolic. He tested this by mounting a chute on a table and letting the ball on it fly off the edge, marking the spot where the ball landed. He showed that any projectile would move in a parabola. This forms the basis of ballistics, a principle used in gunnery and rocketry.

The telescope and the sun-centered Universe

By about 1609, Galileo had worked out the main lines of this new science of uniformly accelerated motion, and told friends that he was preparing a book on it. At this point, he was distracted by hearing of a new and remarkable invention, the telescope. Would a device that made distant objects appear larger show us more of the sky, too? And would not greater knowledge of the sky help to confirm the hypothesis of Copernicus, to which he had been sympathetic for at least a dozen years?

In the early 16th century, the Polish clergyman and astronomer Nicolaus Copernicus (1473-1543) had proposed that the center of what we now know as the Solar System was the Sun; whereas in all ancient cosmologies, the Earth was at the center of the whole Universe. He pointed out that, if we assume the Earth rotates daily about its own axis, and is also in orbit around the Sun, many anomalies in the movements of the planets can be explained quite simply. Copernicus worked out the orbits of the planets in some detail, and showed how his great idea would also enable him to estimate the distances of the planets from the Sun. Fearing ridicule, he delayed the publication of his book until the end of his life, in 1543. In fact, little notice was taken of his ideas for many years. Comprehension of the physical world depended on the centrality and immobility of the Earth. Besides, the Bible occasionally speaks of the Sun moving across the sky, and of the stationary Earth. Gradually, however, the news spread that Copernicus – widely admired for his accurate computations – had put forward this weird theory.

Nearly 30 years after his death, another astronomer took the theory seriously, only to propose a compromise model. In this, all planets except the Moon revolved around the Sun, but the Sun itself, with all the planets in its train, revolved around the Earth. This was the system of Tycho Brahe (1546-1601), a young Danish nobleman who, like all his contemporaries, had been astonished by the sudden appearance of a new star in the sky in 1572 (Galileo later recalled how it was pointed out to him as a boy). Most were concerned with what this phenomenon implied for astrology, but Brahe and a few others examined it further. It failed to show any displacement when observed from different positions – that is, it showed no parallax – as it should if close to the Earth. Brahe decided this must be a true star: not some meteorological wonder, but the first new star to appear in recorded history. As a reward, he was granted the island of Hven and the money to build there his "sky palace", Uraniborg, Europe's first observatory.

The Florentine and other academies

Ever since 1442, when Cosimo de' Medici set up an academy at Florence – called the "Platonic" in imitation of Plato's academy established 2,000 years previously in Athens – academies had proliferated in Italy, until every town with any self-respect had one. They were not schools or colleges, but more like clubs, where like-minded people could get together and debate literary or philosophical questions. Religion was a touchy business, and often politics were, too: in Naples, at one stage, academies were forbidden, as centers of opposition to Spanish rule.

Natural philosophy (science) should have proved a safe alternative subject, but this was not always true. An "Academy of the Secrets of Nature" was formed at Naples in the 1560s, organized by the physicist Giambattista della Porta (1538-1615), who some years earlier had published a book on "Natural Magic". In this, he collected genuine effects – magnetic, optical, chemical – mingled with much fanciful lore. Although della Porta insisted that this magic, however mysterious, was still natural, and did not involve spirits or the supernatural, he was interrogated by the Inquisition and his academy had to be dissolved.

Many of these academies had a somewhat esoteric air. Membership was restricted, and meetings were ceremonial. To counter their own tendency toward the grandiose, the academies might give themselves self-deprecating names. Galileo belonged to one at Padua styled the "Academy of the Sheltered" – a term used for the inmates of a hospital, or almshouse.

▲ A portrait by the Florentine artist Ghirlandaio of the philosopher Marsilio Ficino (1433-1499), second from left. Ficino translated many of the Greek classics into Latin, most notably Plato's dialogs. Chosen to head the new Platonic Academy in Florence, he was a leading figure in the development of Renaissance humanism.

The Academies of Italy

▼ **The Platonic Academy, foremost of Italian academies, was set up by Cosimo de' Medici in Florence in 1442. In this idealized engraving, the members are seen indulging in the debate of philosophical matters, emulating the original Platonic school at Athens.**

▲ **The crest of the Academy of the Lynxes. Named for the animal traditionally held to be the most keen-sighted, the academy was founded in 1603. It specialized in the study of natural history, but broadened its outlook to astronomy on the election of Galileo in 1610.**

The Academy of the Lynxes

In 1603, Marquis Federigo Cesi, then only 18, brought together three of his friends at his palace in Rome, to launch an academy to "penetrate the interior of things and know their causes and the operations of nature which work within them". The company swore to study together; in their single-mindedness they would remain unmarried. Each week they would meet, to report to each other what they had found out. They called themselves the "Academy of the Lynxes", as the lynx was traditionally the most keen-sighted of animals. Cesi's own interests were in natural history; he had a fine collection of minerals and fossils and the only work of his ever published was botanical. Another member produced the first published microscopical observations – of a bee. It was Galileo's success with his telescopic observations that drew the academy to astronomy; they quickly elected him and sponsored some of his books. But, having backed Galileo, the academy could do little to help when Copernicanism was condemned. During the 1620s its activities faded away.

The Academy of the Concrete

Most Italian academies depended on a patron and all were very short-lived. A few carried out some good experimental work in physics and natural history. The Accademia del Cimento ("Academy of the Concrete") brought together by Galileo's pupil, Vincenzio Viviani (1622-1703), at Florence in the late 1650s was important. However it survived less than a decade. Italy's political fragmentation and its relative decline meant that no national institution could emerge to lead the scientific movement.

▼ **This glass spiral thermometer was employed in experimental work carried out by the Accademia del Cimento. It contained alcohol in a vacuum; this expanded and contracted according to the external temperature, and the level was measured by calibrations in the glass.**

Navigation and Mathematics

When, at the end of 15th century, Columbus crossed the Atlantic and Portuguese mariners sailed round Africa into the Indian Ocean, they not only opened the way to new continents, but inaugurated a new science of navigation. Before that, ships had either kept to narrow seas, or used regular winds to bring them to known landfalls. Now that they had to establish their whereabouts in unfamiliar waters, seamen had recourse to the instruments of the astronomers, to find their location from the positions of Sun and stars.

Instruments such as the astronomical astrolabe, quadrant and Jacob's staff were modified for use at sea. Though simpler than the original devices, some mathematical training was needed to handle them properly. The Spaniards developed a system of teaching and certifying pilots to navigate their ships. Books on this new art of compass and celestial navigation appeared, then were translated, pirated, copied and improved. Yet navigation remained a tricky business, and none too easy to put into practice.

Time measurement key to navigation

Finding the longitude was the most difficult problem. An observer was obliged to measure how many degrees east or west of his origin he was. Alternatively, he had to measure the difference in time between events in the sky as he observed them, against the times predicted for those events for a baseline meridian.

Time, then, was the key. Mechanical clocks, driven by falling weights, were a great marvel when they first appeared in northern Italy about 1300. But although they might be good enough to tell the hours for cathedrals, or town halls, they were bulky and not very precise. Clocks driven by springs were introduced in the late 15th century. As they could be much more compact and portable, they became popular with those who could afford them, hung from the neck as a pendant. But they were even less accurate than the others.

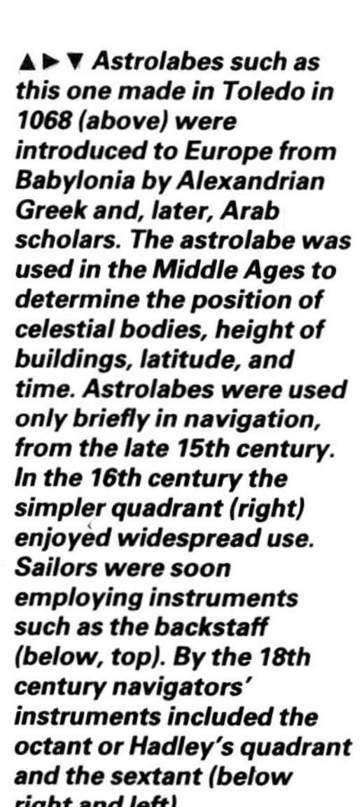

▲▶▼ Astrolabes such as this one made in Toledo in 1068 (above) were introduced to Europe from Babylonia by Alexandrian Greek and, later, Arab scholars. The astrolabe was used in the Middle Ages to determine the position of celestial bodies, height of buildings, latitude, and time. Astrolabes were used only briefly in navigation, from the late 15th century. In the 16th century the simpler quadrant (right) enjoyed widespread use. Sailors were soon employing instruments such as the backstaff (below, top). By the 18th century navigators' instruments included the octant or Hadley's quadrant and the sextant (below right and left).

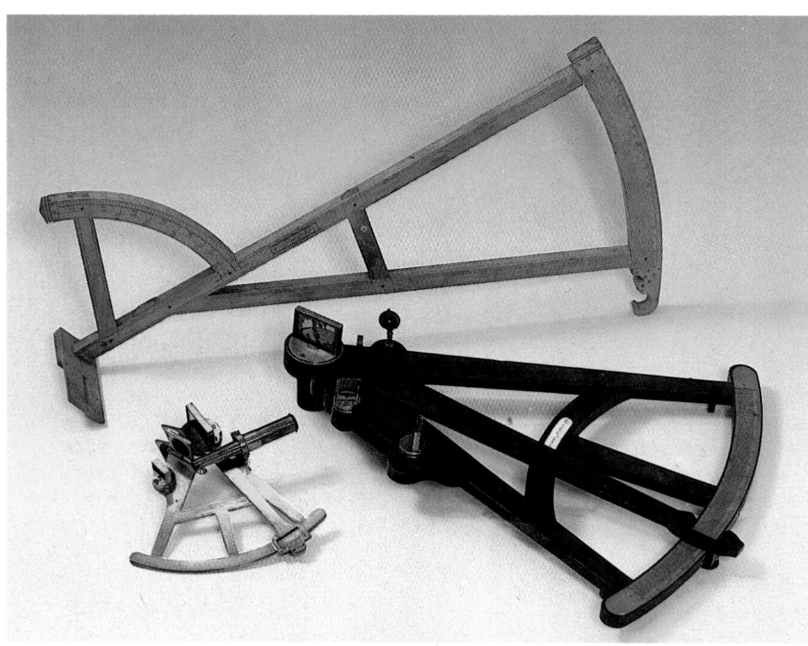

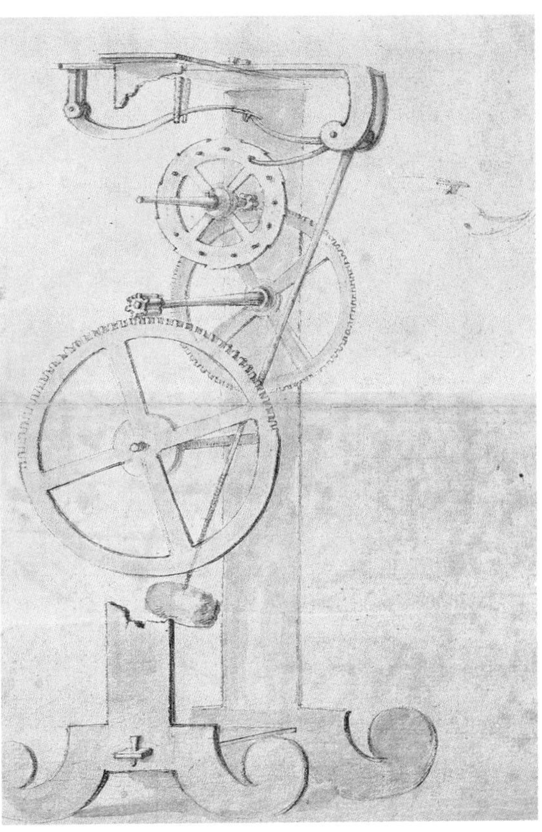

▲ This earliest known picture of spectacles in use was painted by a Tuscan monk in 1352. Compound use of lenses came some 250 years later with the telescope and first compound microscopes.

► A design by Galileo for an escapement and pendulum drawn by his pupil Vincenzio Viviani. Pendulum clocks were to be much more precise than earlier weight- or spring-driven clocks.

*In the last months of his life, Galileo worked on the application of the principle of the pendulum to the regulation of clocks. A pendulum clock was invented in Holland (**➧ page 78**) not long after, in 1656 or 1657. It was designed by Christiaan Huygens (1629-1695), a Dutch physicist who hoped that this more reliable timekeeper would be invaluable at sea. As it turned out, his clocks were too badly shaken about on shipboard to work properly. On land, however, they soon became the norm; older clocks were adjusted to take the pendulum and the new type was soon to be found in every solid middle-class household.*

Surveying and triangulation

On land, too, new mathematical technologies were making their presence felt. Cannon first appeared early in the 14th century in Europe, and by 1500 had evolved to become the major weapon of war. To site, lay, and aim guns required elementary mathematics. To counter artillery, a new style of fortification was developed, in which a series of bastions, like raised gun emplacements, were connected by "curtain" walls. These were spaced and designed so as to allow the widest possible range of fire with the least exposure to the enemy's assault. As land use in the 16th century became more bound to the cash requirements of landowners, exact and detailed surveys of estates became a pressing need. New mathematical techniques were invented, such as triangulation, to make maps and estate plans more precise. These inspired a steady flow of booklets on surveying, cartography and chart-making. Mathematics was also important in gauging (as in Kepler's book on calculating the content of wine and beer barrels).

Lenses, spectacles and the telescope

The first new scientific instrument was the telescope. Yet in its origins it seems to have been an accidental invention. Spectacles were first used in Tuscany, probably by about 1300. By 1500, although still regarded as suitable only for elderly scholars, spectacles were quite common, and there were professional lens-grinders and spectacle-makers.

Three rival claimants for the honor of first combining lenses to create a telescope, all in the Low Countries, applied for patents in 1608. The Dutch spectacle-maker Hans Lippershey, who died in 1619, is usually allowed to have the best claim. There are tantalizing references to earlier instruments giving optical magnification, but there seems to be no reason to believe that anyone actually constructed a telescope before 1608.

*Galileo had proffered his telescope to Venice for use at sea; the captain could identify hostile ships and check their size and armament when they were scarcely visible to the unaided eye. Telescopes did become essential information-gathering devices at sea, and in land warfare, too, by the 18th century (**➧ page 79**). They were also useful in survey work, through the application of telescopic sights to theodolites and other instruments (**➧ page 119**).*

Although Johannes Kepler had never made a telescope for himself, he did suggest in his book "Dioptrice" of 1611, that two convex lenses would be better than Galileo's type, with its concave eyepiece. Only when this was taken up were the astronomers of the 1640s and 1650s able to add to what Galileo had found. Huygens discovered Saturn's ring and its satellite Titan, while others saw the belts on the planet Jupiter.

► The Earth- and Sun-centered systems. In the accepted Ptolemaic system the Earth is encircled by water, air, fire, spheres of the "seven planets" (Moon, Mercury, Venus, Sun, Mars, Jupiter and Saturn) and stars (in 12 zodiacal sections). The new Copernican system has the Sun circled by the six known planets, among them Jupiter with its four moons discovered by Galileo. An engraving of 1661.

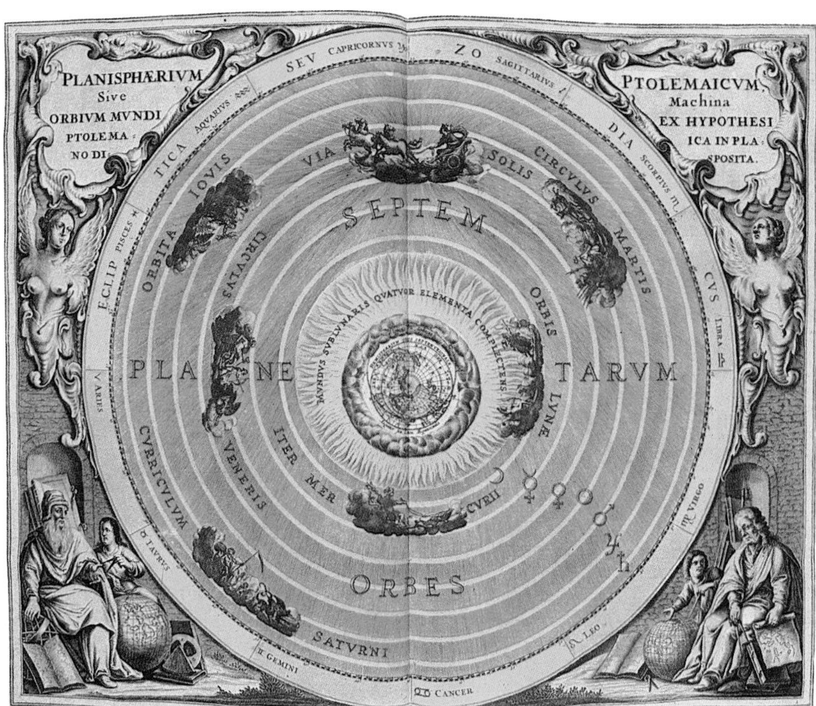

Non parem Pauli gratiam requiro
Veniam Petri neq posco, sed quam
In crucis ligno dederas latroni
Sedulus oro.

▲ Nicolaus Copernicus stands between the Cross and symbols of the new astronomy that he founded.

▼ Tycho Brahe, greatest of pre-telescope astronomers, in his observatory on the island of Hven.

From what Galileo says, he was originally converted to Copernicus's doctrine by a German student at Padua. However, in 1597, he received a copy of a book by a German astronomer a few years younger than himself, Johannes Kepler (1571-1630). There he found fresh proof for the Copernican cosmology. Between the "spheres" of each of the planets (whose radii could only be established on Copernicus's principles), it would be possible, so he claimed, to fit one of the five regular geometrical solids. This underlined the symmetry and simplicity of the new picture of the Universe. Galileo wrote to congratulate Kepler. The book also gained for Kepler the esteem of Brahe, who invited him to help in his work. Thus, after Brahe's death, Kepler inherited Brahe's observations, from which he derived the first two laws of planetary motion, published in 1609. These laws state that the orbits of planets are ellipses with the Sun at one focus; and that a line from a planet to the Sun sweeps out equal areas in equal times.

During those years, while Kepler was hammering out his new revolution in astronomy, Galileo apparently gave the whole subject little thought. Only when another new star blazed in the sky, coinciding in time with a conjunction of Saturn and Jupiter which much excited all the astrologers, did Galileo involve himself. A new star was such a clear challenge to the Aristotelians' world picture, that he could not resist writing a comic dialogue to make fun of their reactions, as they were so sure nothing in the heavens could ever change.

Around 1609, Galileo first heard of a new optical instrument – the telescope – and saw its potential for astronomy. To begin with, he looked at the Moon, whose strange blotches proved to be the surface features of a body that actually appeared to be not so different from the Earth. The dark areas of the Moon he supposed to be seas – and we have retained the name of *maria* (seas), and "ocean" for the largest of them, although there is not a drop of water in them. Elsewhere on the Moon he saw great mountain ranges. However, of all his discoveries, the most dramatic was the sight of unknown "stars", new worlds which humans had never seen before.

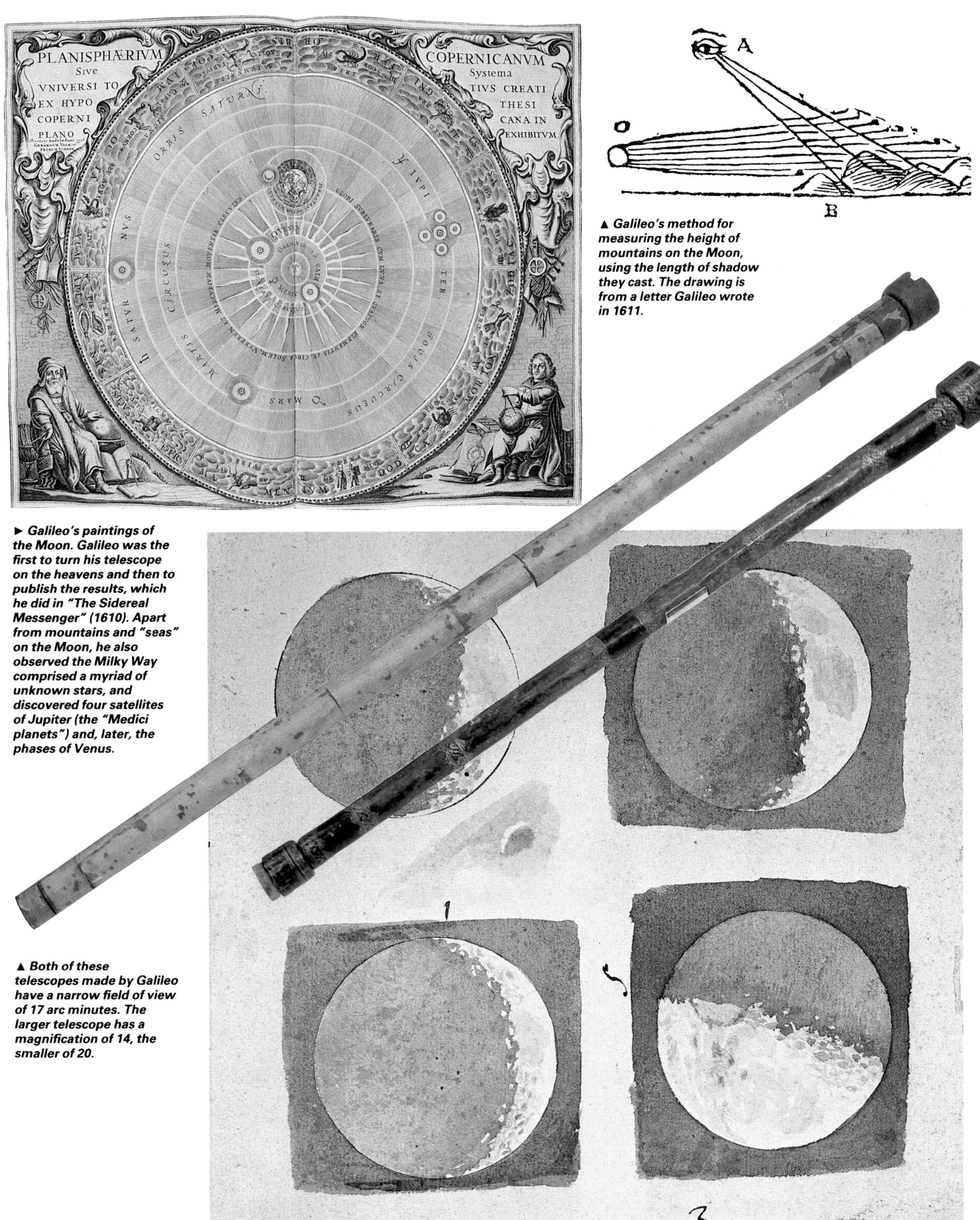

▲ *Galileo's method for measuring the height of mountains on the Moon, using the length of shadow they cast. The drawing is from a letter Galileo wrote in 1611.*

► *Galileo's paintings of the Moon. Galileo was the first to turn his telescope on the heavens and then to publish the results, which he did in "The Sidereal Messenger" (1610). Apart from mountains and "seas" on the Moon, he also observed the Milky Way comprised a myriad of unknown stars, and discovered four satellites of Jupiter (the "Medici planets") and, later, the phases of Venus.*

▲ *Both of these telescopes made by Galileo have a narrow field of view of 17 arc minutes. The larger telescope has a magnification of 14, the smaller of 20.*

Observation of the phases of Venus proved that it traveled round the Sun

Among the new worlds revealed by his telescope, Galileo noticed three next to the planet Jupiter. Over some weeks he followed them, soon adding a fourth, and saw that they now disappeared, then reappeared, always accompanying Jupiter on its path across the sky. Until then it had been held against the Copernican theory that, if the Earth was supposed to be a planet, why should it alone have a moon going round it? Now Jupiter evidently had four. These discoveries were made at the beginning of 1610. In the autumn, the telescope showed him regular phases, like those of the Moon, in the planet Venus. This could only mean that Venus must be traveling round the Sun.

All this astounded Europe. Kepler relates how his friend the courtier, Wackher, called to him from his carriage with the news, within a week of Galileo's publication of his report, *The Sidereal Messenger* (1610). The leading astronomer in Rome, the German Jesuit Christopher Clavius (1537-1612), repeated Galileo's observations, and confirmed them. His was the deciding voice, for many, including some prominent astronomers, insisted they could see nothing, and that Galileo was deceiving himself, or trying to deceive others. Among the doubters was Galileo's friend, the philosopher, Cremonini, who was an uncompromising Aristotelian; he told a mutual friend that "looking through these glasses would make me dizzy". Artists and writers were often Galileo's most enthusiastic supporters.

Now Galileo was invited to be "philosopher and chief mathema-

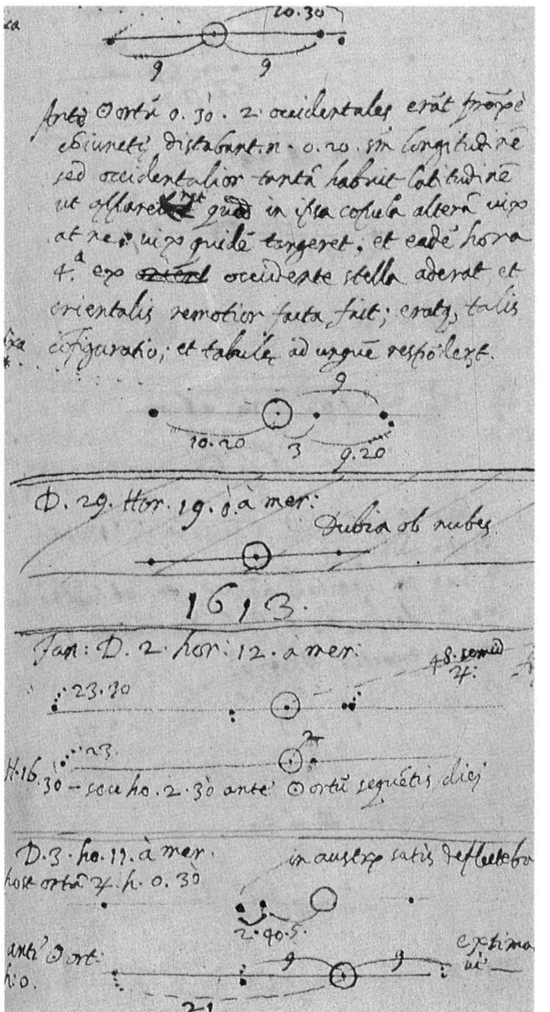

◄ **Galileo's notebook entries for certain nights in 1612 and 1613 relate to the satellites of Jupiter.**

► **Jupiter with Io and Europa. Until Galileo's discovery of four Jovean moons, opponents of the Sun-centered Copernican system could ask: if the Earth is a planet and not the center of the Universe, why should it alone have a moon going round it?**

► **Galileo invented a device called a Jovilabe to calculate the position of Jupiter's satellites. The four largest satellites are represented by the numbered circles.**

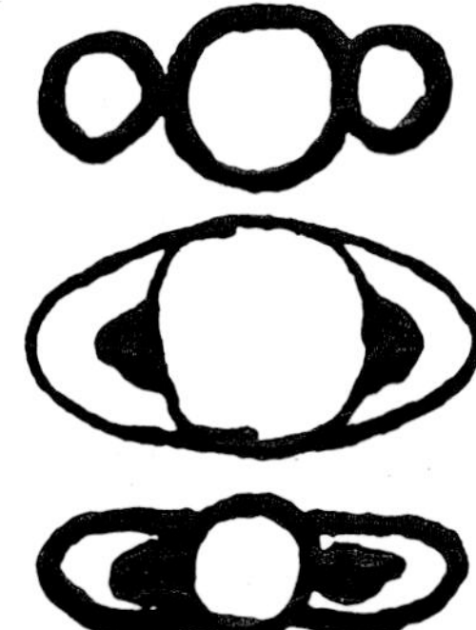

▲ **Galileo was mystified by Saturn's "triple nature" seen in this sketch of 1612. The true nature of the rings, which his telescope was not powerful enough to detect, was revealed in 1667.**

tician" of the Grand Duke of Tuscany, Cosimo II de' Medici. In those days, a place at court was reckoned to be better than any university chair. He visited Rome in near triumph, and was elected to the coterie of science lovers who called themselves the Academy of the Lynxes (◀ page 35).

Others, too, thought they could find new stars in the sky and win fame like Galileo. In 1612, a book came out which described markings or spots on the Sun. The author took them to be satellites; the Sun was consequently a planet with moons like Jupiter, only, in view of the Sun's importance, it had far more of them. Galileo had already observed these sun spots. In some anger he wrote a series of letters about them, in which he presented reasons why they must be on, or at least very close to the Sun, not independent satellites. The book Galileo denounced had appeared under a *nom de plume*; it was, in fact, by Christoph Scheiner (1579-1650), like Clavius a German Jesuit astronomer. In him, Galileo had made a dangerous enemy.

Astrology

▲ *Astrologers cast a horoscope as a child is born (1587).*

Astrology before astronomy

To most people in the Renaissance, astronomy was just the taxing business of facts and figures that underlay the exciting applied science of astrology. In a world where our Earth occupies the central position, the regular movements of Sun, Moon and other planets (as they would have put it) had to have significance for human life. The Sun does clearly govern our whole existence, summer and winter, day and night. The Moon, too, could be imagined to affect all monthly cycles. Why, then, should not all planets have an influence?

In their tracks across the sky, the planets do not diverge more than a few degrees from the Sun's apparent annual path against the background of the stars, the ecliptic. The region of the sky where they are all to be found thus forms a belt (the Zodiac) divided into 12 sections (the signs). These signs were named for the constellations found in each when these divisions were first established.

Between the 7 planets and the 12 signs, almost every eventuality could be accounted for as their intertwined influences mysteriously affected both the behavior of humans and the vagaries of the natural environment, from weather to volcanic eruptions. The various positions, relative to each other, of any two planets could be supposed to produce additional variations.

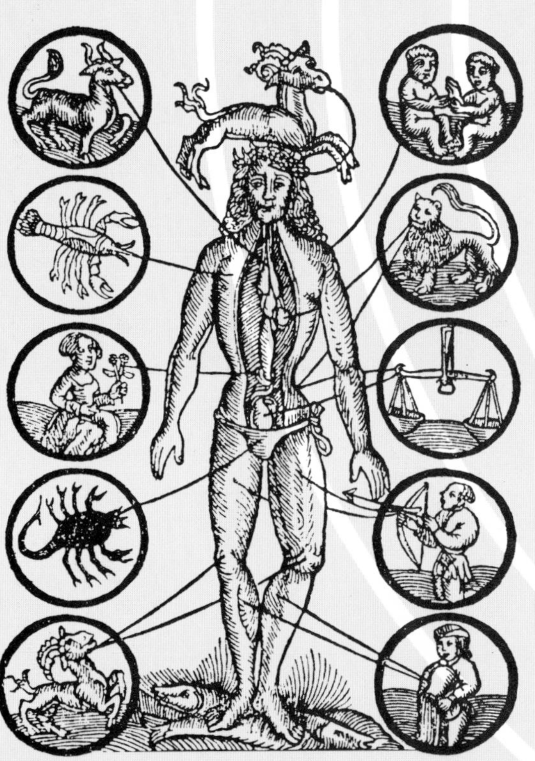

Horoscopes

Forecasts, or horoscopes, were based on the planetary positions at the birth of an individual. Ideally, the moment of conception would be better, but that could not always be fixed with any great precision. Wise parents ensured that their baby's horoscope was at least drawn up by an expert at the instant of birth. Apart from individual fates, the planets were supposed to rule all the different aspects of nature, bringing forth all the general events. Metals, seas, ages of man, days of the week – each had a planet to govern it (that is why they come in sevens). All parts of the human body had a corresponding planet and/or sign, so that any ailment could be interpreted according to the position of celestial bodies governing the part affected, and the course of the disease predicted. Doctors learnt from diagrams that linked each organ to its celestial lord; many such pictures have survived. Human personality, too, was classified by astrological exposition. Our language has retained words like "jovial" and "mercurial".

◄ ***The signs of the zodiac and the parts of the body which they govern. The influence of astrology on diagnosis and treatment was still important in Galileo's day.***

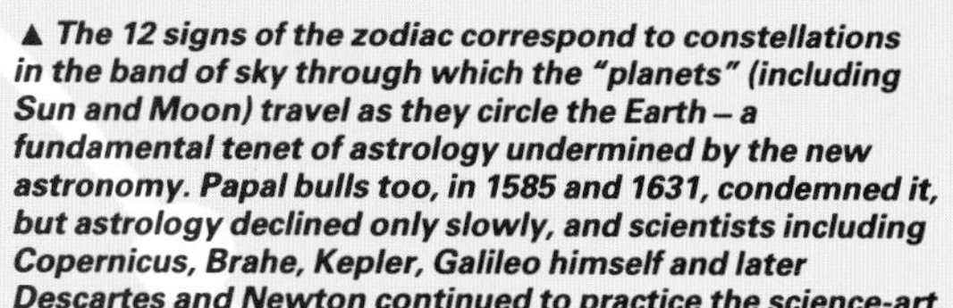

Astrology provided a comforting procedure that offered to show what the future held in store, and was based on accurate, mathematical plotting of the skies. But the whole system was devised in the belief that the planets go round the Earth they influence. What if in fact the Earth also was a planet, and went round the Sun? Would it still make sense? Tycho Brahe and Kepler (◆ page 35) were both active astrologers and accepted the general principles of the art. Even Galileo drew up a horoscope for his daughter Livia when she was born. But the truth of astrology came increasingly to be questioned.

▲ The 12 signs of the zodiac correspond to constellations in the band of sky through which the "planets" (including Sun and Moon) travel as they circle the Earth – a fundamental tenet of astrology undermined by the new astronomy. Papal bulls too, in 1585 and 1631, condemned it, but astrology declined only slowly, and scientists including Copernicus, Brahe, Kepler, Galileo himself and later Descartes and Newton continued to practice the science-art.

▼ This "Month of August" is from a series planned by the court astrologer of the Duke of Ferara and painted about 1470 by Francesco del Cossa. The casting of horoscopes determined the timing of battles, journeys and treaties, and the launching of any important enterprise. Here the corn goddess Ceres rides in triumph; beneath floats the Virgin, the zodiacal sign entered by the Sun in August.

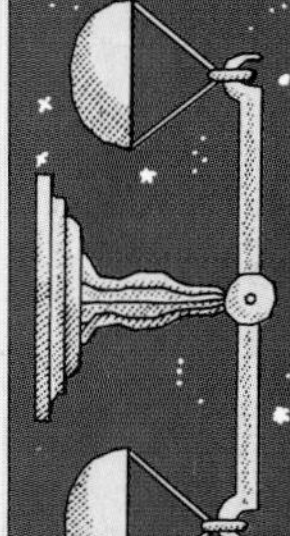

Galileo and the Church

The very success of Galileo began to arouse suspicion and jealousy. Perhaps at Padua he would have been better protected. Since the Medici dynasty in Tuscany was less secure than the Venetian republic, they were more liable to give way to pressure. Cremonini regretted that Galileo had "left the freedom of Padua". Sermons were preached against "Galileists" in Florence; there were complaints that he "defiles the dwelling place of the angels by seeing spots on Sun and Moon, and lessens our hope of Heaven". Perhaps that was the real issue troubling his opponents. For, if Earth was merely a planet and planets so many earths, what was the difference between the Earth and Heaven?

▼ Galileo was tried by the Inquisition in 1633, the year after the "Dialogue" appeared. The pope had granted permission for a work that would favor Aristotle and Ptolemy. Instead, in the everyday idiom of Italian, Galileo demolished the old and argued for the new cosmology. It is said that, after his forced recantation, Galileo muttered (of the Earth), "nevertheless it does indeed move!"

► The title page of the first edition of the "Dialogue on the Two Chief Systems of the World". The figures are Aristotle, Ptolemy and Copernicus, but inside the book it is the ideas of Copernicus which triumph. The crest is that of Galileo's patron Cosimo, Grand Duke of Tuscany, and "linceo" after Galileo's name refers to his membership of the Academy of Lynxes.

Galileo tried to defend himself by maintaining that it had always been permitted to interpret Scripture as allegory. But then he was accused of wishing to explain the Bible in ways other than those of the Roman Catholic church, a dangerous idea when the Protestant Reformation was a major threat. Galileo went to Rome and persuaded the authorities that he was no heretic. But he was bluntly told that Copernicanism was contrary to sound doctrine, and must not be taught or defended in writing.

Back in Florence, Galileo gave up astronomy for a while. Aroused by the appearance of three comets over the winter of 1618-1619, he wrote to criticize the account of them published by another Jesuit astronomer, Grassi. His own explanation was more wrong than Grassi's, in fact, but in the controversy that followed he did get a chance to expound his underlying ideas of the right way to comprehend nature in mathematical and atomistic terms. Encouraged by the acclaim for *The Assayer*, his wittiest book, Galileo hoped he could now return to astronomy and present the case for Copernicanism. Permission was granted, on condition that he put the case for the other side, too. Perhaps he did not see that Pope Urban VIII, who was now personally involved, meant that the book should come down firmly for the old cosmology. Any alternative would have to look dubious, or at best a convenient basis for calculation.

Instead, Galileo wrote a vigorous argumentative book, *Dialogue on the Two Chief Systems of the World* (1632), in the form of discussions between three characters, two of them named for friends, former students of his, who had died young – Sagredo and Salviati. The third, Simplicio, was the spokesman for Aristotle. Their debate is set in Sagredo's house in Venice, where the Florentine Salviati speaks for Galileo. Salviati pulls to pieces Aristotle's conviction that celestial bodies never change, refutes physical arguments against the Earth's mobility, and seeks to prove that the observer cannot perceive motions which he shares. Then he shows how the Sun is the obvious center for all planetary movements. Finally, there is a discussion of the only idea in the *Dialogue* which Galileo had not before made public – his theory of the tides. He portrays these as movements of the sea influenced by the different motions of the Earth, like water sloshing about inside a boat.

If Galileo thought that his book would clinch the debate, he was mistaken. It was rash of him to make his assault on so powerful traditional belief, and the storm soon broke over his head. Despite his age and illness he was obliged to go to Rome and abjure openly and absolutely the heresy that "the Sun is in the center of the world and immovable, and the Earth is not the center". At first, he thought that would be the end of the matter, but a public recantation was not enough. For a while he recuperated with the sympathetic archbishop of Siena, but that was cut short and he was condemned to spend the rest of his life restricted to his country house at Arcetri. Even a request to go into Florence for medical treatment the next year was refused, with the warning that "he should refrain from handing in supplications or he will be taken back to the jail of the Holy Office." Later the Pope relented, and allowed him these medical visits.

Then, in 1634, his daughter Virginia died; she was only 33. Galileo was deeply scarred by his loss. Sick and in despair he wrote to a friend that he was suffering from "a sadness and immense melancholy, complete loss of appetite, I am hateful to myself and cannot sleep, continually I hear my darling little daughter calling me..."

The Church in a New Age

Reformation and the Council of Trent

Ever since Luther nailed up his theses attacking the theology and conduct of the Church, at Wittenberg, Germany, in 1517, western Europe had been torn by the conflict between Protestant reformers and a Catholic Church at first on the defensive. In Protestant lands, the triumph of reform subordinated the Church to the secular state and suppressed its monasteries. Wherever the Church was able to withstand these attacks, bitter struggles broke out. The Thirty Years War devastated Germany and drew in France, Sweden, Denmark and Poland, not coming to an end until 1648. In response to the challenge of the Reformation, the Catholic Church reorganized itself completely at the Council of Trent. Almost everything came within the Council's purview; our present calendar was introduced there. The Council (1545-1563) also strictly defined the true doctrine of the Church.

▼ From his dungeon in Naples, the Dominican friar Tommaso Campanella wrote a defense of Galileo against accusations that his scientific theories were contrary to the Bible: this was when Galileo went to Rome to plead his case in 1616. Campanella called for freedom of inquiry and opposed the subordination of science to theology. His tract was eventually published on Protestant territory, in Germany. In the circumstances, it cannot have done Galileo much good. In 1628 Campanella was transferred to the Roman Inquisition, which treated him well, and he was even consulted on astrological questions by Pope Urban.

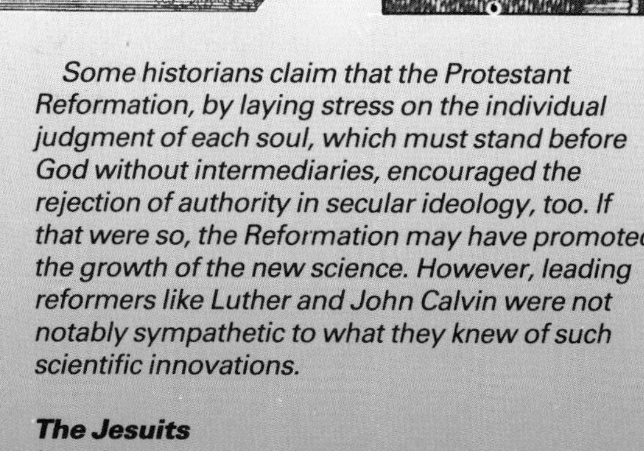

Some historians claim that the Protestant Reformation, by laying stress on the individual judgment of each soul, which must stand before God without intermediaries, encouraged the rejection of authority in secular ideology, too. If that were so, the Reformation may have promoted the growth of the new science. However, leading reformers like Luther and John Calvin were not notably sympathetic to what they knew of such scientific innovations.

The Jesuits

In order to combat the enemies of the Church – with renewed devotion and reasoned argument rather than force – Saint Ignatius Loyola founded the Society of Jesus in 1539. Education was to be the main field of Jesuit activity. Schools and colleges run by the society adopted new ideas (but not Galileo's), and mathematics was prominent in their curriculum. They wished to offer a modern training to equip their pupils both practically as well as spiritually. Sometimes they even sought to replace less committed institutions. Thus, in Madrid the Academy of Mathematics was suppressed, because the Jesuits held they could, and should, teach the same subjects better.

A later generation of Jesuits played an important part in the diffusion of science not only in Europe, but through their missions all over the world. It was they who brought Galileo's telescopic discoveries to China (although they naturally tried to fit them into Brahe's Earth-centered system).

▲ **Jesuit astronomer Christoph Scheiner observes sunspots by projecting the Sun's image onto a screen. As opponents of Gailieo and the new cosmology, the Jesuits are often depicted as enemies of science. But there were good Jesuit astronomers, among them Scheiner himself.**

◄ **The ceremonial costume of a man condemned to burn by the Inquisition. In the 15th century some 2,000 people were burned under Spanish grand inquisitor Torquemada. The Roman Inquisition which tried Galileo was set up in 1542 to combat Protestantism. Although the Inquisition was installed in Venice, the Venetian republic was determined that it should not harass its own subjects, at least not those whom the republic found useful. One of Galileo's friends, the physicist Paolo Sarpi was such a vigorous opponent of papal authority that he was excommunicated. However, his standpoint was so attractive to the republic's government that he was well protected.**

Inquisition and heresy

Like other religions, Christianity has always had its reformers. In the Middle Ages, an institution was set up to investigate potentially dangerous teaching. It was officially called the Holy Office – unofficially, the Inquisition. In Spain, the Inquisition flourished as a means of keeping an eye on the many Jews and Moslems who had been forcibly converted to Christianity. When the Reformation began, the Spanish Inquisition was used against it in territories under Spanish rule, which included the Low Countries and southern Italy. In theory, all Catholic countries possessed branches of the Inquisition, but most made sure that the local branch did not interfere with their own courts.

Bruno and Campanella

Two contemporary Italian thinkers who suffered from the thought control of the day were caught up in the debate over the Sun-centered world. One, Giordano Bruno (1548-1600), was a Dominican friar who conceived a new pantheist religion, which he imagined to be the most ancient religion – that of Egypt – resurrected. He saw that Copernicus's system, in which the fixed stars were at rest, could imply that the universe had no boundaries. In that case, might it not be infinite? Should not a Creator of infinite power create an infinite Universe, in which all the stars were so many suns, with their own planets? As he fled from Italy, and traveled through Switzerland, France, England, Germany and Bohemia, his new religion outraged all parties. Returning to Italy in 1591, Bruno was arrested in Venice, extradited to Rome as a renegade friar, and after years of imprisonment burnt at the stake in 1600. He was put to death – as a dreadful heretic, not for supporting Copernicus – but his case must have prejudiced the papacy against the whole concept of a world with the Sun at its center.

Another Dominican, Tommaso Campanella (1568-1639), had been won over to the call for a new philosophy that would derive from sense evidence only. He studied briefly with Galileo at Padua in the 1590s, was later imprisoned under suspicion of heresy and tortured. On release, he was confined to his friary. There, astrological predictions, with sermons about a new age about to be born, made him the mentor of a revolutionary circle. Betrayed, Campanella spent nearly 30 years in prison in Naples – without ever being tried. He acted as if mad – and was tortured to make him confess he was only pretending. Yet, at the worst, he never ran out of paper, and wrote voluminously, including an account of his communist utopia, "The City of the Sun". Some of his writings were smuggled abroad and published.

Campanella defended Galileo in 1616, and when Galileo's "Dialogue" was published in 1632, he wrote to say what a wonderful book it was. "If we were together in the country for a year", he declared, "great things would be achieved... These innovations of ancient truths, of new worlds, new stars, new systems, new knowledge of nature, are the beginning of a new age."

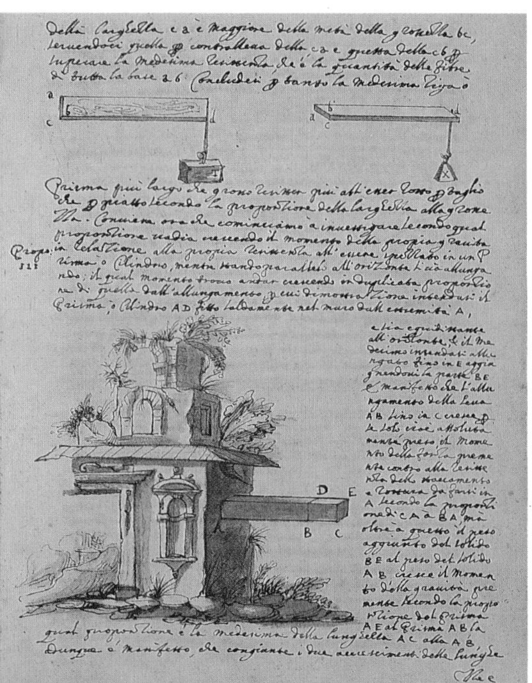

▼ **After he went blind, Galileo continued to carry out scientific investigations. He also kept up a considerable correspondence with other scientists and received distinguished visitors, such as the English republican poet John Milton.**

▲ **In his "Discourses upon Two New Sciences" (1638) Galileo studied resistance to fractures in beams then went on to outline his observations on the vacuum, acoustics, light, the pendulum, and a geometrical presentation of his dynamics.**

Working to the end

Yet Galileo was not silenced. Soon he began at last to write up for publication his work on the new science of accelerated motion, and the motion of projectiles, which followed logically from it. In 1637 he went blind. Nevertheless, he found assistants to write out his material. Galileo boldly gave it the form of the *Dialogue*, a discussion between the same three characters, likewise spread over four "days". These *Discourses upon Two New Sciences* open with studies on the resistance of beams to fracture, then on cohesion and resistance to the formation of a vacuum; there are also telling observations on acoustics, light and the pendulum. He goes on to set out his dynamics in straightforward geometrical form. As publication in Italy was forbidden him, the book had to be smuggled out; the *Discourses upon Two New Sciences* were published in Holland in 1638.

Even after that, Galileo went on working. In 1641 he began to think of the possible application of the pendulum to timekeeping. A drawing (◀ page 37) by Galileo's last pupil, Vincenzio Viviani, depicts a kind of clock mechanism using this principle. He continued to correspond with other scientists, and in his last years, Galileo became something of a tourist attraction, with whom foreign visitors could commiserate, when they could get permission to see him. Among them were the English philosopher Thomas Hobbes and the poet John Milton.

Perhaps only the last of his visitors, Evangelista Torricelli (1608-1647), who as a student had written to Galileo in 1632 that he had been converted by his *Dialogue*, was truly able to learn from him and develop his ideas further, so as to console the old man's last weeks. Torricelli, who invented the barometer in 1643, arrived in October of 1641. Galileo died early in January the following year.

William Harvey 1578-1657

Galileo and Harvey...Europe's leading medical faculty ...Galen's physiology...A physician in London... Experiments on the heart..."The Motion of the Heart" ...The English Civil War...The "Generation of Animals" ...The puzzle of respiration...The microscope confirms Harvey's work...PERSPECTIVE...Centers of medical science...Medical treatment...Science in a divided nation...Health, epidemics and superstition

▲ **Harvey in about 1650. His proof that the blood circulates ushered in a new age of medicine and biology. Harvey's methods included meticulous observation of experimental dissection, and interpretation of quantitative data.**

While Galileo was teaching mathematics at Padua, a young Englishman came to the university to study medicine. His name was William Harvey, and he was to become a founder of modern physiology with his discovery of the circulation of the blood, some 20 years after he left Padua in 1602. Animal physiology considers how animals stay alive and how they take in their energy requirements, so an understanding of the system of energy distribution is an essential first step. It follows that an understanding of the flow of blood round the bod is of basic importance.

Although there is no-reason to suppose that Harvey attended any classes given by Galileo or even spoke to him, Harvey's work as a founder of modern physiology does resemble Galileo's work in physics. Both rejected what they saw as the dead hand of the past. Yet both respected their predecessors in the classical world and maintained that they themselves were closer to the spirit of rational enquiry which is the true heritage of Aristotle than were those who just repeated whatever an ancient authority said. However, the comparison with Galileo reveals differences as well as similarities.

▲ **Galen's famous vivisection of a pig, as portrayed in the 1550 edition of his works. In Harvey's youth, the Greek physician to Roman emperors was still the great authority in anatomy and physiology. Harvey followed Galen in using vivisection, but effectively demolished Galenic ideas on the role and activity of the blood.**

A revolution in life sciences

The Scientific Revolution concentrated on physics, and the great debate over rival world systems in astronomy. But great things were done in the life sciences too. The biology, like the physics, of the Middle Ages tended to take too much ancient learning on trust. By the 16th century the medical schools of the European universities provided the nearest thing to a scientific education. Anatomists like Andreas Vesalius and Realdo Colombo reformed teaching methods to ensure that their students saw for themselves all the organs of the human body. However, they were disinclined to tamper with the physiology of the ancient Greeks.

William Harvey was dissatisfied with current explanations of the function of heart and blood. He adopted Aristotle's view that the operations of organs in life are to be inferred from their forms; and like Aristotle he hoped to understand human physiology by comparing a large number of different animal species. Through his discovery of the circulation of the blood Harvey demonstrated how experiment in controlled circumstances could trace the body's most active component, blood, through the motion of its most rapidly moving member, the heart. The discovery opened up research into the human vascorespiratory system. But an understanding of the chemical exchanges as blood travels round the body required knowledge of the biochemical role of oxygen, and so awaited the interest of Antoine Lavoisier.

William Harvey – His Life, Work and Times

Harvey and Galileo

Harvey founded his work on the Aristotelian assumption that the shape, structure and size of each organ of the animal body indicated its purpose and function in the whole system. These principles appeared to make better sense in the life sciences (where they were not abandoned until the time of Darwin, ➧ page 149) and Harvey was much closer to Aristotle than Galileo was. Indeed, Harvey once told a young enquirer to read Aristotle first and foremost: modern writers on medicine, he said, were mere "shit-breeches" by comparison.

Galileo and Harvey alike insisted that observations made by our senses carried more weight than deductions drawn from theoretical assumptions, although both were guided more than they realized by their own assumptions. Both often started by noticing where current explanations seemed confused, or inconsistent. They tried to see and describe just what was going on, then think out a clear and testable hypothesis to account for what they believed they had observed. Finally, they predicted what would follow should their hypothesis prove to be true. Both made use of mathematical argument, and both thought in terms of time – how *long* does an action take? However, where Galileo stressed the logical certainty of geometrical reasoning, being convinced that there were fundamental mathematical laws of nature, Harvey, a more conservative thinker, often used traditional, qualitative language. He represents the observational, one-step-at-a-time side of the Scientific Revolution of the 16th and 17th centuries.

A student at Padua

Harvey's family were Kentish farmers, who had done well enough to move into commerce; his father was a Folkestone merchant, a man of means who became mayor of the town. William was born there on 1 April 1578, the eldest of a large family. (All his six brothers survived to adult life, an unusual occurrence in those times, and stuck together, eventually moving to London to become prosperous merchants trading with the Levant.) At the age of ten, William entered the King's School, Canterbury, and from there went to Cambridge University with a medical scholarship in 1593. He entered Caius College, which had been reorganized not long before and was the best place in England for medical studies. For really first-class medical training, however, he would have to go abroad. Of all the medical faculties in Europe, Padua was then outstanding with a well-established tradition of original research in anatomy. So there Harvey went at the age of 21. At Padua he became "Councillor of the English Nation" – the university was nominally run by the student organization, whose council included delegates of all the main nationalities among the students.

Among his teachers the most important was Girolamo Fabrizi, or Hieronymus Fabricius of Aquapendente (1537-1619), whose work on comparative anatomy took up this branch of science at the point where Aristotle had left off. From 1600 onward, Fabricius produced a number of books which derived from his lectures and research over many years. They represented his attempt to advance physiology and zoology on traditional foundations. It was while Harvey was attending his classes and carrying out dissections in the anatomy theater that Fabricius must have been writing his pamphlet on the valves – or, as he called them, the "sluice-gates" – in the veins. Fabricius had discovered most of these valves, but concluded wrongly that they regulated the outward flow of venous blood, to prevent the limbs receiving more than they could absorb at any one time.

Historical Background

■ **1558-1603** Under rule of Elizabeth I, Protestantism reinstated in England

■ **1577-1580** Francis Dra
became first Englishm
circumnavigate the w

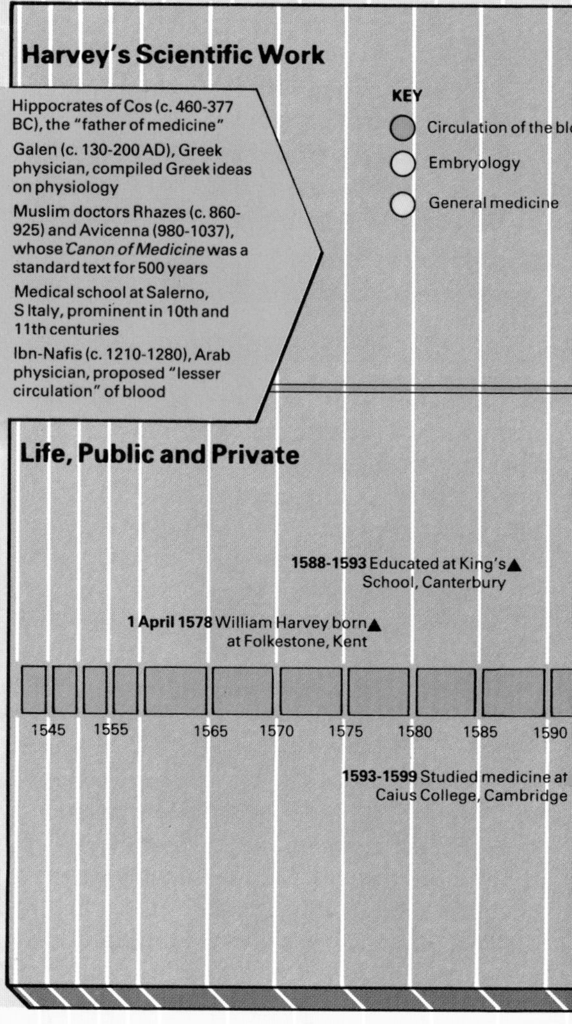

Scientific Background

1559 Matteo Realdo Colombo (1516-1559), Italian anatomist, in *On Anatomy* claimed blood passes from one side of the heart to the other only through the lungs

1543 Andreas Vesalius (1514-1564), Flemish anatomist, published *On the Structure of the Human Body*; reformed teaching and practice of anatomy

1561 Gabriel Fallopius (1523-1562), anatomy professor at Padua, described function of Fallopian tubes

1493-1541 P. A. Paracelsus, Swiss physician and alchemist; prefaced his lectures at Basle by publicly burning the works of Avicenna and Galen

1540 1550 1560 1565 1570 1575 1580 1585 159

1620 Pilgrim Fathers landed in North America

1625, 1630 Outbreaks of plague in England

1649 Ex-king Charles I of England (Harvey's patient) executed; Oliver Cromwell set up republican Commonwealth, invaded and subdued Ireland and (1650) Scotland

1665, 1666 Great Plague and Great Fire of London

...y Edict of Nantes, at end of ...n wars of religion, ...enots and Catholics given ...rights

1642-1646, 1647-1648 English Civil Wars, or Puritan Revolution

1660 English monarchy restored

From 1610 Worked on the heart and on circulation of blood, in a long series of meticulously observed dissections and experiments, on a variety of animals

1628 Published *The Motion of the Heart and Blood in Animals* (printed at Frankfurt)

From 1628 For most of the rest of his life, worked on reproduction, embryology; dissected animals from royal deer parks of Windsor and Hampton; developed the epigenetic approach

1651 Published his work on embryology, *The Generation of Animals*

3 June 1657 Died at Roehampton, Surrey, near London

1629-1633 Traveled in Europe, chiefly with James Stewart, later Duke of Richmond

1647 Returned to London to live with his brothers

1604 Married Elizabeth Browne

1625 Appointed physician to King Charles I

1642 Left London with King Charles I, with whom he stayed during the Civil War

...**602** Returned to London

1618 Appointed physician to King James I of England

1605 1610 1615 1620 1625 1630 1635 1640 1645 1650 1655 1660 1665 1670

...**-1602** At Padua University, ...me time as Galileo

1645 In Oxford, made warden of Merton College

1607 Elected fellow of Royal College of Physicians

1636 Went to Germany and Italy with a diplomatic embassy to Emperor Ferdinand II

1654 Elected president of Royal College of Physicians but refused the honor

1609-1648 Was physician at St Bartholomew's Hospital, London

1610-1623 Lectured on surgery for Royal College of Physicians

1674-1676 Anton van Leeuwenhoek (1632-1723), Dutch microscopist, described bacteria and protozoa; also red blood corpuscles, capillaries, sperm cells

1622 Gaspare Aselli (1581-1626), Italian anatomist, discovered that the lacteals carry products of digestion to body

1648 In his *Ortus medicinae*, J. B. van Helmont (1579-1644), Flemish chemist and physician, attributed physiological changes to chemical causes

1665 Robert Hooke (1635-1703), English scientist, published his *Micrographia*, microscopic observations using a compound microscope

1621 Hieronymus (Girolamo) Fabricius (1537-1619), Italian physician and teacher of Harvey, published work on embryology

1661 Marcello Malpighi (1628-1694), Italian physician and biologist, using microscope, discovered capillaries in lung of frog

1620 Francis Bacon (1561-1626), English philosopher (and patient of Harvey), proposed inductive scientific method, in place of *a priori* reasoning

1637 Réné Descartes (1596-1650), French philosopher, used circulation of blood as the major element in his mechanistic treatment of physiology

1661 *Sceptical Chymist* published by Robert Boyle (1627-1691), Irish-born natural philosopher; his 1684-1691 medical books stressed medicine as a science

1780s Antoine Lavoisier (1743-1794), French chemist, demonstrated the task that oxygen performs in the body

1828-1837 Theory of epigenesis rehabilitated by K. E. von Baer (1792-1876), Estonian biologist, a founder of modern embryology

1605 1610 1615 1620 1625 1630 1635 1640

Traditional medicine and teaching methods had already been challenged in the 15th century by Paduan professors such as Vesalius and Colombo

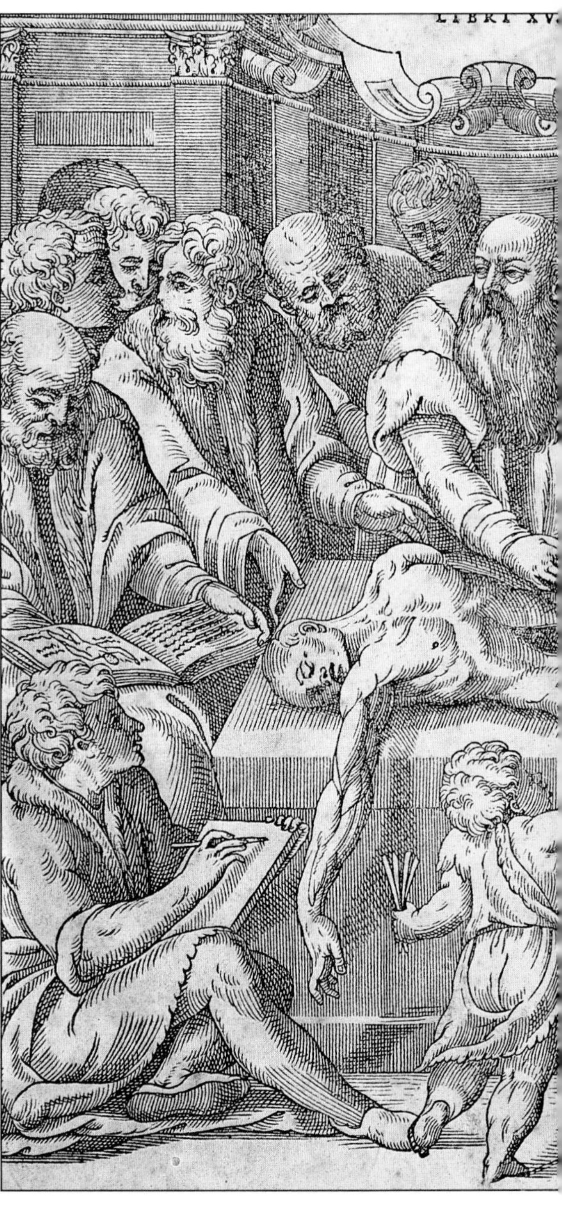

Galen's views are challenged

For all the improvements that had been made in the way anatomy was taught, the physiology which explained how the dead body on the table had functioned in life was effectively still that of Galen. This Greek, who lived in the 2nd century AD, was physician to the emperors and gladiators of ancient Rome. By seeming to sum up centuries of research and controversy, his work came to be regarded as the encyclopedia of anatomy and physiology, the only base for sound and exact medical instruction. He was the authority on the subject – as Aristotle was for science in general.

To Galen, the difference in color between blood in the veins and blood in the arteries indicated a real qualitative difference; they were of two distinct kinds. Food was converted in the stomach into a fluid the body could use, called chyle, and then to blood in the liver. So the liver was the beginning and fountainhead of all the veins. Through these the whole body was nourished with blood, like irrigation channels feeding fields. The heart received some of this blood, which was then transmitted further to the lungs.

Some blood passed from the right side of the heart to the left, where the aorta (the main trunk of the arterial system) arises. Galen supposed that, since the heart looks like a single organ, some blood must seep through pores in the septum (the partition between the two sides). There it supposedly was refined by the heart's heat, and mixed with air that came down the pulmonary vein from the lungs. This revitalized blood formed a "vital spirit" which bestowed its vitality on the rest of the body through the arteries. In the brain, some might be still further refined into "animal spirits", which ran through the nerves and tendons. The lungs, then, served to supply air to the heart, as an ingredient of arterial blood, and also ventilated it, keeping it at the right temperature. Some kind of exhaust was produced by this process, "sooty vapors", which passed along the vessels leading from heart to lungs, and so were discharged into the air.

▼ *An early, and quite realistic, printed depiction of the viscera (heart, lungs, liver, stomach, spleen and bladder are labeled) in Reisch's "Pearls of Wisdom" (Basle, 1508).*

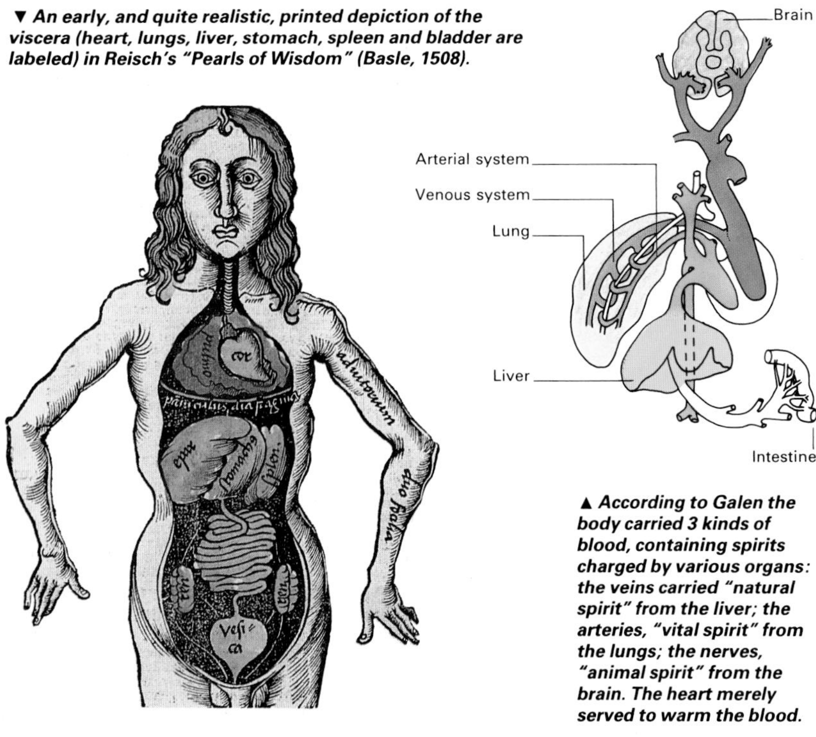

▲ *According to Galen the body carried 3 kinds of blood, containing spirits charged by various organs: the veins carried "natural spirit" from the liver; the arteries, "vital spirit" from the lungs; the nerves, "animal spirit" from the brain. The heart merely served to warm the blood.*

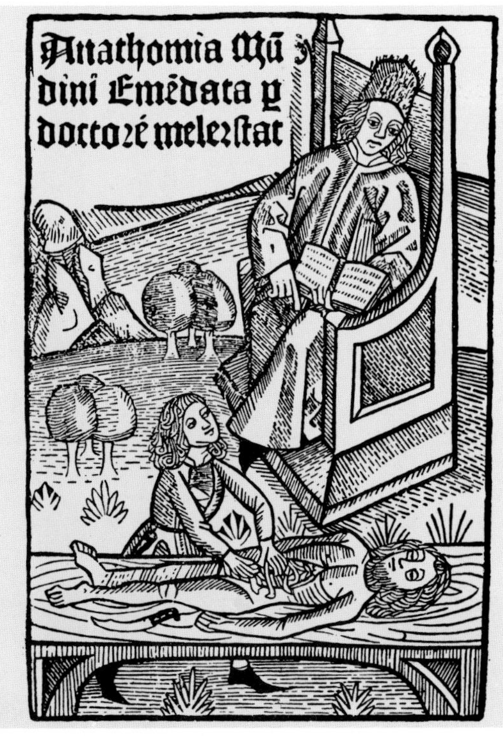

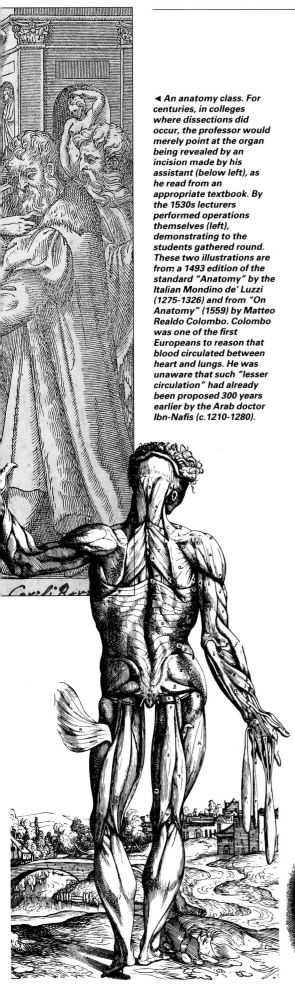

◄ An anatomy class. For centuries, in colleges where dissections did occur, the professor would merely point at the organ being revealed by an incision made by his assistant (below left), as he read from an appropriate textbook. By the 1530s lecturers performed operations themselves (left), demonstrating to the students gathered round. These two illustrations are from a 1493 edition of the standard "Anatomy" by the Italian Mondino de' Luzzi (1275-1326) and from "On Anatomy" (1559) by Matteo Realdo Colombo. Colombo was one of the first Europeans to reason that blood circulated between heart and lungs. He was unaware that such "lesser circulation" had already been proposed 300 years earlier by the Arab doctor Ibn-Nafis (c.1210-1280).

Perhaps the warmth of the expired breath might be the source of the "sooty vapors" idea; indeed, in those days before toothpaste, perhaps the sootiness was not a figment of the imagination. Heart, aorta and all the arteries were also supposed to have in their sturdy coats a "pulsific power", that enabled them to expand. As they did so, they drew into themselves the blood from adjacent areas.

This model of the relationship between blood, heart and lungs had already been challenged by the time Harvey went to Padua. The 16th-century anatomists had been worried by their inability to find any pores, or passages in the septum between the two sides of the heart. Some began to suspect there might well be another route, absurd as it seemed, by which blood entered the heart, left it and then returned again. If air came from the lungs to ventilate the heart, or vapors of exhaust moved from heart to lungs, then the pulmonary vein linking heart to lungs ought to contain them. So, in a living animal, when that vein is cut, air ought to escape. However, when so cut, the vein proved to be full of blood. Could not blood then be flowing from lungs to heart? And perhaps it also traveled by the pulmonary artery from the other side of the heart, to the lungs?

Among the anatomists who held to this opinion, the most important was Matteo Realdo Colombo (1516-1559), who had been professor of anatomy and surgery at Padua, but subsequently moved to Pisa and then to Rome. His idea was not widely accepted, perhaps because he had been a mere surgeon and held in low regard by "proper" doctors, and perhaps because he quarreled with his predecessor, Vesalius, as well as his successor, Fallopius. Nevertheless, the theory had been published in Colombo's textbook of anatomy of 1559, even if it had not become part of normal instruction. Harvey certainly knew of the theory, and knew, too, that Colombo had proposed that the active phase of the heart's movement is not its dilation (diastole), but its contraction, when it expels the blood it contains into the lungs and through the aorta into the rest of the body.

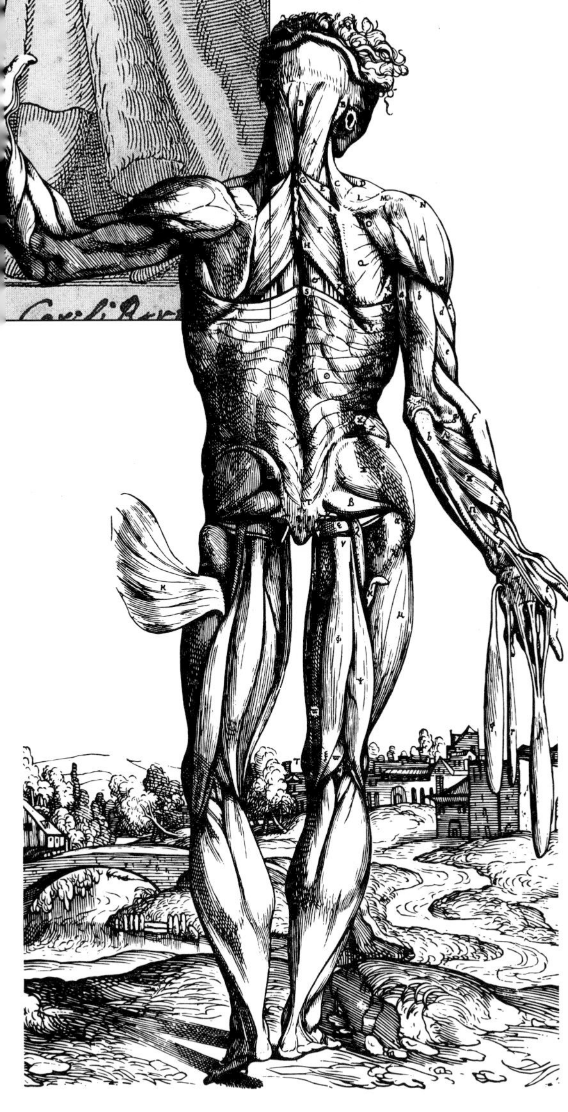

◄▼► Outstanding among the innovators was Andreas Vesalius (1514-1564), who was professor at Padua. Vesalius carried out some unprecedentedly scrupulous dissections and used the latest in artistic techniques and printing for the more than 200 woodcuts in his "On the Structure of the Human Body" (1543). Originally himself a Galenist, Vesalius became a leading figure in the revolt against Galen's teachings.

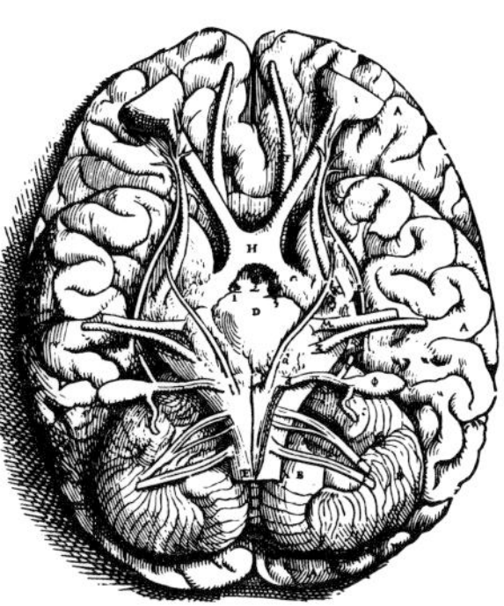

Medical Schools and Hospitals

The medical schools of the Renaissance

*The great medical schools of a few universities already dominated the practice of medicine in the later Middle Ages. Of these, the old rivals, Padua and Bologna in Italy and Montpellier in southern France, were reckoned to be the best. Philosophy was the basis of their teaching, for physicians were keen to be treated as scholars, not mere tradesmen. Often they studied mathematics and astronomy, too, in order to become astrologers. Almost all accepted that the planets and zodiacal signs affected the diseases from which each organ might suffer. Like philosophy, medicine and anatomy were taught in a traditional manner and depended on the exposition of standard texts of ancient and medieval learning – what the ancients knew was always best. The Renaissance revival actually emphasized this by providing a more detailed knowledge of ancient authors on medicine, as good texts of Hippocrates and Galen (**♦ pages 12, 49**) were printed and medieval Arabic additions were pruned away – often unwisely. So by 1550, although doctors knew better what their great guides had said, they did not necessarily know more about the workings of the human body.*

The teaching of anatomy

Romans and Moslems had shared a revulsion to touching dead bodies, let alone cutting them up. Even the ancient Greeks had only carried out dissection of human bodies occasionally (chiefly in Egypt under the Ptolemies, where the technology of mummification may have eased the way). Galen, living under Roman rule, probably never dissected any human cadavers. In medieval Europe, dissection began as postmortems, when the cause of death was uncertain but poison was suspected. Although medical faculties might officially oblige students to learn practical anatomy through watching dissections as a feature of their course, these regulations were not always observed.

*From about 1530 onward, anatomy teaching was reformed, and properly integrated into medical studies. The lecturers began to perform operations for themselves, demonstrating to their students who stood around them, asking them to look and, if necessary, feel for themselves. Andreas Vesalius, professor at Padua, was the most outstanding of the innovators (**♦ pages 52-53**). As they proceeded anatomists gradually came to realize that Galen's data – taken from dogs, pigs and apes – were often unsatisfactory. They would correct him, and even discover new organs, still sometimes named for those who first described them – such as the Fallopian tubes, for the Italian anatomist Gabriel Fallopius (1523-1562).*

Sometimes it was quite hard to obtain corpses to dissect, as families frequently objected. Vesalius tells us of robbing gallows: in one instance, the body, which had been swinging in the wind for weeks, came to pieces in his hands. A medical student at Montpellier in the 1550s recalled how he and his friends had stolen the body of a fellow student from a monastery cloister and sneaked it back through a gap in the town wall. Unluckily, the hole was not quite big enough, so the nose was badly damaged.

▲ *The courtyard of Padua University depicted in the 1601 gazette. The timetable featured Galileo and Girolamo Fabricius (inset) whose lectures in the anatomy theater (top left) Harvey attended. Fabricius studied valves in veins and pioneered embryology.*

► *Inside the crowded Hôtel-Dieu in Paris, in about 1500. Nuns care for the sick and sew the dead into shrouds.*

▲ *Thomas Linacre founded London's College of Physicians.*

In practice, anatomists often used the remains of patients, or friends, or their own kin. Eventually, medical schools persuaded kings to let them have executed criminals for dissection. There were so many offenses for which capital punishment was decreed that the shortage of material ceased.

Colleges of physicians

Where medical schools existed, the faculty provided health services for the town. Elsewhere, university-trained physicians gathered together to form colleges (groups of colleagues, not places of education), which had official support. Their task was to regulate medicine: they examined candidates and forbade practice by the unqualified. In return, town councils might designate some as municipal physicians, with a duty to treat the local poor and advise the council on health matters, especially when plague threatened. These colleges also tried to control the surgeons and apothecaries. They were viewed by the physicians as mere agents, who had not had their academic training. In Italy, medical associations began in the Middle Ages, then slowly spread northward. The English physician Thomas Linacre (?1460-1524) followed this model when he obtained from Henry VIII a charter for the College of Physicians in 1518. This empowered the physicians to allow only doctors with a satisfactory knowledge of Galen to practice in, or near, London, and to destroy drugs for sale which did not come up to their standards. To us, however, the medicine of these official physicians was scarcely less founded on folklore than the remedies of the wise women of the countryside, who in nearly every village dispensed age-old herbal remedies to their fellow peasants.

Hospitals

In a few places, hospitals were already attached to universities in Harvey's day. Some Italian professors had begun to supplement their lectures by going the rounds and discoursing on illness by the patient's bedside. For the most part, however, hospitals were still hospices, places of refuge where the destitute sick could live out their last days in a little comfort. There were hospitals for the regular sick as there were foundling hospitals, old people's almshouses, hospitals for the insane, plague hospitals and leper hospitals. These last were also intended to isolate the disease and save others from infection. Even if clean blankets and food were supplied, no cure was available.

As a rule, all these institutions began as charitable foundations of the church. In Paris, the Hôtel-Dieu (God's Hostel) was literally built onto the wall of the cathedral of Nôtre Dame. Such, too, were the two London hospitals, St Bartholomew's in the city, and St Thomas's across the River Thames. St Bartholomew's was, indeed, obliged to take in every sick person who was judged curable, and the inmates had to gather in the hospital's great hall once a week for the doctor to prescribe for them. When patients did recover, the hospital could take the credit (▸ page 57). Nevertheless, people did not go into hospitals to be healed, and seldom expected to leave alive.

Between giving lectures and work as a hospital physician, Harvey began his experiments, dissections and reasoning on the heart and blood circulation

Return to England: the action of the heart

Harvey went back to England in the summer of 1602 as a doctor after graduating from Padua. He settled in London, where he soon applied for admission to the College of Physicians. In 1604 he made a good marriage with Elizabeth Browne, daughter of King James I's physician. In 1607 he was duly elected fellow of the college, after an examination in which he impressed all his examiners, not least his father-in-law. Dr Browne wrote that "I did never in my life know any man anything near his years that was any way match with him in all points of good learning, but especially his profession of Physician" (but that was in a reference for a job, so perhaps it should not be taken too literally). At all events, Harvey seems to have been an ambitious young man, determined to make his mark. Two years later he was appointed physician to St Bartholomew's, the only hospital within the City of London, where Harvey remained for most of his life. For the College of Physicians he always felt great loyalty, and was keen to strengthen its members' sentiments of comradeship. He served on its council for many years, and donated large sums to provide it with a library. Although the college was for physicians, it tried to retain control over surgeons and apothecaries (chemists). One way was to provide lectures on anatomy for surgeons, given by a leading physician. Harvey was appointed to this post in 1616. He was supposed to keep to an existing stardard textbook, but it is clear from his lecture notes that he introduced new ideas and criticized old ones. These notes show the way in which his mind was working. Harvey must have been an enjoyable lecturer to listen to: the notes reveal that same sly, sardonic humor that others recalled in anecdotes about him.

To begin with, Harvey felt he had to sort out the action and role of the heart. To Aristotle, the heart was the most important organ, center of the blood system and also center of the emotions (which is why we talk of a "broken heart"). Galen, who knew much more anatomy than Aristotle, had demoted the heart. In his view the liver was primary for venous blood, while the seat of all thought was in the brain. He carried out a famous vivisectional experiment on a pig, depicted in editions of Galen's works published in 1541 and 1550. This demonstrated that the voice is controlled by nerves from the brain, in this case the laryngeal nerves. So the power of speech, that essential of human thought, must also be directed by the brain, not the heart.

Harvey was an Aristotelian in making the heart the monarch of the body, but a Galenist in his use of vivisection to see what happened when some particular organ could not function. However, he now introduced a method which neither Aristotle, nor Galen, nor anyone else had conceived. The heartbeat is so rapid that it is very hard to follow what is happening. Harvey thought of using the dying heart, moving more and more slowly, particularly in less warm-blooded animals, like snakes. The complex rapid action would then break down into a succession of simple movements. This idea of simplifying the complex in order to understand what was happening became the key to all future scientific research. As Harvey explained, the firing of a musket seems complicated but we know that it consists of a sequence of simple movements, from pressure on the trigger to striking the flint, which sends a spark into the gunpowder. Now he could trace the sequence for the heart. He saw that the contraction (systole) is the active phase: the arteries do not pulsate independently, but dilate as the blood moves into them. He could also confirm Colombo's finding that blood crosses the heart only by way of the lungs, whose soft and

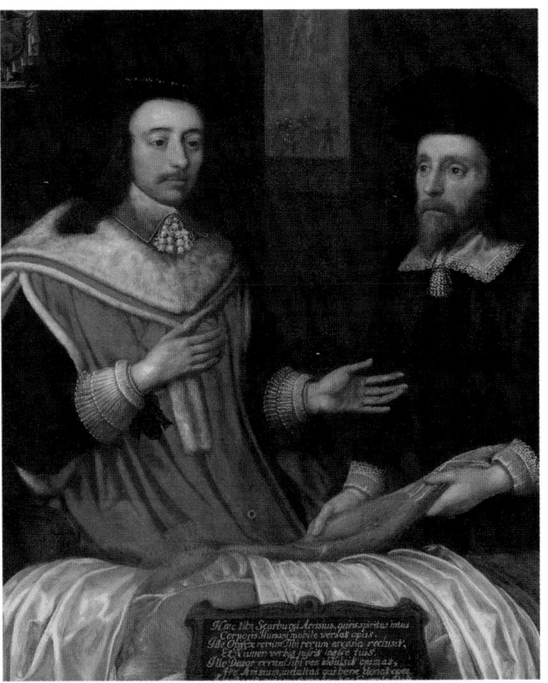

▲ *Dissection of peripheral nerves by Charles Scarburgh, Harvey's successor as lecturer at the Barber-Surgeons, a precursor of the London College of Surgeons.*

► *A page from Harvey's notebooks, in which the physician recorded the results of his dissections of the blood circulation system and also wrote up his lecture notes.*

▲ *The diploma of Doctor of Arts and Medicine granted to Harvey by the University of Padua on 25 April 1602. Harvey had completed some three years' study at Europe's most outstanding center of medical teaching and research. This opening spread depicts, in the top oval, the device chosen by Harvey when elected "Councillor of the English Nation" on the rector's council.*

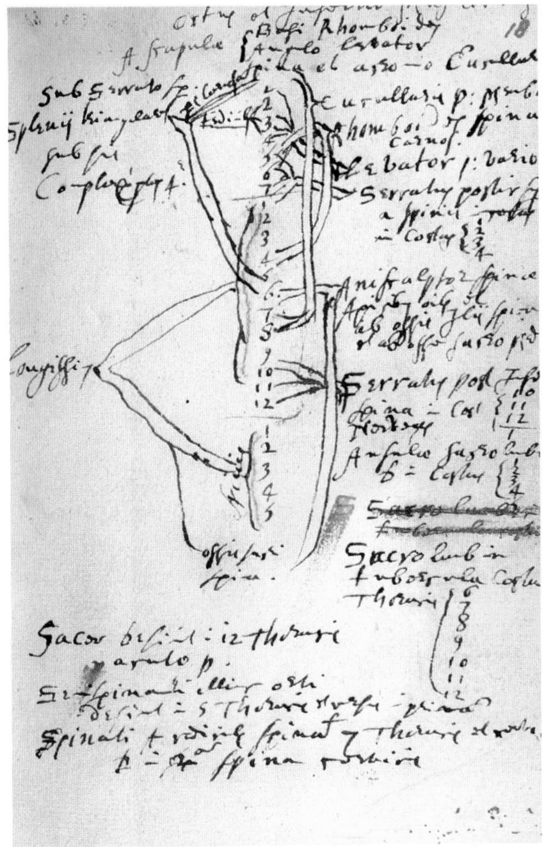

spongy tissue does not offer the resistance presented by the tough impermeable septum of the heart. So blood moves from the heart to the lungs and back again to the heart – whose contraction provides the impulse to drive it into the lungs from one side, and into the aorta from the other. Harvey also realized – as by no means everyone did – that this impulse squeezes out the heart's entire contents.

Blood enters the heart by the largest of the veins, the two venae cavae which converge as they enter. Valves in the heart must prevent blood from moving backward, so there is a one-way flow. On its way outward, the blood enters the biggest and toughest artery. Harvey observed that these valves of the heart, and the large size of these vessels, must have purpose. As a good Aristotelian he accepted that the shape, size and position of organs showed how they fulfilled their tasks. Then he began to consider the system from the time aspect: not just how much blood was ejected at each pulse, but how much would leave the heart in a given period, given the number of heartbeats each minute. Although his measurements look rough and ready to modern eyes, they were good enough to show that more blood passes through the heart in a short space of time than the whole body contains, far more than the intake of food could possibly supply. Toward the end of his life, he told the chemist Robert Boyle (1627-1691) that it was the valves in the veins which guided him to the next phase in his thinking. He always stressed that "so provident a cause as Nature" would not produce organs "without design"..Harvey came to the conclusion that one-way flow cannot last long, unless the flow is closed. There must be a perpetual return, "as it were, in a circule", which would preserve the amount of blood in the body.

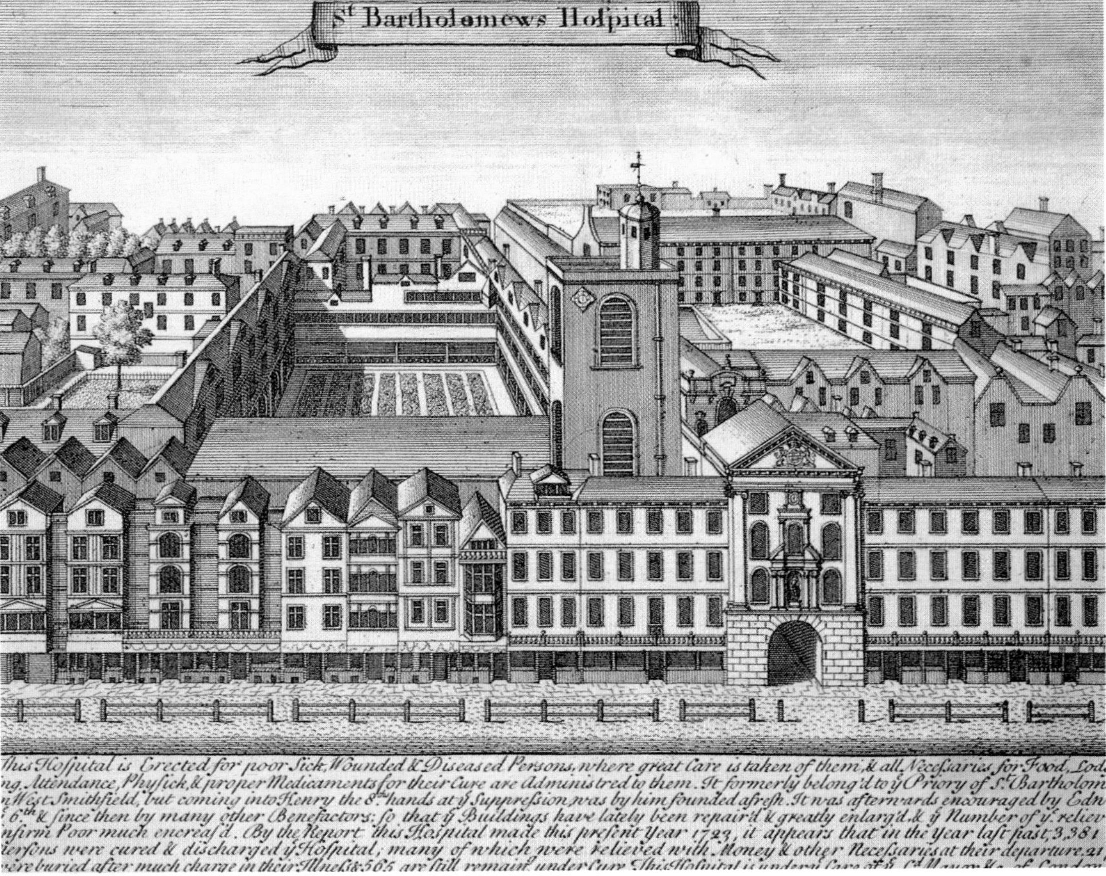

► St Bartholomew's Hospital, London, where Harvey was physician from 1609 to 1648. The print records that in 1723 of 4,163 patients 3,381 were discharged cured.

Kill or Cure?

When William Harvey died in 1657, medical techniques had hardly changed for 2,000 years, and they remained ineffectual. The only useful suggestions the doctor could make were plenty of healthy exercise, rest in illness, and a moderate, sober diet – an approach identified with Hippocrates, the supposed founder of medical science (◆ page 12). What we now recognize as vitamin deficiency diseases were rife among the masses of the population, while gout and stones in the bladder seem to have been common afflictions of the upper classes, to which the Harveys belonged. Harvey himself suffered from gout, and could do little to help himself (he tried sitting on the roof until chilled, and then putting his feet in a pail of water, perhaps imagining that he was giving his circulation a useful shock).

A few ointments were known, which had a soothing effect and reduced pain somewhat. Mostly, the physician tried forcibly to rid the body of superfluous humors (◆ page 65). Bleeding was the favored method: quantities of blood were removed by nicking a vein, or raising a blister (cupping). This must have hastened many a patient's end, as he or she might lose pints of blood in a few days, when already weakened.

Then there were violent purges and laxatives, and enemas. Emetics induced vomiting, regarded as healthy, since patients were throwing up whatever ill humors were causing the trouble. In a culture which knew that cleanliness was, in theory, a good thing, but in practice did not worry too much about it, antiseptics were little used. Even if some drugs were applied to induce drowsiness, they were not employed to reduce pain.

Surgical operations were always desperate measures. Apart from "cutting for the stone" in the bladder, surgeons cut out bullets, or splinters caused by gunshot, and amputated shattered limbs which had become gangrenous. Only when patients were already undergoing such acute pain that even the agony of the operation might not be greater, and death was anyway the sole alternative, would the surgeon's knife and saw be called for.

◄ In the absence of anesthetics, it was a lucky amputee who got himself knocked out by the bearded man with the bandaged fist (left).

▲ To rid the body of an imbalance of humors, bloodletting, often after a bath, was favored, usually by nicking an arm vein (woodcut of 1513).

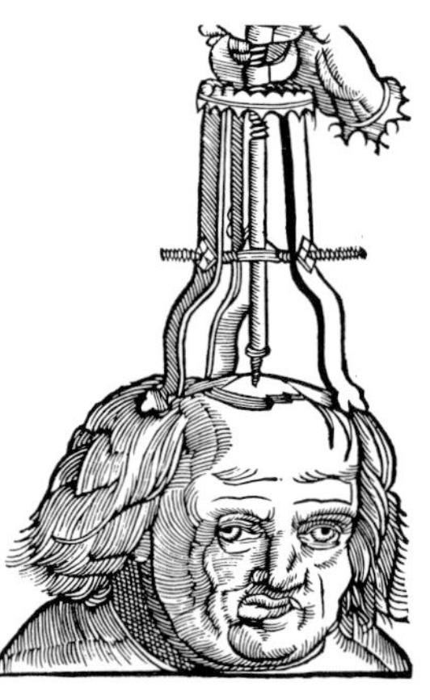

► Trephination, or trepanning, an ancient if drastic practice for relieving pressure within the skull and curing insanity. This woodcut dates from 1517, as does the one above, left.

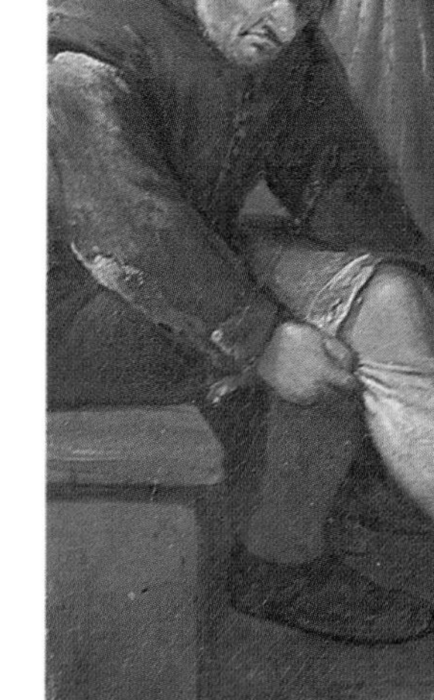

▼ The apothecary approaches, the patient doubts the enema's efficacy, the servant defends her mistress's modesty, and the maid is fed up with cleaning the commode, say the verses that accompany this 17th century French engraving.

In the mid-16th century, the Parisian surgeon Ambroise Paré (1517-1590) introduced an ointment to assuage pain and prevent the onset of gangrene, and techniques to save the patient from bleeding to death after the operation, as well as halting harmful practices. Few comparable improvements were made for a long time after.

In several countries, official pharmacopoeias were printed, once the guilds of apothecaries had been given formal recognition and apothecaries' education had become supervised by the physicians. As most of their drugs required many ingredients and complicated processes of manufacture, perhaps the power of suggestion, based on their elaboration, may have had some effect. For the rest, superstition still played a role. Robert Boyle claimed that Harvey tried successfully an old remedy for a wen on the neck – stroking by the cold hand of a corpse. More often, Harvey was the kind of physician who pursued the best Hippocratic tradition, and would not meddle too much with the ways of nature.

▲ "The Village Barber" by Egbert van Heemskerk (1634-1704). Sprains and chin stubble, cuts and haircuts all await the barber's attention.

► The great Ambroise Paré started as a barber's apprentice, then became a wound-dresser at the Hôtel-Dieu and on the battlefield. Paré reintroduced the ancients' use of ligatures to stop bleeding, and ceased the use of burning oil to treat gunshot wounds, and of cauterizing irons. His "A Universal Surgery" (1561) included many innovations.

Harvey's experiments demonstrating circulation of the blood are easily tested on oneself

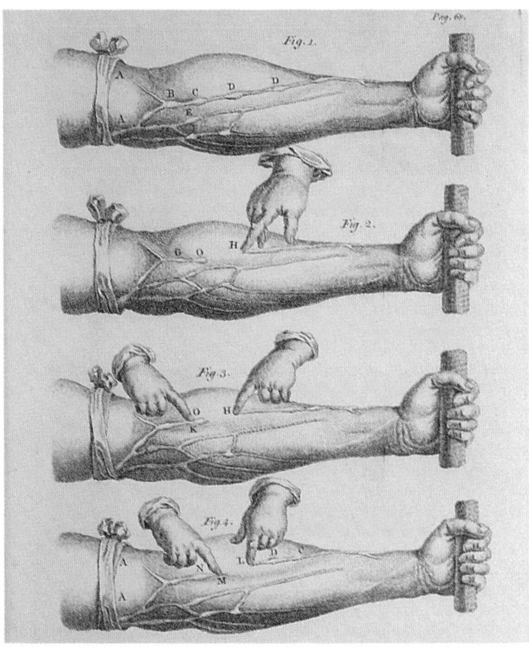

***►* A false-color X-ray of arteries arising from the aorta above the heart to supply the face and neck (carotids) and arms (subclavian arteries).**

***◄* Harvey's demonstration that the blood circulates in one direction round the body. The arm is bound above the elbow, but not too tightly, so that just the veins are affected and stand out. The valves show as swellings. Pressure causes blood to empty out of the vein between the pressure-point and the next valve toward the heart. From "The Motion of the Heart and Blood in Animals" (1628).**

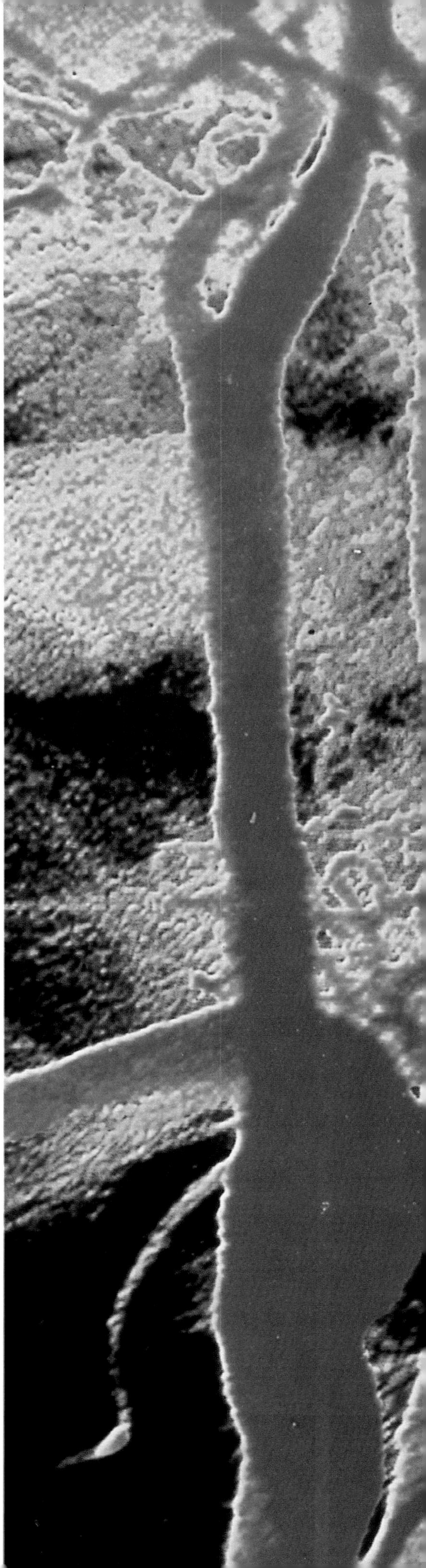

Demonstrations of circulation

Harvey was quite clear that the reality of the circulation of the blood could only be tested by experiment. From the hypothesis of a circulation, three phenomena should follow, breaking the circulation into its three main stages. First, blood must be driven uninterruptedly across the heart from venae cavae to arteries. To check this, he opened up a snake, and squeezed the vena cava so as to block the passage of blood. The heart then became "whiter in color" and smaller in size. It beat slowly until the pressure was removed, when it soon recovered. If the aorta was compressed, the arteries began to swell between the heart and the point of constriction, so the heart became "distended, turned purple to livid in color, and eventually was so suppressed with blood you would believe it about to choke."

The second expectation was that the blood would be driven into the arteries. For that, a simple experiment was sufficient. If you tie a bandage very tightly round your arm, hardly any pulse can be felt at the wrist. There is a throbbing at the bandage, as if – in Harvey's words – the blood was "trying to burst throught an impediment to its passage and re-open the channel." If the bandage is just somewhat loosened, the passage is cleared: the hands look full of blood, and the subject feels the blood run ("pins and needles"). But the veins are still blocked, which is why the hand retains its redder appearance. For Harvey's third supposition was that, to complete the circle, the veins must constantly bring the blood back to the heart from all over the body. Here, too, an experiment is easy to perform. Just tie the arm moderately, so as to block the veins, but not the arteries: the little swellings produced by the valves in the veins stand out, particularly in those who suffer from varicose veins. A finger can push blood back as far as the next valve outward, but no further. Behind the finger there is empty space, blood cannot get through, unless a little escapes under the pressing finger. Inward, toward the heart, blood can be swept along by the finger without much difficulty, and without any such empty section appearing after it. Thus blood moves in an unceasing circuit, caused solely by the heart's pulsation, whose purpose is to drive the blood round the body.

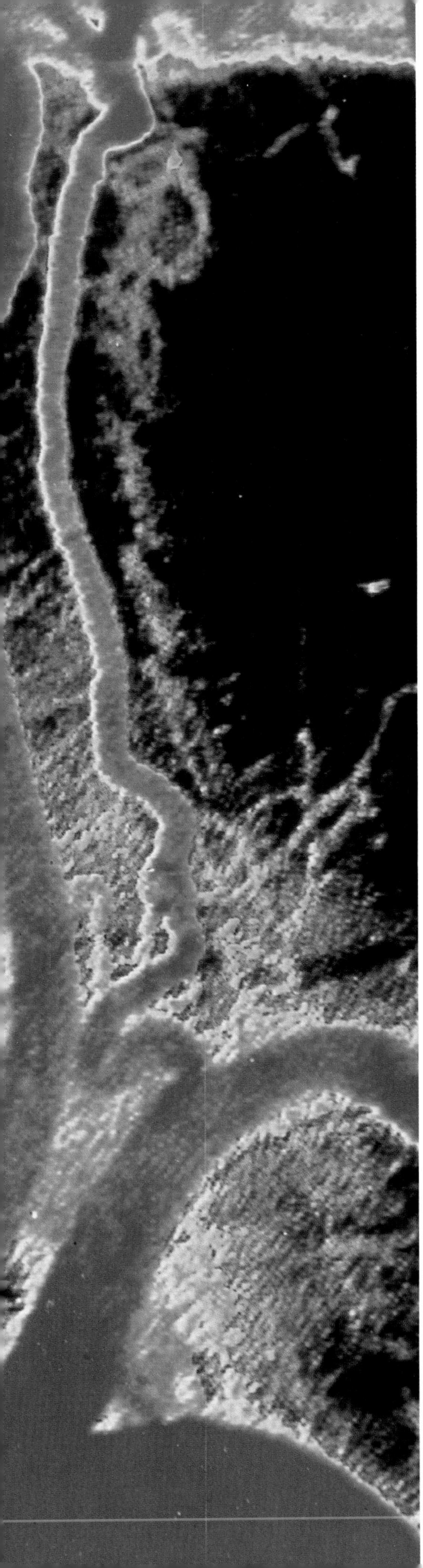

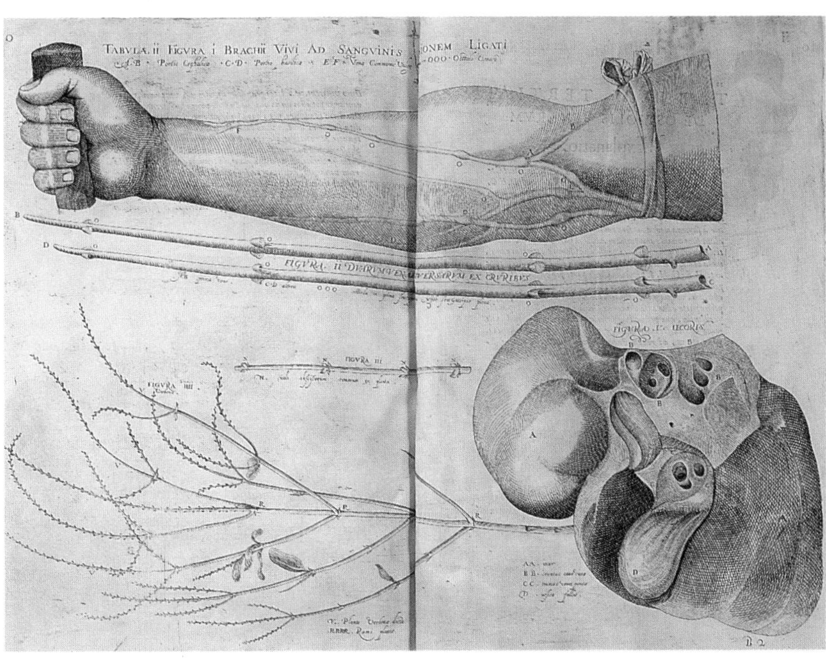

▲ Illustrations from Fabricius's "On the Valves of the Veins", the book in which Harvey's teacher misinterpreted the function of "sluice-gates".

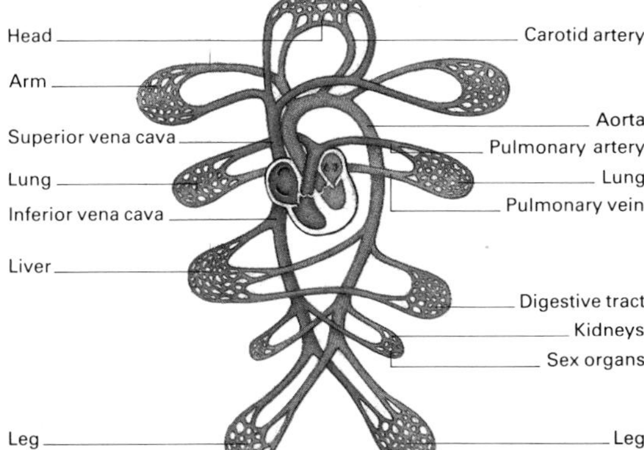

► Blood flows through the pulmonary artery to the lungs. It returns, oxygenated, to the left side of the heart to be pumped in arteries to organs and muscles, then back via capillaries and veins to the heart's right side.

▼ Studying the heart of a frog through his microscope, Malpighi in 1661 discovered the capillaries (below) postulated by Harvey.

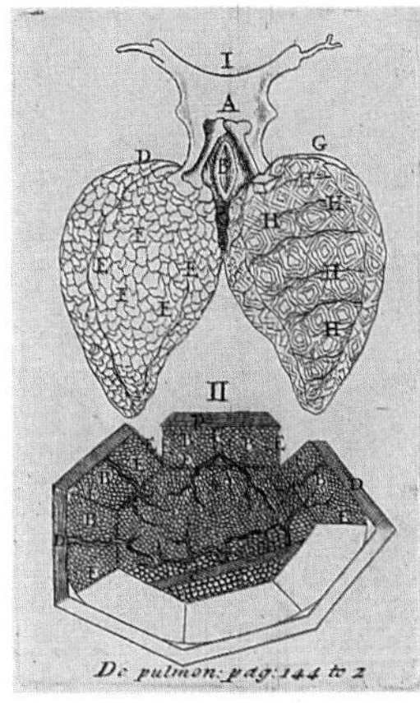

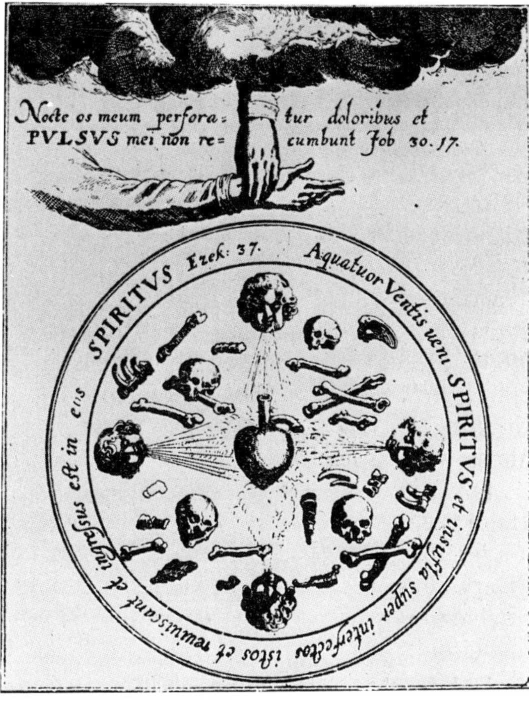

► In his treatise on the pulse, Robert Fludd accepted Harvey's views on the heart, and incorporated them into his own mystical world view.

Puritans and Royalists

Visions of a better world

Harvey survived the English puritan leader, Oliver Cromwell, by a few months. Many have seen a connection between the rise in England of science and of puritan attitudes. They argue that the puritans believed laymen had their own religious calling in the world, one that was justified by worldly success reinvested for use.

The utilitarian aspect of their doctrine led puritans to think that wealth was good so long as it was employed, and increased, for the public good. Immediately before, and during, the Civil War between the parliamentary party and King Charles I which broke out in 1642, many puritan preachers published their visions of a better future when, they prophesied, charity and piety would be manifested by a reformed attitude to material goods. This would be reflected, for instance, in the invention of new and useful medicines. Once the victory of the Parliament was in the offing, many publicists called for a new and fitting respect for trades and crafts, and honor for manual skill. Knowledge had to be applied – spiritual knowledge to personal salvation, and natural knowledge to invention, particularly to medicine to relieve the sufferings of humanity.

In other countries affected by Calvinist ideology, similar attitudes prevailed. This was true of Scotland, and above all in the Netherlands, which was puritan in outlook and the most progressive and democratic society of the time: the Dutch were distrusted and envied in other countries, but everywhere they also had their admirers.

Political allegiance and science

All the same, in England, puritans had no monopoly of enthusiasm for science. Harvey himself was a royalist, and was entrusted with the care of King Charles's young sons at the battle of Edgehill in the Civil War. The young medical men whom he knew at Oxford had all cast in their lot with the king. The College of Physicians remained a center of interest in new ideas throughout the period of puritan rule, but without becoming puritan in political sympathies. Attempts have been made to assess the scientific concern of the two parties by determining the relative proportions of parliamentarians to royalists in the early Royal Society, founded soon after the restoration of King Charles II in 1660. It is clear that scientists and technologists were overwhelmingly drawn from the same upper-middle and professional classes that provided the prominent parliamentarians. The statistics are complicated by the way people could change sides. John Wilkins (1614-1672) and Seth Ward (1617-1689), whom the parliamentary commissioners had appointed respectively to the headship of a college and a professorial chair in Oxford, nevertheless defended the university against major change. In the 1660s, both were very active in the Royal Society. Finally, Wilkins and Ward were able to reconcile themselves with the monarchy: Charles II made both of them bishops.

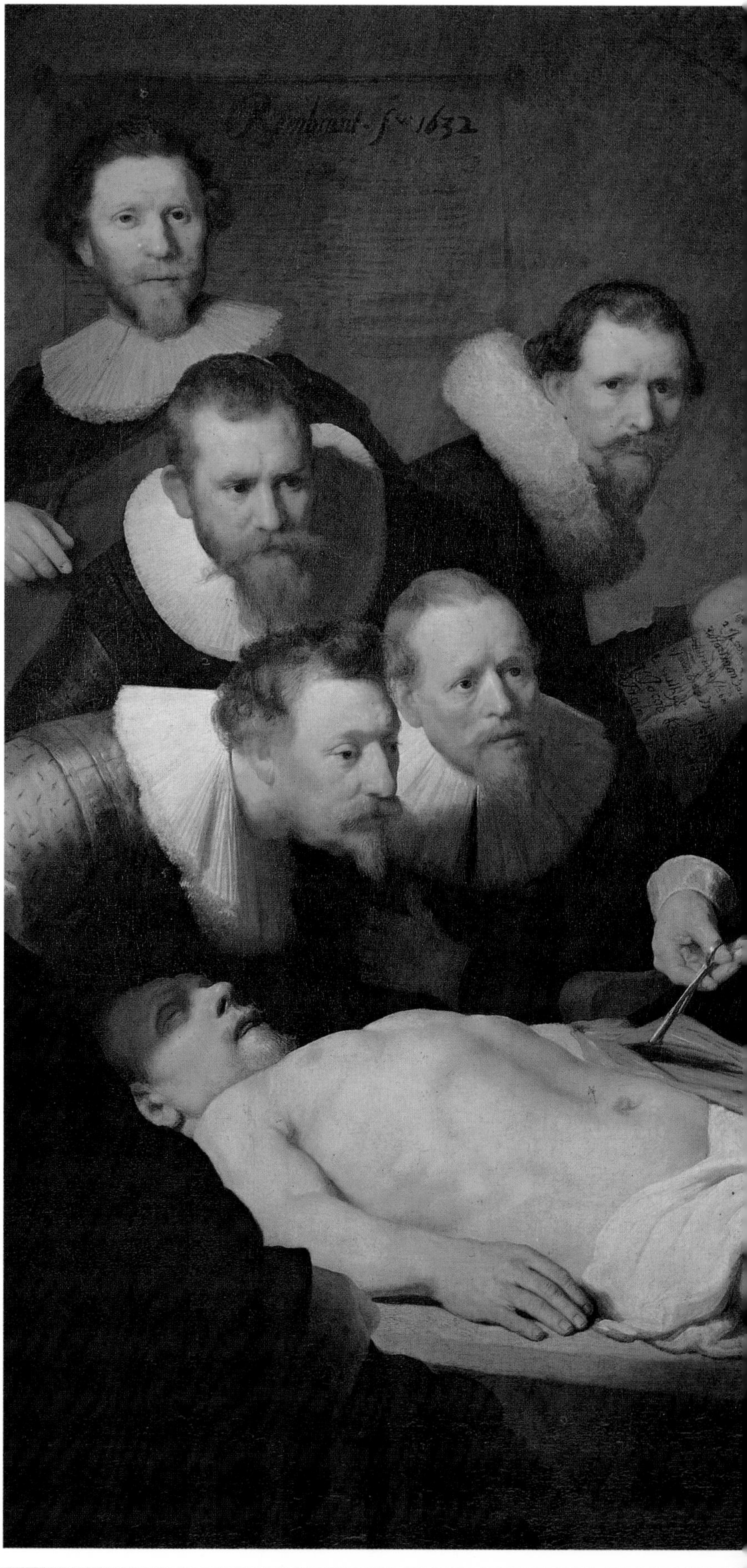

► Rembrandt's "The Anatomy Lesson of Doctor Tulp" was commissioned by the Amsterdam College of Surgeons and painted in 1632. The somber clothes are a reminder that, in the Low Countries especially, puritans sought to apply science to alleviate suffering through medicine.

Harvey publishes his discovery

Harvey's great discovery was published in 1628, in a book entitled *The Motion of the Heart and Blood in Animals*. By the 1620s, Harvey had become an eminent figure on the medical scene. King James I sometimes consulted him, and he was a regular physician to Charles I, who succeeded James in 1625. Harvey dedicated his book on the heart to King Charles, whom he saw as the "heart" of his kingdom. He had many other well-known patients, among them the philosopher Francis Bacon (1561-1626) (though Harvey does not seem to have had a high opinion either of Bacon, or of his approach to science). The 17th-century antiquary John Aubrey's description of Harvey accords with an early portrait painted at about this time: "not tall but of the lowest stature, round faced, olivaster like wainscot in complexion, little eye, round, very black, full of spirit; his hair was black as a raven." Words like "olivaster" (olive-ish) and "wainscot" (the wood paneling then often used to line the main rooms of a house) suggest that Harvey's complexion was quite dark. (Whether this came from a Mediterranean ancestor, we do not know).

Years after, Harvey used to complain that his medical practice fell off after his book came out; some people thought he was too clever and unconventional to be a good doctor. Although he had demonstrated his experiments to his colleagues, some time passed before his views were generally accepted. Indeed, of those who did believe him, many did so for what Harvey would see as the wrong reasons. Robert Fludd (1574-1637) a leading London doctor, decided he could fit the idea of the circulation into his mystical picture of the world. From the opposite side, the French philosopher René Descartes (1596-1650) felt it could be the key to his mechanistic cosmology and biology, in which everything about the animal body, or the human body apart from its immortal soul, worked like a machine. But he supposed that the great heat of the heart makes the blood expand as it enters, so that it overflows into the lungs and arteries, like milk boiling over. That makes the diastole the main phase, which is precisely what Harvey had begun by disproving.

Neither a mechanist in biology, nor a mystic, Harvey held to the view that life itself possesses powers not be found in the rest of nature. Originally, he may have thought that these powers resided in the heart itself, and were given anew to the blood at each circulation. As time went by he shifted somewhat, stressing that the blood is in the body from the start, and that it is the life source – a very ancient idea, which is not just Aristotelian, but biblical. Soul and blood are then the same, not that the blood contains spirit, but that it *is* spirit.

There was another route of investigation which Harvey thought could reveal the function of heart and blood. According to a view then traditional, all creatures stand on a ladder of nature; humans are the most perfect on Earth, then mammals, birds, reptiles, and so on. The lower levels lack some of the organs which the higher ones possess. Thus fish have no lungs, and simpler hearts than mammals. If growth is a process of development, then it is rather like an ascent of the ladder by each individual (a process of perfection as Aristotle and Harvey would have put it). Early stages in development, particularly embryonic stages, can show how the animal operates with some organs either missing, or simplified. In either case, it is as if one factor is removed, and we can observe the consequences of its absence. So, as the liver appears in an embryo only after heart and blood are already active, we deduce that the liver can hardly be the supplier of the veins.

Humors, Plague and Witchcraft

The four humors

During the 17th century, physicians still interpreted the workings of the human frame according to the doctrine of the four humors, or body fluids, inherited from the early days of Greek medicine. (Indeed, the "humors" were not finally abandoned until the advent of modern biochemistry in the 19th century.) If these four fluids passed through the system in just the right proportions, all was well and a person was then in good humor. If any one of them was either in excess or deficient, the imbalance manifested itself as disease.

The four humors were blood, phlegm, yellow bile ("chole"), and black bile ("melaina chole"). As the balance of the four humors affected the psychological state of the individual no less than the physiological one, people could be phlegmatic, sanguine (or cheerful and optimistic), choleric (easily angered), or melancholy – words we have kept long after these medical categories have been rejected. No one ever tried either to calculate how much of each fluid the average body contained, or to isolate any of the fluids. Presumably, they derived originally from simple observation of the excreta, vomit and mucus discharged in illness.

Each humor had its own center in the body: black bile in the spleen, yellow bile in the gallbladder, blood in the liver, and phlegm in the head. Each also had its own season. Thus, phlegm prevailed in the winter, "when people spit and blow from their noses the most phlegmatic mucus", as a popular Hippocratic lecture explained.

The plague

This model may have been plausible for when individuals fell ill, but what if an epidemic swept through a whole region, with thousands of men, women and children falling dead? Plague was widespread: there was an outbreak while Harvey was at Cambridge, and at least two while he was Royal Physician (in 1625 and 1630). The doctors did have a natural explanation, as distinct from the pious belief that all plagues were punishments sent by God. They held that something – maybe a comet, or a planetary combination, or some local accident – had corrupted the air, which then gave off a vapor, called a "miasma", imperceptible save in its effects, and thus uncontrollable.

Although physicians knew they were all but helpless against pestilence, they could hope for mitigation through sanitary measures. When King Charles' Council asked the College of Physicians' advice in the outbreak of 1630, they suggested the filth of London was at least aggravating the trouble: let the Council remedy the overcrowding of the poor, the rubbish thrown into the streets, and the seepage of sewage.

***►** In a world where much was unexplained (not least the recurrent plagues), witch-hunts were common. In 1647 the English "Witchfinder General" Matthew Hopkins was himself hanged as a wizard, within months of publishing an account of his hunt for witches.*

***▲◄** Death triumphs in the 1625 plague. Doctors believed a plague "miasma" flourished in crowded, filthy towns but did not suspect rats and fleas were carriers.*

***►** The workings of the human frame were understood through the ancient Greeks' doctrine of the four humors, or body fluids, that rule the body – phlegm, blood, yellow bile (chole) and black bile (melaina chole). Galen linked temperaments to each: phlegmatic, sanguine, choleric and melancholic, as in these medieval manuscripts.*

THE
Difcovery of VVitches:
IN
Anfwer to feverall QUERIES,
LATELY
Delivered to the Judges of Affize for the County of NORFOLK.
And now publifhed
By MATTHEVV HOPKINS, Witch-finder.
FOR
The Benefit of the whole KINGDOME.

EXOD. 22. 18.
Thou fhalt not fuffer a witch to live.

may . 18
LONDON,
Printed for R. Royfton, at the Angell in Ivie Lane.
M. DC. XLVII.

In that, the London physicians were right, even if they did not understand the reason why; they knew nothing of the carriers of disease or of bacteria. If the famous plague of 1665 was the last (and one of the worst) in England, medicine had still not come up with better methods to cope with the terrible outbreak at Marseille 60 years later. For the individual, the best thing was to run away as soon as plague was suspected – but in that way, too, it often spread.

Strict sanitary regulations might have helped, but were seldom enforced. The impact of plague may have been limited by quarantine laws, introduced by Venice, which obliged those who arrived by sea to wait 40 days before landing, to see if they would develop plague or not. Other cities followed suit. To obviate the inconvenience to travelers, a health certificate, signed by an official in the coast town along the road, could assure the authorities that the traveler had come from a place free of plague. Harvey was held up at Treviso on his way to Venice in 1636 because his certificate was found inadequate. While he waited, the only accomodation offered him was the plague hospital – where he reasonably feared he might catch the contagion – or a garden shed.

The work of the Devil

If, despite all these precautions, plagues still slaughtered thousands, and made their way from town to town, people doubted doctors' explanations, and fell back on old beliefs and superstitions. In fear of the intangible, they readily looked for witches and wizards as the cause: in the early 17th century, the Devil was seen to be at work everywhere, suborning humans to his service.

While Galileo was busy writing his "Dialogue on the Two Chief Systems of the World" in 1630-1632, a major epidemic ravaged Italy. One of Venice's best known landmarks, the church of Santa Maria della Salute ("of good health") commemorates its ending. In Milan, dreadful rumors whispered that diabolic agents had smeared the walls with plague poison: those suspected were tortured until they confessed, and then executed horribly. Those years just before the victory of rational ideas, with the acceptance of the ideas of Galileo, Harvey and their contemporaries, were indeed the worst of the witch mania, which flared up in an insecure world before finally dying down altogether. Likely enough, more people – mostly women – were put to death as witches than for heresy. The astronomer Johannes Kepler (1571-1630) once had to dash home to the small town in the Black Forest where he was born, to rescue his mother from an accusation that she was a witch who had cast spells on her neighbors. This witch-hunt was less hysterical in England and the Netherlands, where none were executed after 1600, than it was in Germany, Poland and Scandinavia. In England a turning point was perhaps the famous case of the Pendle witches. Those fortunate enough to be sent to London for full investigation were inspected by midwives advised by Harvey. In the event, they were set free; their accuser soon confessed he had made up the whole story.

Harvey was no campaigner, though other anecdotes imply his disbelief in witchcraft. He might declare the innocence of anyone he was asked to examine, but would not go out of his way to defend those charged, or attack the very idea of witchcraft, as a few brave doctors did, notably the Dutch physician Jan Wier (1515-1588). In the end, the witch craze was not halted by some cause célèbre: in the new, rational climate this strange and terrible persecution simply withered away.

"All things from an egg": without the microscope, Harvey took embryological studies to the limits possible

The generation of animals

Harvey's quest for information provided by the imperfect or undeveloped, involved him in embryological research, and an inquiry into how individuals produce fresh individuals which inherit their main characteristics. The first step to an answer to this question was to trace the process of growth before birth. Like his predecessors, Harvey found it hard to satisfy himself what was going on. Fabricius had made a start, and had adopted Aristotle's procedure of opening one of a clutch of eggs each day, and his description had been more detailed and exact than Aristotle's. Harvey gave still more details, and corrected both on several points through his more careful observation. Now that he was King Charles's physician, the king could help him extend the research to mammalian embryology, by allowing him to dissect pregnant does from among the deer at Windsor Great Park.

Harvey continued these investigations intermittently through the 1630s, in time taken from his duties (he had moved to the royal palace at Whitehall, and in 1636 crossed much of western Europe as doctor to a major embassy). On one occasion, Harvey's role at court gave him a direct glimpse of the motion of the human heart, with the arrival of a young nobleman who had been left with a hole in his chest after a bad fall. What was thought to be a lung could be seen through the hole whenever his dressing was changed. Immediately Harvey saw that he was actually looking at a heart in action, and could watch how it leapt up in its systole (contraction).

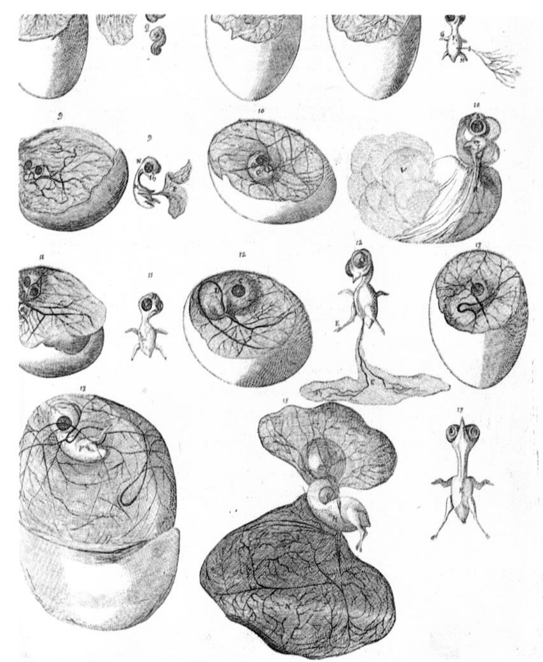

▲ The development of a chick from a work by Fabricius, who prompted Harvey's own studies.

▼ Harvey demonstrates his experiments on deer to Charles I, with the future Charles II looking on.

▲ *Anton van Leeuwenhoek, Dutch pioneer microscopist.*

Attachment to the royal court was to have its drawbacks for Harvey. When the English Civil War broke out in 1642, the king set up his headquarters in Oxford, as London was decidedly for Parliament. Harvey left London with him. When Whitehall palace was ransacked, his researches on insects and many other papers were destroyed. Charles made him head of Merton College at Oxford for a year or so, and several young army surgeons and doctors gathered round him – the nearest he ever came to having a research group. With the defeat of the royalists and the surrender of Oxford, Harvey lost his post, his friends were dispersed, he was soon forbidden to attend the king, and a massive fine was imposed on him.

His brothers, too, were heavily fined, but as merchants they would find it easier to recoup their losses. Harvey had to live off his inheritance and his shrunken practice. He was widowed toward the end of the Civil War, and never had his own home again. Instead, he lodged with one or other of his brothers and fell back on research as a refuge from the world. One of his young colleagues persuaded him to publish his unfinished work, *The Generation of Animals*, in 1651.

Although overall Harvey kept to the main principles of Aristotle and Fabricius in this field, he was able to improve on them in many ways. For instance, he realized that Fabricius had been wrong to suppose that the future chick is formed from the white of an egg. Instead, he saw that the chick grows from what looks like a tiny scab at first; he identified what is indeed to the naked eye the earliest appearance of the embryo. In search of something comparable in mammals, he decided that all creatures must grow from a simple undifferentiated point of pulsating blood. This he called the "primordium", the first principle of future life. Harvey was looking in the right place, and the successive development of organs which he then tried to trace – the process of differentiation – has today kept the name Harvey gave it: epigenesis. Grubs, he thought, provide a good example of early development: he remarked how embryo birds and mammals resemble insect grubs at an early stage in their development. Just as insect grubs come from eggs, so the same should apply to the origin of all animals. Hence the phrase "all things from an egg" became the motto of his book. As an introductory poem declares,

"Both Hen and Housewife are so matched,
That her Son born is only her Son hatched,
And when her teeming hopes have prosperous been,
Yet to conceive is but to lay within."

Although this little rhyme is a caricature of Harvey's quite sophisticated concept, it does make the point that all individual lives start out in a wonderfully similar manner, however different the final forms.

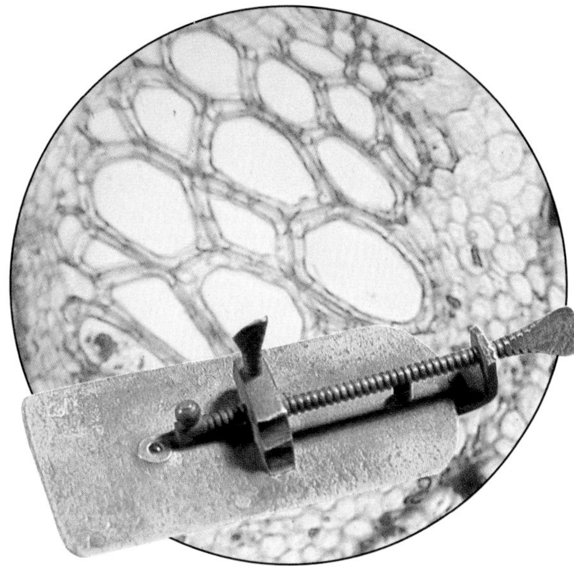

▲ ***A microscope built and used by Leeuwenhoek and a photomicrograph of a fern taken using the spherical lens from one of his microscopes. The magnifying power of one of these instruments is typically × 125. The construction consists basically of two brass plates between which the lens is held, with an adjustable screw on which the specimen is mounted. Leeuwenhoek recorded his observations of capillaries, red blood corpuscles, sperm cells and, in 1674-1676 bacteria and protozoa (little animalcules), which he was the first to observe.***

▼ ***Animalcules, including microscopic animals and sperm, based on Leeuwenhoek's drawings.***

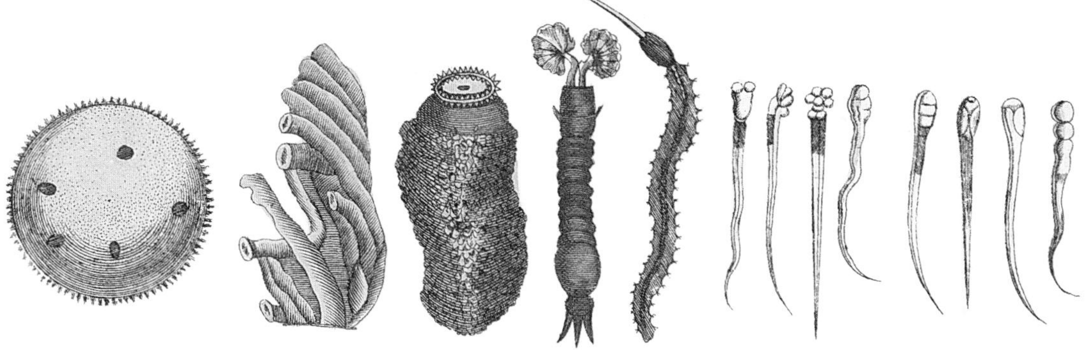

It took 200 years before evidence from improved microscopes vindicated Harvey's idea of epigenetic development

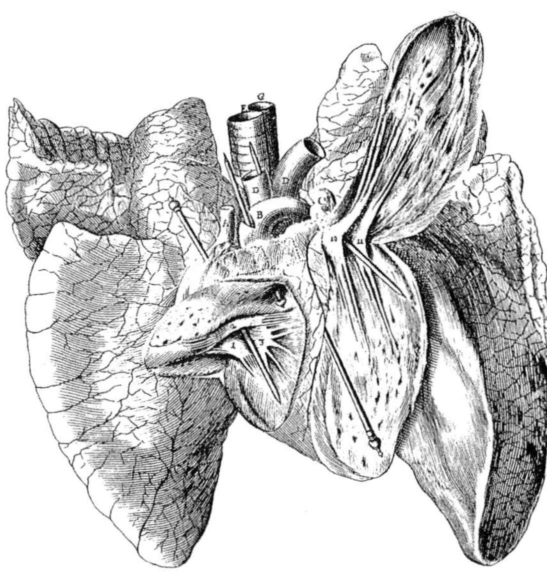

▲ ***The heart, from a work by René Descartes, whose views on biology were mechanistic. He believed the heart's dilation was caused by vaporization of blood within it.***

► ***Marcello Malpighi, whose discovery of capillaries clinched the case for Harvey's ideas.***

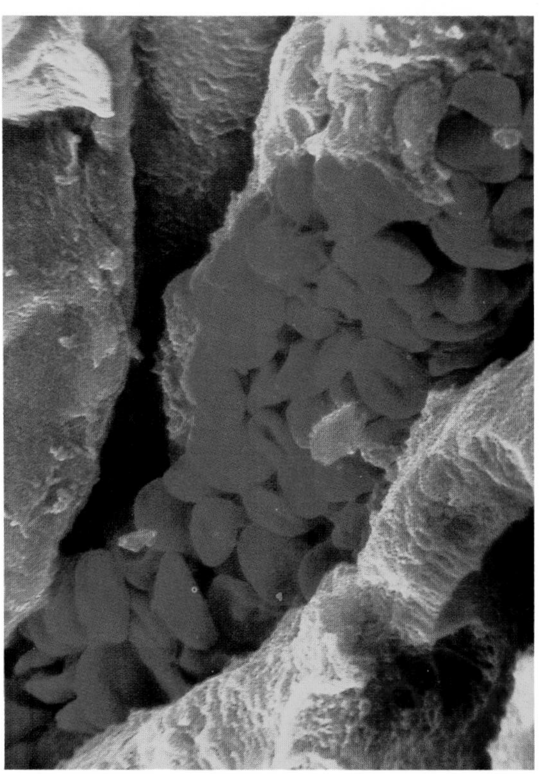

▼ ***An electron micrograph of red blood cells in a capillary. Advances in microscopy were to become central to progress in medical science.***

Harvey in old age

After publication of *The Generation of Animals*, Harvey stayed on in London or nearby, ailing, but revered by his medical colleagues. He died in London in 1657. An entertaining account of his old age appears among the *Brief Lives* of the gossip writer John Aubrey, who talked with him in the 1650s, and much admired him. Aubrey depicts an eccentric but lovable old fellow with a quizzical, sardonic outlook, no longer obliged to conform, and with few illusions. Man, said Harvey, is but a great mischievous baboon.

Harvey's heritage

In the same way that Copernicus (◀ page 34) had been able to envisage the existence of a Sun-centered Universe without using a telescope, Harvey discovered the circulation of the blood and the site of the early development of the embryo without the aid of a microscope. The telescope added powerful arguments for Copernicus's system, and the microscope served Harvey's doctrine in a similar way. Just four years after Harvey's death, the Italian physiologist Marcello Malpighi (1628-1694) found the very small blood vessels, that enable the blood to reach the veins from the arteries.

In embryology, sperm was first observed through a microscope 20 years after Harvey's death, by the brilliant Dutch amateur, Anton van Leeuwenhoek (1632-1723). His compatriot, Regnier De Graaf (1641-1673), found what seemed to be the mammalian egg at about the same time. In fact what De Graaf saw were the follicles which still bear his name, little sacs which contain the actual egg cells, like tiny peas in their pods. For a time, it seemed that he had confirmed Harvey's theory. But the microscope revealed that organs were already present at earlier stages than Harvey had seen them. So his account came to be regarded as inadequate and out of date. It was only in the 19th century that better microscopes showed the genuine egg cell, and the earliest growth after fertilization, when the embryo is a bunch of unspecialized cells, looking rather like a raspberry. Harvey's key idea of epigenetic development had been finally vindicated.

Harvey did, however, bequeath a set of problems to his successors which their microscopes could not solve. The action of the blood remained one such. Clearly, venous blood differs a little from arterial blood, and the change must take place in the lungs. But how? Physiologists could, like Stephen Hales (1677-1761), refine and extend Harvey's crude estimates of quantities, even observe the fine structure of the lungs and red cells in blood, as Malpighi did, and agree that respiration was the key; but they could not work out what was taken in from the air. At Oxford once, while a friend was talking of his adventures exploring Egyptian pyramids, Harvey questioned how he had managed to breathe inside the pyramid, for we "never breathe the same air twice, but still new air is required to a new respiration, the nourishing juice of it being spent in each expiration." What was this nourishing juice?

The answer would have to come from a chemistry of life, something in which Harvey had never shown much interest. His successors at Oxford tried to make chemical sense of Harvey's, and their own, discoveries. But their chemistry relied on old, ill-defined ideas which went back to alchemy. All their theorizing failed to make progress, and the problem had to wait for the discovery of oxygen 100 years later, after which Antoine Lavoisier (1743-1794) could establish the task which oxygen performed in the body.

Isaac Newton 1642-1727

Successor to Galileo...Halley visits Newton at Cambridge...Why an ellipse?...Newton's laws of motion...The universal force of gravitation... Descartes' "vortices"...and Newton's rebuttal of them ...Earth bulges and comets...Investigating light...The public man...PERSPECTIVE...The publishing of science... Advances in instrumentation...Newton's impact on poets and artists...The Age of Newton and Locke

"Nature, and Nature's Laws lay hid in Night:
God said, *Let Newton be!* and All was *Light*".

Alexander Pope's couplet is probably the most famous tribute by a major poet to a major scientist. Like most tributes, it conceals as much as it reveals. Modern science was not created by one sudden leap forward, but advanced via a series of steps – some big, some small; some helpful, some not.

In particular, the investigation of how and why objects move in the way they do forms a clear line of development from Copernicus, through Kepler and Galileo to their successors. One might say that Galileo and Newton span the history of this inquiry which lay at the base of the Scientific Revolution; for Isaac Newton was born a few months after the death of Galileo. In those days, when little in the way of scientific instrumentation existed (◀ page 36), motion was something that could be measured relatively easily. At the same time, the theory of motion was sufficiently complex that developing it stretched the best European minds for over a century. Measurement and theory together were just simple enough for quantitative results to be obtained even within the limitations of the 17th century. So the study of motion provided the first breakthrough into modern science, and it dragged other branches of science in its wake.

"Newtonian science"

The "Scientific Revolution" was so important to the development of mankind that modern historians honor the phrase with initial capital letters. The new way of seeing the world that it introduced first tentatively surfaced with the publication of Copernicus's work in 1543. It reached its triumphal acceptance with the appearance in 1687 of Isaac Newton's "Principia".

In this, Newton summed up the partial and fragmented work of his predecessors and contemporaries – Copernicus, Galileo, Kepler, Descartes, Huygens, Hooke and many others. He added major insights of his own to the advances they had made, and so provided not only a new picture of the world around us, but also a clearly-defined approach, based on observation, quantitative experiments and mathematical theory.

The influence of the "Principia" was backed up some years later by the publication of Newton's "Opticks". Between them, these two books laid the foundations for the development of "Newtonian Science" in the 18th century. Though particularly important for astronomy and physics, Newton's ideas were to influence activities in all fields of science.

In the early 20th century, Einstein fundamentally challenged Newton's ideas on space and time, but we still use many of Newton's concepts and methods. In a very real sense, modern scientists are "Newtonian" scientists.

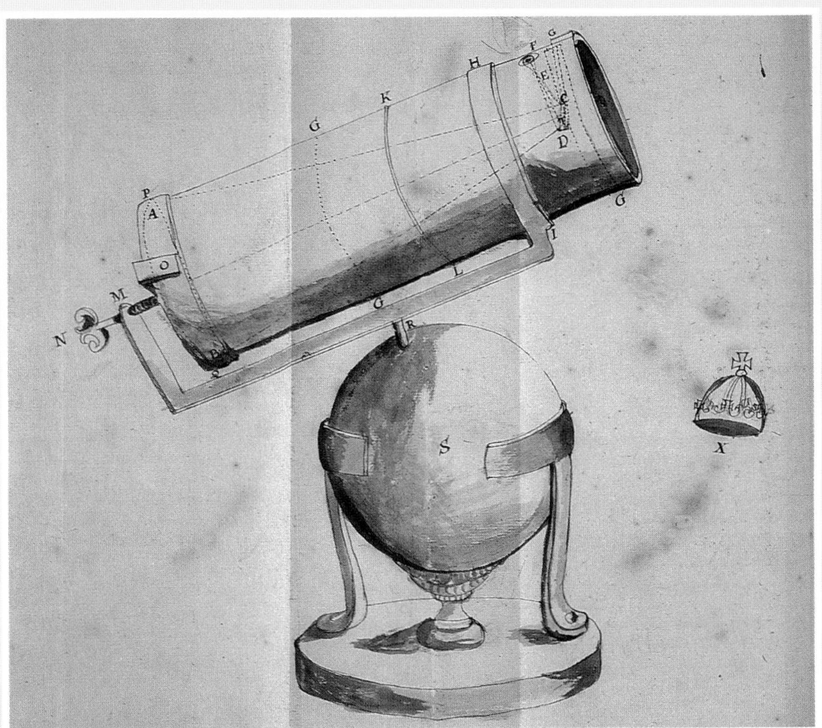

◀ Isaac Newton's own drawing of the reflecting telescope he invented. Newton sent a telescope to the Royal Society in London, to which he was elected a fellow.

▲ Newton aged 59, 15 years after the "Principia" appeared. Godfrey Kneller's portrait reflects the successful civil servant and politician, rather than the absent-minded scientist.

Isaac Newton – His Life, Work and Times

Newton's synthesis

The work of Copernicus, Kepler and Galileo was taken up not by Newton alone, but by a number of his contemporaries. Newton's great achievement was to bring together all the significant steps made up to his time, and to synthesize them into a single, basically simple, picture. This sounds less dramatic than the claim poetically put forward by Alexander Pope, but, to his contemporaries, Newton epitomized the great power of the new science.

According to this view, each piece of scientific knowledge should be built on previous observation and experiment, and should in turn, provide the foundation for further research. Unlike other knowledge, scientific knowledge was therefore cumulative, building on what had gone before in much the same way that a house was built of bricks. When Newton said, "If I have seen further it is by standing on the shoulders of giants," this was not just a modest disclaimer. He was underlining his belief in the way that the new science operated – by building on previous work. Most scientists today still approach their research in this same spirit. The success of Newton's synthesis lies not only in its creation of a new picture of the world, but also in that it finally ensured the acceptability of the scientific approach.

Understanding the Universe

Isaac Newton was born on Christmas day 1642 in the manor house at Woolsthorpe in rural Lincolnshire. His father had died two months earlier, and Newton, a rather lonely boy, was brought up mainly by his grandmother. It became evident that he had little inclination toward agriculture, so he was dispatched to Cambridge University as a not particularly distinguished entrant in 1661.

Newton is an outstanding example of the value of long vacations. During the years 1665-1666 he stayed at home, because Cambridge was then being visited by the plague. For 18 months he concentrated on the problems which were to make him famous – relating to the calculus (or "fluxions"), gravitation and light – and advanced a long way in their solution. Although he said little of this work, his reputation at Cambridge grew, and, in 1669, he was appointed professor of mathematics. Newton then remained quietly at Cambridge until after the publication of his great work, the *Principia*, in 1687.

In the summer of 1684 Edmond Halley visited Isaac Newton at Cambridge. Newton by that time had been professor of mathematics for 15 years. Although some knew him as a highly intelligent man with interesting ideas, he had published little research and was far from being famous. Halley (1656-1742), on the contrary, was already well known both at home and abroad. He had come to the fore as a scientist while still an undergraduate at Oxford. He had taken time off from his studies to sail to St Helena in the South Atlantic, so that he could measure the positions of stars in the southern heavens. These first detailed measurements were designed to complement observations of northern stars then being made at the newly founded Royal Observatory in Greenwich. With these he was able to show that certain stars had moved perceptibly against their background since Ptolemy's time. Halley had subsequently become an official of the Royal Society (newly founded, like the Royal Observatory), and so was in contact with most of the leading scientists of his day. One of these correspondents was Newton, and, when the two first met in Cambridge, their conversation naturally turned to that great unsolved problem of the day – the motion of the planets about the Sun.

Historical Background

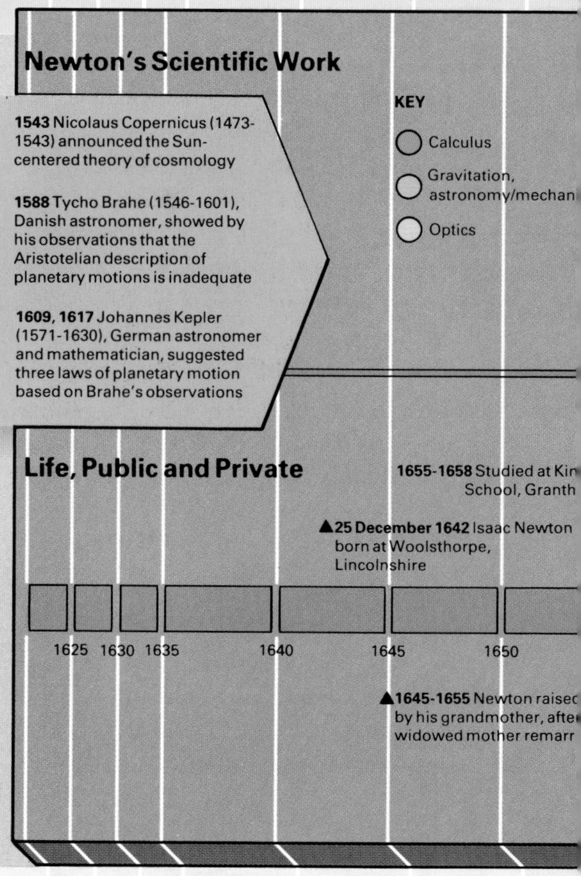

1642-1646, 1647-1648 English Civil Wars or Puritan Revolution
Ex-king Charles I of England executed; Oliver Cromwell s
up republican Commonwea
Newton's Scientific Work
KEY
Calculus
Gravitation, astronomy/mechan
Optics
1543 Nicolaus Copernicus (1473-1543) announced the Sun-centered theory of cosmology
1588 Tycho Brahe (1546-1601), Danish astronomer, showed by his observations that the Aristotelian description of planetary motions is inadequate
1609, 1617 Johannes Kepler (1571-1630), German astronomer and mathematician, suggested three laws of planetary motion based on Brahe's observations
Life, Public and Private
1655-1658 Studied at Kin
School, Granth
25 December 1642 Isaac Newton born at Woolsthorpe, Lincolnshire
1625 1630 1635 1640 1645 1650
1645-1655 Newton raisec by his grandmother, afte widowed mother remarr

Scientific Background

1644 Réné Descartes (159 1650), French mathematic and philosopher, publishe works on motion of bodie fluids

1642 Blaise Pascal (1623-1662), French mathematician and physicist, invented an adding machine; studied hydraulics, probability theory; confirmed (1646) atmospheric air has weight

1620 Francis Bacon (1561-1626), English philosopher, proposed inductive scientific method, in place of *a priori* reasoning

1654 Otto von Guericke (160 1686), German physici inventor of vacuum pum showed 16 horses unable separate 2 halves of evacuated hollow sphe

1638 Galileo Galilei (1564-1642), Italian mathematician and physicist, published work on laws of how bodies fall

1643 Evangelista Torricelli (1647), Italian physicist and mathematician, invented th mercury barometer

1657 Accademia del Cim Florence, founded, th scientific research ins

1620 1625 1630 1635 1640 1645 1650

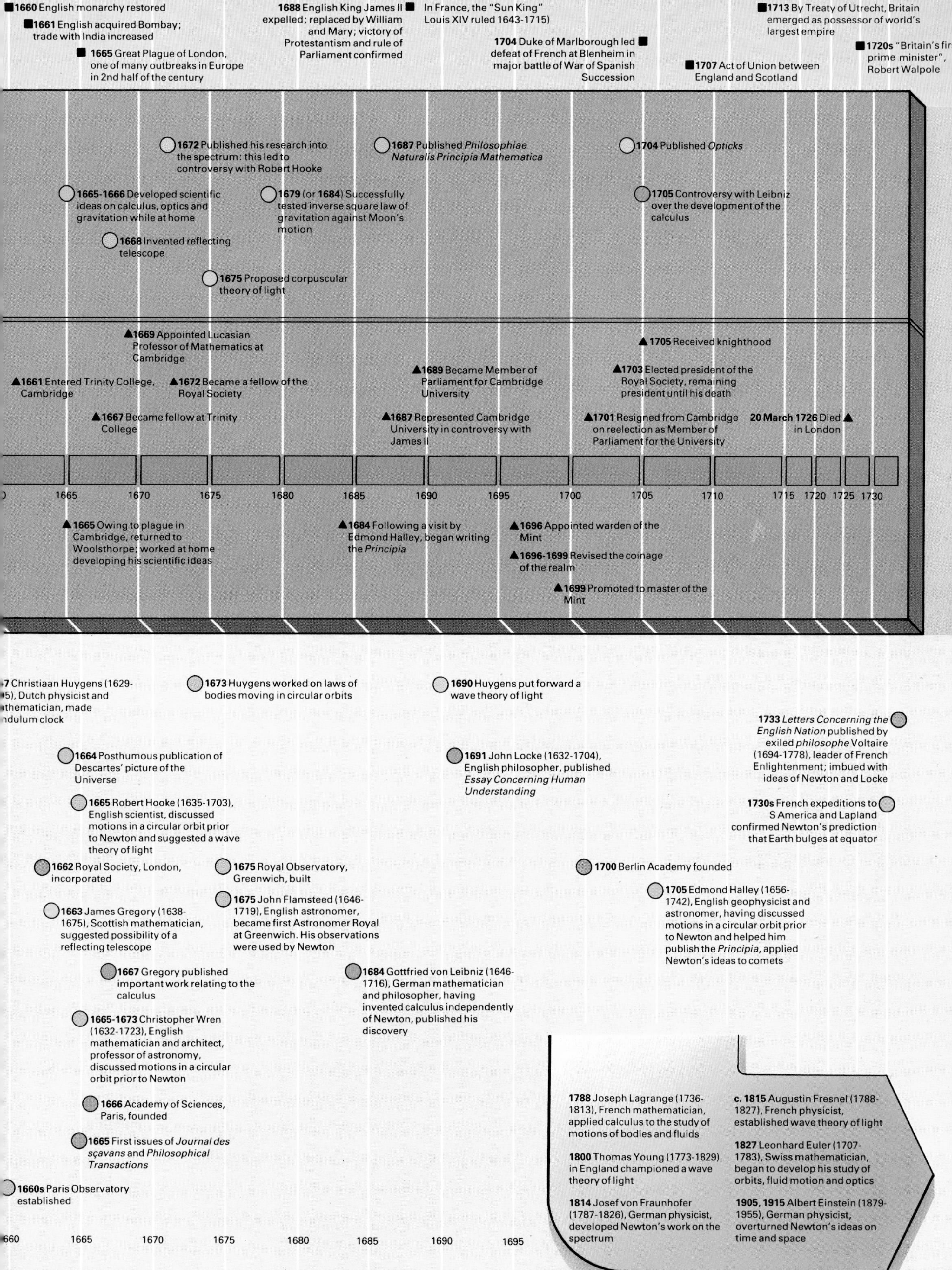

1660 English monarchy restored
1661 English acquired Bombay; trade with India increased
1665 Great Plague of London, one of many outbreaks in Europe in 2nd half of the century
1688 English King James II expelled; replaced by William and Mary; victory of Protestantism and rule of Parliament confirmed
In France, the "Sun King" Louis XIV ruled 1643-1715)
1704 Duke of Marlborough led defeat of French at Blenheim in major battle of War of Spanish Succession
1707 Act of Union between England and Scotland
1713 By Treaty of Utrecht, Britain emerged as possessor of world's largest empire
1720s "Britain's first prime minister", Robert Walpole
1665-1666 Developed scientific ideas on calculus, optics and gravitation while at home
1668 Invented reflecting telescope
1672 Published his research into the spectrum: this led to controversy with Robert Hooke
1675 Proposed corpuscular theory of light
1679 (or 1684) Successfully tested inverse square law of gravitation against Moon's motion
1687 Published Philosophiae Naturalis Principia Mathematica
1704 Published Opticks
1705 Controversy with Leibniz over the development of the calculus
1661 Entered Trinity College, Cambridge
1667 Became fellow at Trinity College
1669 Appointed Lucasian Professor of Mathematics at Cambridge
1672 Became a fellow of the Royal Society
1687 Represented Cambridge University in controversy with James II
1689 Became Member of Parliament for Cambridge University
1701 Resigned from Cambridge on reelection as Member of Parliament for the University
1703 Elected president of the Royal Society, remaining president until his death
1705 Received knighthood
20 March 1726 Died in London
1665
1670
1675
1680
1685
1690
1695
1700
1705
1710
1715
1720
1725
1730
1665 Owing to plague in Cambridge, returned to Woolsthorpe; worked at home developing his scientific ideas
1684 Following a visit by Edmond Halley, began writing the Principia
1696 Appointed warden of the Mint
1696-1699 Revised the coinage of the realm
1699 Promoted to master of the Mint
7 Christiaan Huygens (1629-5), Dutch physicist and athematician, made ndulum clock
1664 Posthumous publication of Descartes' picture of the Universe
1665 Robert Hooke (1635-1703), English scientist, discussed motions in a circular orbit prior to Newton and suggested a wave theory of light
1662 Royal Society, London, incorporated
1663 James Gregory (1638-1675), Scottish mathematician, suggested possibility of a reflecting telescope
1667 Gregory published important work relating to the calculus
1665-1673 Christopher Wren (1632-1723), English mathematician and architect, professor of astronomy, discussed motions in a circular orbit prior to Newton
1666 Academy of Sciences, Paris, founded
1665 First issues of Journal des sçavans and Philosophical Transactions
1660s Paris Observatory established
1673 Huygens worked on laws of bodies moving in circular orbits
1675 Royal Observatory, Greenwich, built
1675 John Flamsteed (1646-1719), English astronomer, became first Astronomer Royal at Greenwich. His observations were used by Newton
1684 Gottfried von Leibniz (1646-1716), German mathematician and philosopher, having invented calculus independently of Newton, published his discovery
1690 Huygens put forward a wave theory of light
1691 John Locke (1632-1704), English philosopher, published Essay Concerning Human Understanding
1700 Berlin Academy founded
1705 Edmond Halley (1656-1742), English geophysicist and astronomer, having discussed motions in a circular orbit prior to Newton and helped him publish the Principia, applied Newton's ideas to comets
1733 Letters Concerning the English Nation published by exiled philosophe Voltaire (1694-1778), leader of French Enlightenment; imbued with ideas of Newton and Locke
1730s French expeditions to S America and Lapland confirmed Newton's prediction that Earth bulges at equator
660
1665
1670
1675
1680
1685
1690
1695
1788 Joseph Lagrange (1736-1813), French mathematician, applied calculus to the study of motions of bodies and fluids
1800 Thomas Young (1773-1829) in England championed a wave theory of light
1814 Josef von Fraunhofer (1787-1826), German physicist, developed Newton's work on the spectrum
c. 1815 Augustin Fresnel (1788-1827), French physicist, established wave theory of light
1827 Leonhard Euler (1707-1783), Swiss mathematician, began to develop his study of orbits, fluid motion and optics
1905, 1915 Albert Einstein (1879-1955), German physicist, overturned Newton's ideas on time and space

Copernicus had supposed that planetary motion could be represented by combining circles together in various ways. Kepler had shown this was untrue: planets follow elliptical paths round the Sun. The question that faced his successors was – why an ellipse? The French mathematician and philosopher René Descartes (1596-1650) thought he could provide an explanation, but it was descriptive rather than quantitative. The Dutchman, Christiaan Huygens (1629-1695), first provided a quantitative estimate of the force required to move an object in a circle. His work was taken up by three Englishmen, who applied it to the case of a planet moving in a circle round the Sun. These were Christopher Wren (1632-1723), normally thought of now as an architect but then an accepted mathematician, Robert Hooke (1635-1703), like Halley an official of the Royal Society, and Halley himself. Their results showed that planets would move in a circle round the Sun if the Sun exerted a force of attraction on them which fell off with distance from the Sun as the inverse of the square of the distance. In other words, if one planet is twice the distance of another from the Sun, it will experience only a quarter of the attractive force felt by the inner planet; if it is at three times the distance, the force will be one-ninth, and so on.

This result was interesting, but did not seem to contribute much to an understanding of why the planets move in ellipses, not circles. The next advance was elicited by a question that Halley put to Newton, not expecting an answer. Newton's reply was recounted later (by which time he was knighted):

"The Doctor [Halley] asked him what he thought the curve would be that would be described by the planets supposing the force of attraction towards the Sun to be reciprocal to the square of their distance from it. Sir Isaac replied immediately that it would be an ellipse, the Doctor struck with joy and amazement asked him how he knew it.

▲ Christopher Wren was knighted for his work as an architect, but he was also known as a mathematician and astronomer. He was president of the Royal Society in 1680-1682.

▼ The Royal Observatory at Greenwich (left) was designed by Sir Christopher Wren and built in 1675. Astronomical measurements were made at the top from the octagon room (right).

▼ Johannes Kepler applied mathematics to the study of planetary orbits. Kepler was Tycho Brahe's assistant, and he inherited Tycho's careful observations of the orbit of Mars. Kepler based his laws of planetary motion entirely on observation. Newton derived them theoretically by using his new ideas on gravity and laws of motion.

► Newton's father, a farmer, died shortly before Newton was born. The family lived in Woolsthorpe Manor in Lincolnshire. When Newton was three, his mother remarried and left him at Woolsthorpe with his grandmother. Newton returned here during the plague years to lay the foundations of his future work.

'Why,' saith he, 'I have calculated it,' whereupon Dr Halley asked him for his calculation without any further delay. Sir Isaac looked among his papers but could not find it, but he promised him to renew it, and then to send it him."

Newton sat down to work on the problem anew. But he found that additional queries and new problems arose, so that his final conclusions were only published (at Halley's expense) in 1687. With the passage of time, his original short proof had expanded into a volume, known ever since as the *Principia*. The full title, in translation from the Latin, is "The Mathematical Principles of Natural Philosophy". In the 17th century, natural philosophy had a somewhat similar meaning to our word "physics". So Newton's choice of title proclaims that he intended to explain the physical world in terms of mathematics. Here he was very clearly following in the steps of Copernicus, Kepler and Galileo, all of whom had argued that the Universe must be understood quantitatively, that is, mathematically.

The three books of the *Principia* laid out in logical sequence how Newton saw the world. The first book explained basic principles. There were, he thought, three basic laws that governed the way in which objects moved. The first law laid down that an object not being pushed or pulled by a force would either sit still, or move in a straight line. The second law defined what happens when a force is applied to such an object. It accelerates in the direction of the force: for a given force, the amount of acceleration depends on the mass of the object. (Thus, a light stone is easier to throw than a heavy one.)

The major breakthrough in the "Principia" was to show the universality of the force of gravitation.

The third and last of the laws of motion proposed by Newton in the *Principia* says that, if you push or pull an object, it will push or pull you to an equal extent. For example, if you stand on a floor, it pushes you upward by an amount that equals your weight. If it pushed less, you would fall through the floor; if it pushed more, you would be thrown into the air. Some aspects of these laws had been already recognized by Newton's predecessors and contemporaries, but the group together has always been known as Newton's laws of motion.

The other major topic in Book I was gravitation. Newton now developed a detailed specification of what the force keeping planets in orbit round the Sun must be like. He had already shown early on that a force decreasing as the square of the distance could produce elliptical, as well as circular orbits. But he was also able to show that the pull exerted by the Sun must depend on the masses of both the Sun and the planet concerned. From this new idea of gravity and his three laws of motion, Newton was able to derive theoretically Kepler's laws of planetary motion, which had previously been based entirely on observation (◀ pages 38, 72). Indeed, he went further: he was able to show that there must be slight deviations from Kepler's laws because the planets all had different masses.

This was a major step forward but, even so, was not the most important breakthrough in Book I. Questions about how bodies moved had always been split into two parts – the first concerned the motion of the planets; the second concerned the motion of bodies on Earth, and especially why all objects fall toward the Earth's surface. Aristotle saw these as two entirely separate questions. Galileo realized that they overlapped, but did not progress very far in uniting them. Newton showed they were different aspects of a single basic theme – the universal force of gravitation. Every piece of matter in the Universe attracts every other piece. The Earth's gravitation attracts the Moon and keeps it in orbit; in just the same way, the Earth attracts a stone, making it fall to the surface. This is the point behind the well-known (and possibly true) story of Newton watching an apple fall from a tree in the orchard of his home. He realized that the force attracting the apple was the same force that attracted the Moon: the same laws applied to both bodies. An admirer explained: "In the year 1666 he retired again from Cambridge to his mother in Lincolnshire and whilst he was musing in the garden it came into his thought that the power of gravity (which brought an apple from the tree to the ground) was not limited to a certain distance from the Earth but that this power must extend much farther than was usually thought. Why not as high as the Moon, said he to himself, and if so that must influence her motion and perhaps retain her in her orbit, whereupon he fell a-calculating what would be the effect of that supposition."

Newton realized that it was necessary to distinguish between the mass of an object and its weight. On Earth, the distinction is seldom important, but in space the difference is vital. Astronauts may have zero weight in space, but they still retain their mass. A careful distinction between words which are often used interchangeably in everyday life was characteristic of Newton. He realized that scientific language must be precise. This comes over particularly well in his discussion of time and space. These are basic to the question of motion: speed is measured in, say, kilometers per hour – that is, in terms of the space covered during a given period of time. Newton queried whether a measurement of motion could be absolute, or only be relative, for it is usually measured against some kind of background.

Correspondence between scientists

Seventeenth-century scientists communicated ideas by writing letters to each other, a development made possible by the introduction of reasonably regular and reliable postal services in Europe during the 16th and 17th centuries. The interchange of letters was not private in the modern sense. They were used to convey news of research carried out not only by the writer, but also by his circle of acquaintances. Similarly, the recipient was expected to pass on the information to all his own scientific contacts.

Scientific societies and their journals

These informal groups of scientists developed into the first scientific societies. In 1662 the Royal Society of London was incorporated, from just such casual beginnings. It was followed by the Academy of Sciences in Paris (◀ page 94). Although information continued to be exchanged by correspondence and by visits at home and abroad, there quickly arose a need for a more permanent record. So the new societies began to publish printed journals reporting new research. Newspapers, similar in principle to our own daily papers, had begun to appear in the early 17th century. Scientific journals were printed and distributed along the lines established by newspapers. They could be made available to far more people than a handwritten letter, and lasted longer, so that later generations of scientists could refer back to earlier work.

One of the first scientific journals, the "Philosophical Transactions" of the Royal Society, still survives today. (The word "philosophical" was used here in the same way as in the title of Newton's "Principia" or "Principles of Natural Philosophy", to mean "scientific".) The first issue of the corresponding French journal, "Le Journal des sçavans", appeared in 1665 a few months before the "Philosophical Transactions", and over the next few years other similar journals were started elsewhere in western Europe. They typically differed from the "Philosophical Transactions" in one important respect. Whereas it was mainly concerned with science they often included other topics, such as literature or law. In this, the Royal Society pointed the way toward the future, for the trend since the 17th century has clearly been in favor of more specialized journals. In fact, specialist medical journals split off at an early date, and by the latter part of the 18th century, journals dealing only with single subjects, such as biology, geology and chemistry, had appeared. This increasing specialization reflects the development of science itself, with knowledge accumulating so that no individual scientist can absorb it all.

The early scientific journals were all published in western Europe, which remains a major geographical focus for science publishing today. But journals soon began to appear farther afield – for example, in Russia and North America during the 18th century. Most of these were associated with scientific societies, but commercial publishers became interested.

Societies and Journals

In the 17th century, the common language of ach country was becoming increasingly popular ɔr communication (Newton wrote the "Principia" 1687) in Latin, but the "Opticks" (1704) in English). arly journals contained a mix of articles in Latin nd the vernacular. The introduction of a variety of ational languages obviously made ommunication between scientists more difficult, ut the changeover had the great advantage that eople such as instrument makers, who had no lassical background, could still participate in cience. Since World War II, English has become asily the commonest language of international scientific communication. A typical article in a modern journal begins with the title, author's name, author's institutional affiliation, abstract of the article's contents and continues through the main text before ending with a detailed list of references. Most journals employ referees who scan each article for errors and imbalances: many current journals reject an appreciable proportion of the submitted articles. Nevertheless, the central role of scientific journals, as the means of conveying new research results quickly to a wide audience, continues today as it has for the past three centuries.

▼ The frontispiece of Thomas Sprat's "History of the Royal Society" shows its first president Viscount Brouncker, founder Charles II, and spiritual father Francis Bacon (1561-1626). From 1665 the Society's correspondence with scientists abroad was the basis of its "Philosophical Transactions". In 1887, this journal was divided into two specialized series, A (mathematical and physical) and B (biological).

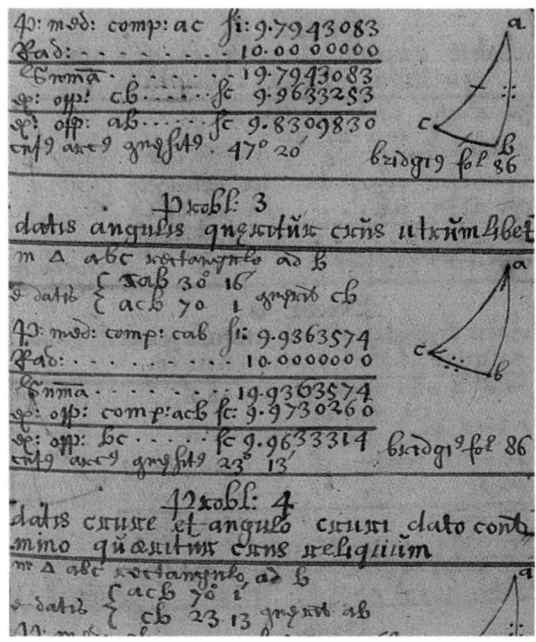

▲ ***A manuscript page of a book on trigonometry by Newton, who developed the binomial theorem, and fluxions, an early form of calculus. His terminology was later abandoned for that of Gottfried von Leibniz (1646-1716), codiscoverer of the calculus.***

▼ ***A typical page of the "Principia" shows how Newton dealt with orbits by applying the geometry of the ancient Greeks. His book was authorized for publication by the Royal Society's president – at that time Samuel Pepys, the famous diarist.***

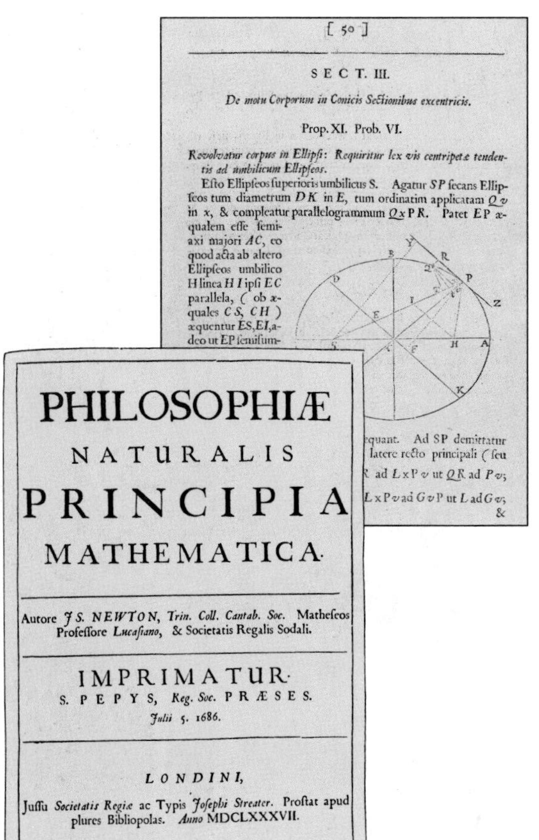

[50]

SECT. III.

De motu Corporum in Conicis Sectionibus excentricis.

Prop. XI. Prob. VI.

Revolvatur corpus in Ellipsi: Requiritur lex vis centripetæ tendentis ad umbilicum Ellipseos.

Esto Ellipseos superioris umbilicus S. Agatur SP secans Ellipseos tum diametrum DK in E, tum ordinatim applicatam Qv in x, & compleatur parallelogrammum QxPR. Patet EP æqualem esse semiaxi majori AC, eo quod acta ab altero Ellipseos umbilico H linea HI ipsi EC parallela, (ob æquales CS, CH) æquentur ES, EI, adeo ut EP semisum-

...quant. Ad SP demittatur latere recto principali (seu ... ad LxPv ut QR ad Pv; LxPv ad GvP ut L ad Gv; &

PHILOSOPHIÆ

NATURALIS

PRINCIPIA

MATHEMATICA.

Autore *J. S. NEWTON*, *Trin. Coll. Cantab. Soc.* Matheseos Professore *Lucasiano*, & Societatis Regalis Sodali.

IMPRIMATUR.

S. PEPYS, *Reg. Soc.* PRÆSES.

Julii 5. 1686.

LONDINI,

Jussu *Societatis Regiæ* ac Typis *Josephi Streater*. Prostat apud plures Bibliopolas. *Anno* MDCLXXXVII.

Relative and absolute motion

The speed of a carriage is usually measured relative to the road surface; but the road, itself, is in motion due to the Earth's rotation; the Earth, in turn, is moving about the Sun; the Sun is moving relative to other stars, and so on. Newton recognized this problem, but a careful analysis led him to believe that motion such as the Earth's rotation could be detected absolutely – that is, without reference to any other object in the Universe. It was this belief in the existence of absolute space and time that Einstein discarded in the 20th century with his relativity theory (➧ page 233). But, as Einstein commented, the remarkable thing was that Newton realized there was a problem here. No one else in the 17th century analyzed motion in this way.

Newtonian and Cartesian world views

Having established his own picture of how the world worked, Newton turned, in the second book of the *Principia*, to demolishing the most important of the competing world views then in existence. By the latter part of the 17th century, the old Aristotelian picture had been discarded by common consent, but there was no general agreement on what should replace it. The Roman Catholic church insisted on retaining a Universe in which the Earth was stationary. The most popular view among scientists was at variance with this, for many of them accepted a model put forward in the 1640s by René Descartes. Descartes supposed that the Universe was filled with a transparent fluid, which he called the "ether" (the same term Aristotle had used for a supposed transparent solid filling the Universe). This fluid was in continual eddying motion: Descartes envisaged it as a series of whirlpools (or "vortices"). Opaque matter floated in this transparent fluid and was caught up by the whirlpools. Most of it came to be concentrated at the center of each whirlpool, but smaller amounts remained in circulation round the center. Descartes now identified one of these whirlpools as forming the Solar System. The Sun at the center contained most of the material, but smaller pieces – the planets – moved round it. Occasionally, material from an adjacent whirlpool would be thrown into our own, and we would see it as a comet.

This Cartesian Universe, as it was called, attributed the motion of the Earth entirely to the fluid ether. Indeed, Descartes argued that the Earth was stationery in his picture, because it did not move relative to the nearby ether. Hence, it was possible to have both a stationary Earth and a heliocentric system. This argument, though ingenious, did not appeal greatly to the Roman Catholic hierarchy. In fact, the Cartesian picture created additional theological problems. It envisaged an infinite Universe filled with planetary systems similar to our own. Since, presumably, these other systems contained life, what did this imply for the uniqueness of man? Or, again, in an infinite Universe, where, physically, could Heaven be situated?

Descartes' system explained the orbits of the planets in a general sense. A whirlpool spins faster near its center that at its periphery (as does the vortex that forms after removing the plug from a bath full of water). Correspondingly, the inner planets move more quickly round the Sun than the outer planets. But Descartes never made his picture quantitative. In Book II of the *Principia*, Newton took up the very difficult challenge of studying mathematically how fluids move. He showed that a whirlpool motion did not satisfy the actual quantitative observations of planetary motion. In consequence, Descartes' picture could not explain the Universe in which we live.

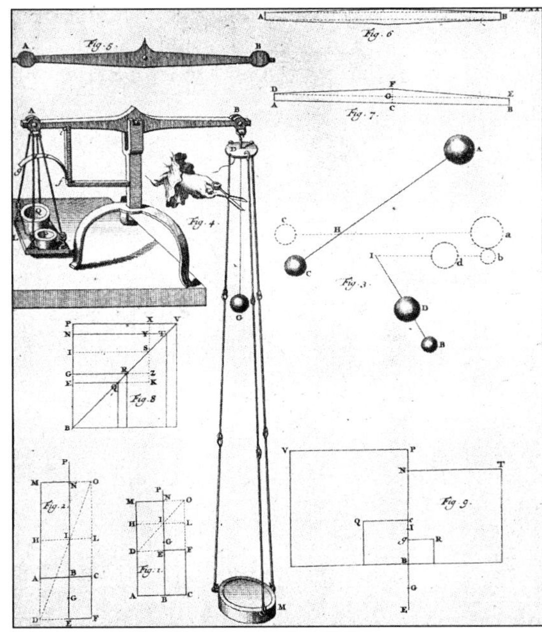

▲ Newton's ideas were soon accepted in the Netherlands. This Dutch experiment used scales to test predictions concerning falling bodies.

◄ Launching rockets involves all of Newton's laws of motion.

▼ René Descartes was an important figure in 17th-century science. His idea that the Universe was filled with whirlpools of ether was accepted by many contemporaries. Newton attacked some of Descartes' work in building his own picture of the Universe.

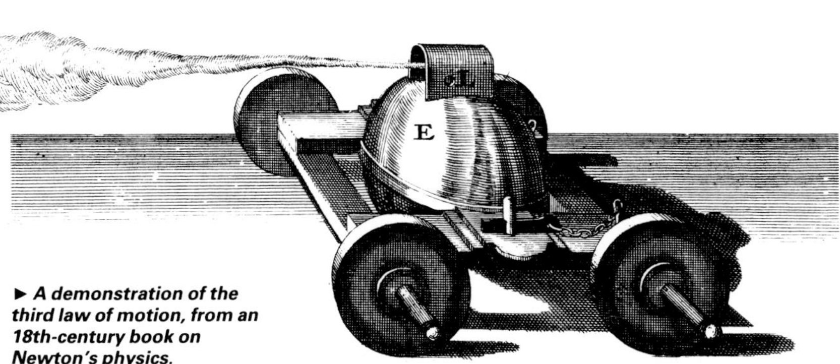

► A demonstration of the third law of motion, from an 18th-century book on Newton's physics.

Scientific Instrumentation

The microscope and telescope

Sophisticated scientific instruments were being built well before the 17th century, as witness the small, delicately-wrought astrolabes of earlier centuries (♦ pages 16, 36), or the large-scale astronomical instruments of the 16th-century Danish astronomer, Tycho Brahe (♦ page 38). What was new in the 17th century was the sudden appearance of a range of instrumentation which derived from the scientific interests of the period. Instruments based on the same general principles are still in use today. Two of the new instruments – the microscope and the telescope – involved properties of light: their invention was an important stimulus for Newton's work on optics. Both seem to have originated almost accidentally in the Netherlands around the year 1600 (♦ pages 37, 39, 67), and knowledge of these discoveries diffused through western Europe.

The compound microscope was so called because it contained at least two lenses. This distinguished it from the simple microscope – a single lens of the traditional "burning glass" type. The combination of two lenses should, in principle, produce a clearer image. In practice, the single lens was able to compete with the compound microscope for much of the 17th century. For example, the Dutch naturalist, Anton van Leeuwenhoek (1632-1723) taught himself to make excellent simple microscopes, and used them for a number of important observations (♦ page 67). Perhaps the most striking was his discovery that ordinary water is full of tiny animals (now called protozoa) which are invisible to the naked eye. Simple microscopes were sometimes called "flea glasses" because they were used in particular for looking at insects.

A simple hand lens could sometimes compete with an apparently better optical system mainly because of the defects of 17th-century lenses. The glass used was typically of very low standard by comparison with today's, often containing air bubbles and specks of dirt. More importantly, the combination of lenses distorted the image and produced colored halos, so reducing their advantage over a single lens. Similar problems affecting the early telescope were more serious, because a simple lens could not be used as a substitute for looking at distant objects. Consequently, the desire to improve the performance of the telescope – as Newton did with his reflecting telescope of 1668 – and microscope stimulated many 17th-century scientists to think about optical problems (♦ page 180).

The thermometer and barometer

Another range of new instrumentation derived from the contemporary interest in the properties of the atmosphere and of gases in general. The air thermometer invented by Galileo consisted of a glass bulb with a long, thin glass tube attached to it. This was inverted over a trough containing water in such a way that some of the water was sucked up the tube. When the air in the bulb was heated, it expanded and drove the water down the tube. So the height of the water in the tube provided a measure of the temperature of the bulb.

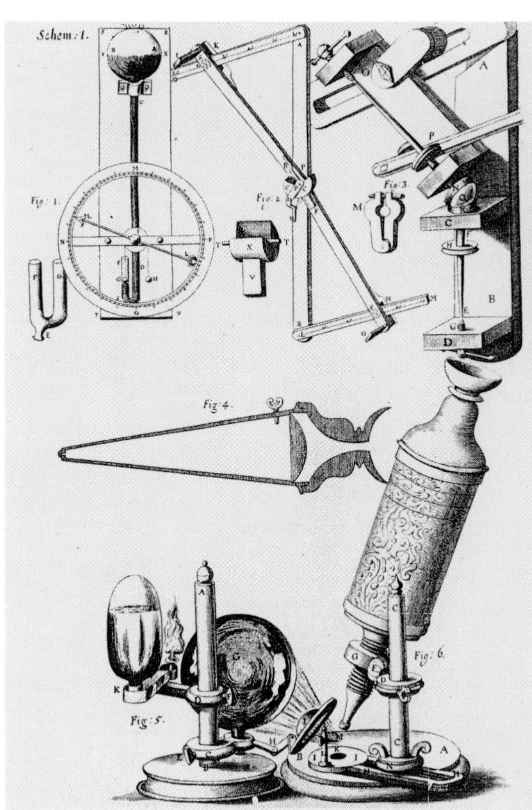

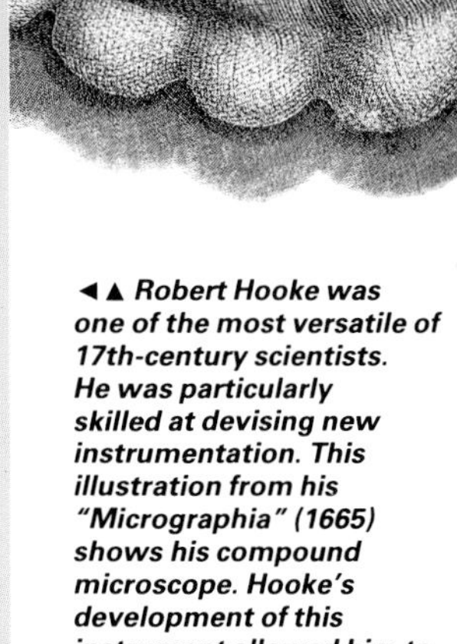

◄▲ Robert Hooke was one of the most versatile of 17th-century scientists. He was particularly skilled at devising new instrumentation. This illustration from his "Micrographia" (1665) shows his compound microscope. Hooke's development of this instrument allowed him to make the first detailed observations of such objects as this louse (attached to a hair).

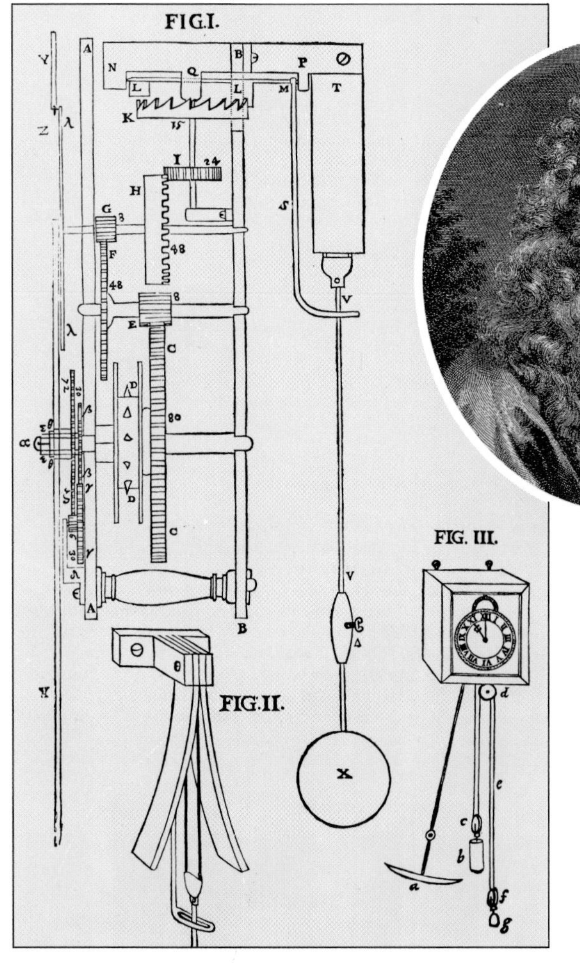

◄▲ The Dutch scientist, Christiaan Huygens, worked on many of the same problems as Newton. He investigated the forces on bodies moving in circular paths and applied his results to the pendulum clock; these studies in his book, "Horologium Oscillatorium" (1673) were the most significant step forward since Galileo. He also helped improve the design of telescopes and discovered that Saturn has rings. He stimulated Leibniz's interest in the calculus.

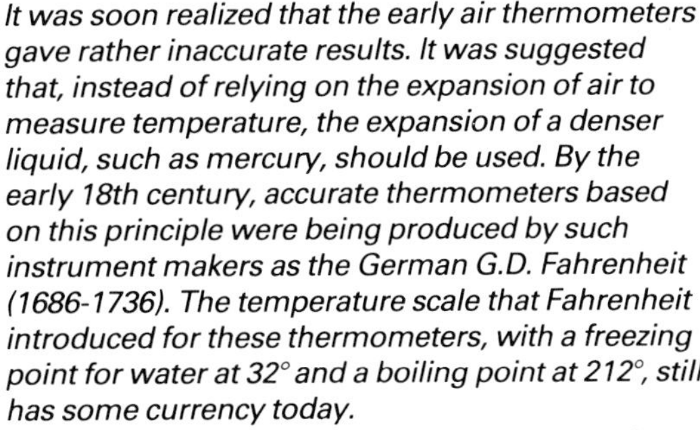

It was soon realized that the early air thermometers gave rather inaccurate results. It was suggested that, instead of relying on the expansion of air to measure temperature, the expansion of a denser liquid, such as mercury, should be used. By the early 18th century, accurate thermometers based on this principle were being produced by such instrument makers as the German G.D. Fahrenheit (1686-1736). The temperature scale that Fahrenheit introduced for these thermometers, with a freezing point for water at 32° and a boiling point at 212°, still has some currency today.

Mercury was also used for experiments on the pressure of the atmosphere, which led to the development of the barometer. Galileo's pupil Evangelista Torricelli (1608-1647), realized that an apparatus rather similar to the air thermometer could be made to measure pressure. In this case, the space at the top of the glass tube was left empty – instead of air, there was a vacuum. The height of the mercury in the tube then depended on the pressure exerted by the atmosphere on the mercury in the trough round the bottom end of the tube. In France, Blaise Pascal (1623-1662) confirmed that it was really atmospheric pressure that was being measured by having a "Torricellian" barometer carried to the top of a mountain in central France. The level of the mercury in the tube fell because the pressure of the atmosphere decreased with height.

Clocks

The high-precision scientific instrument of the 17th century was the clock. Clocks had originally appeared in Europe as devices for sounding bells in church towers: the word "clock" comes from the same root as the French cloche or "bell". These early mechanisms were very inaccurate, often losing or gaining many minutes a day. What was needed was clearly some way of regulating the rate at which the mechanism worked. In the latter part of the 17th century, Christiaan Huygens, following Galileo's suggestion, used a pendulum for this purpose, because the time taken for each swing of a pendulum remained fairly constant, day-in, day-out. Introduction of the pendulum clock, in much the same form as the grandfather clocks which are still in use, immediately increased the accuracy of time-keeping by a factor of a thousand. Since measurement of longitude requires an accurate knowledge of time, this meant that mapping of the Earth soon also became far more accurate than before (for seafarers, the equivalent breakthrough came with the invention of the chronometer by John Harrison in 1735). The importance of this advance is reflected in the high status assigned to clockmakers by the early 18th century. For example, the leading English clockmaker of the day, Thomas Tompion, was accorded the same honor as Newton of being buried in Westminster Abbey.

The Reflecting Telescope

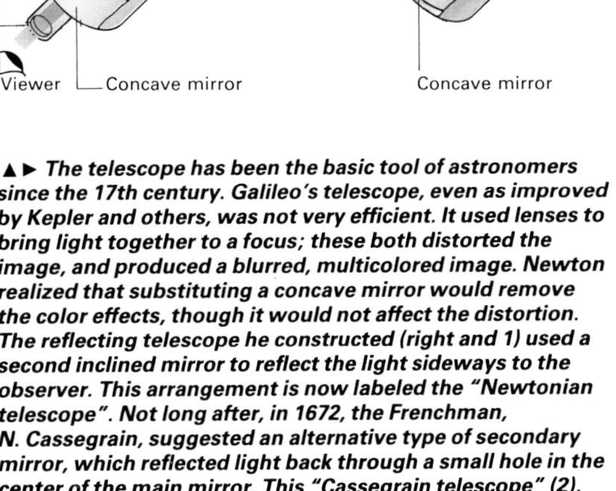

▲► The telescope has been the basic tool of astronomers since the 17th century. Galileo's telescope, even as improved by Kepler and others, was not very efficient. It used lenses to bring light together to a focus; these both distorted the image, and produced a blurred, multicolored image. Newton realized that substituting a concave mirror would remove the color effects, though it would not affect the distortion. The reflecting telescope he constructed (right and 1) used a second inclined mirror to reflect the light sideways to the observer. This arrangement is now labeled the "Newtonian telescope". Not long after, in 1672, the Frenchman, N. Cassegrain, suggested an alternative type of secondary mirror, which reflected light back through a small hole in the center of the main mirror. This "Cassegrain telescope" (2), along with the Newtonian, is commonly used today.

◄ A portable barometer made about 1705. Evangelista Torricelli invented the mercury barometer in 1643. Blaise Pascal's experiments in 1646 confirmed that the atmosphere had weight.

Newton's predictions concerning cometary orbits and the shape of the Earth were impressively confirmed, but only after his death

Predictions and tests

In the first two books of the *Principia*, Newton laid down his own ideas concerning the workings of the Universe, and showed that they fitted the observations, whereas other theories did not. However, the important point about a theory for Newton was that it should not only represent known data: it should also predict new and, preferably, unexpected results. In the final book of the *Principia*, he turned his attention to the predictions that stemmed from his ideas. Two of these proved to be especially important.

In the first place, Newton showed that a planet is held together because of the gravitational forces between all its constituent pieces; these forces acted in such a way that the planet would necessarily be spherical. However, this was only true if the planet was not spinning round. Any rotation of the planet would tend to make the equator bulge out: the faster the rotation, the bigger the bulge. Since the Earth was obviously rotating, it could not be a perfect sphere as Aristotle and everyone since had supposed. From the known size, mass and spin rate of the Earth, Newton was able to calculate, to within 1 percent of the present accepted value, just how much the Earth should deviate from being a sphere.

Newton's second important advance concerned the most mysterious objects then known to astronomy – the comets. From time immemorial, comets had been objects of dread, for they seemed to appear and disappear without obvious cause or reason. All the leading astronomers of the 17th century tried to account for comets, without much

◄▲ *Newton's contemporaries were struck by his prediction that the Earth is flattened, not spherical. Results derived from the mapping of France suggested he was wrong. Only after his death was the prediction confirmed.*

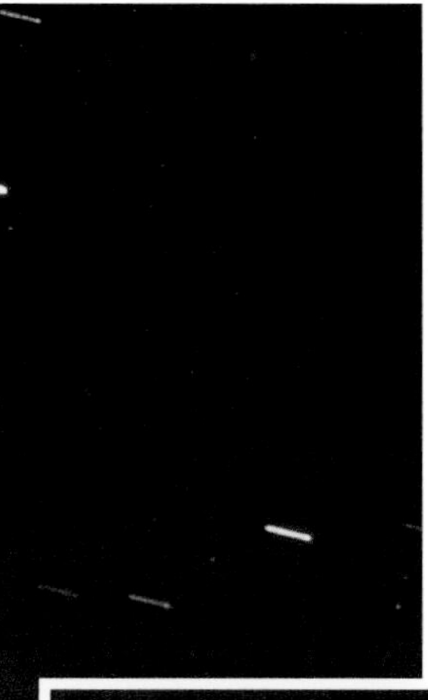

◄▼ *Edmond Halley, when secretary of the Royal Society, played a major part in ensuring that the "Principia" was published. Applying Newton's theory to the study of comets, he found that one returned at regular intervals. This comet ("Halley's comet") was seen in AD 66, at the same time when Jerusalem was destroyed (bottom).*

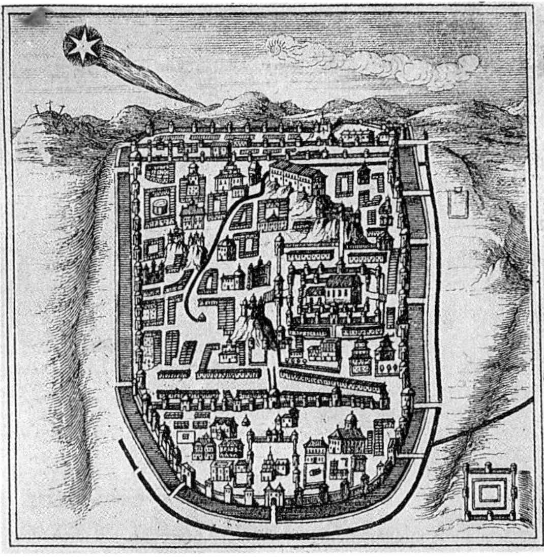

success. Now Newton found he could predict what paths comets should follow through the Solar System. He pointed out that the near-circular orbits of the planets round the Sun were only one possibility – objects could follow highly elongated orbits round the Sun, perhaps leaving the planetary Solar System altogether. Newton showed that comets were in this latter type of orbit: their paths were therefore as predictable and understandable as any planetary orbit.

These two predictions were especially important because they provided crucial tests of Newton's ideas. During the latter part of the 17th century, considerable effort was expended in France on the accurate mapping of that country. As a part of this activity, measurements were made of what distance on the ground corresponded to one degree of latitude. On a perfectly spherical Earth, each degree of latitude from equator to pole would correspond to exactly the same length on the ground. If the Earth is not a sphere, each degree of latitude will differ in length. Hence, measuring ground lengths at different latitudes enables the shape of the Earth to be determined. The French looked at the results from mapping their country and decided that they certainly suggested a non-spherical Earth. Unfortunately, they seemed to indicate an Earth that bulged at the poles and was constricted at the equator – the opposite of Newton's prediction!

Edmond Halley continued to be as interested in the *Principia* after its publication as he had been before. He took up, in particular, Newton's ideas on comets and tried to test them against existing observations. In looking through records of old cometary observations, he found that comets seen in 1531 and 1607 together with the one he had observed himself in 1682 seemed to be following a similar path round the Sun. He proposed that these were actually three appearances of a single comet which orbited the Sun approximately every 76 years. Consequently, he predicted that the comet would be seen again, following the same path, in the middle of the 18th century. Because of the great significance his contemporaries attached to comets, this was an especially striking prediction. Unfortunately, confirmation entailed a long wait: the return of "Halley's comet", as it became known, would occur – if it did – when Halley, Newton and their generation were dead.

As it happened, the controversy over the shape of the Earth was also not finally resolved until after Newton's death. The result was that, during his lifetime, he saw full acceptance of his ideas by colleagues in Britain, but only slow and often partial acceptance elsewhere. The breakthrough which led to universal belief in Newtonian ideas was again due to a French initiative. French scientists realized that measuring the shape of the Earth from two adjacent degrees of latitude in France was not the most accurate way, since big differences in length would only show up for well-separated latitudes. Two French expeditions were therefore specially dispatched in the 1730s, one to South America and one to Lapland, to make new sets of measurements. The results showed not only that Newton was right in declaring the Earth to bulge at the equator, but that his quantitative estimate of the size of this bulge was about right. As the writer Voltaire, one of Newton's strongest supporters in France and his chief popularizer, commented sardonically to a French scientist involved in these measurements: "You have found after tedious effort, what Newton found without leaving his home." When, in 1758, Halley's comet returned as predicted, acceptance of the Newtonian picture of the world became total.

Newton and the Muse

Science and the educated layman

Newton's "Principia" was published by the Royal Society, whose president had delegated powers to approve material for publication. So it is that the title page of the "Principia" bears the names not only of Newton, but also of Samuel Pepys, who was president at the time. There was nothing unusual in an enthusiastic amateur occupying this post: what was required was a genuine interest in science and good standing in society. Pepys' diary records his interest. In August 1666 he was on the roof of his house till one o'clock in the morning, "looking on the Moon and Jupiter, with this twelve-foote glasse and another of six foote, and the sights mighty pleasant, and one of the glasses I will buy, it being very useful."

A genius admired

The new picture of the world that Newton constructed influenced the thought of his contemporaries and successors at a variety of levels. At the simplest level, they showed their great admiration of his genius. James Thomson in "The Seasons", one of the best-known English poems of the 18th century, refers to

"Newton, pure Intelligence, whom God
To mortals lent, to trace his boundless works
From laws sublimely simple."

This adulation peaked at the time of Newton's death, when a number of commemorative poems were written. Some have a common theme: Newton's soul leaves the Earth, voyages through the Solar System and then onward into space, en route surveying with approval the various celestial bodies obeying his laws of motion and gravitation.

Imagining the Universe

At a deeper level, writers wondered what impact Newtonian ideas might have on their beliefs. For example, Newton had demonstrated what his predecessors had only guessed at – that the Universe was likely to be infinite in extent. This raised in an acute form not only the theological question of where Heaven might be situated, but also the personal question of whether the Universe could even be imagined. As Edward Young wondered in his "Night Thoughts" in the mid-18th century:

"Where ends this mighty building? Where begin
The suburbs of Creation? Where, the wall
Whose battlements look o'er into the vale
Of non-existence? Nothing's strange abode!"

Newton had not excluded God from his Universe. This was something reserved for his successors, especially those in France, to claim they could do. Newton, on the contrary, invoked the intervention of God: for example, for the initial placement of the planets in their orbits round the Sun. The use of the word "attraction" in talking of gravitation was also often felt to have Christian overtones. Indeed, one of Newton's contemporaries even saw the workings of the British monarchy in terms of an attraction which he equated to mutual love. He compared ministers of state dancing attendance on the king with the planets moving round the Sun.

"Opticks" and the rainbow

Yet it was Newton's "Opticks" which ultimately made most impact on 18th-century writers; for here was a book they could read in English and understand for themselves, and which, moreover,

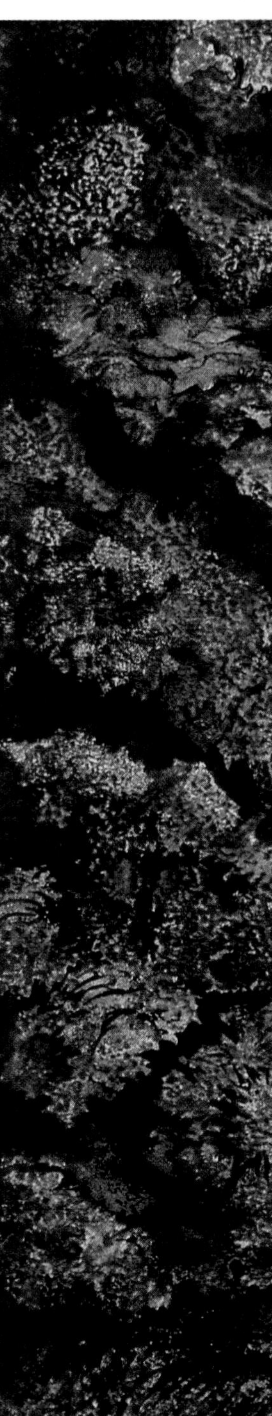

▲ Newton as pictured by the English artist and poet, William Blake. Although Blake recognized Newton's genius, he deplored the mechanistic view of the world that stemmed from Newton's ideas.

◄ Newton's explanation of the rainbow captured the imagination of many 18th-century artists and writers. Among these was Joseph Wright of Derby who painted this picture.

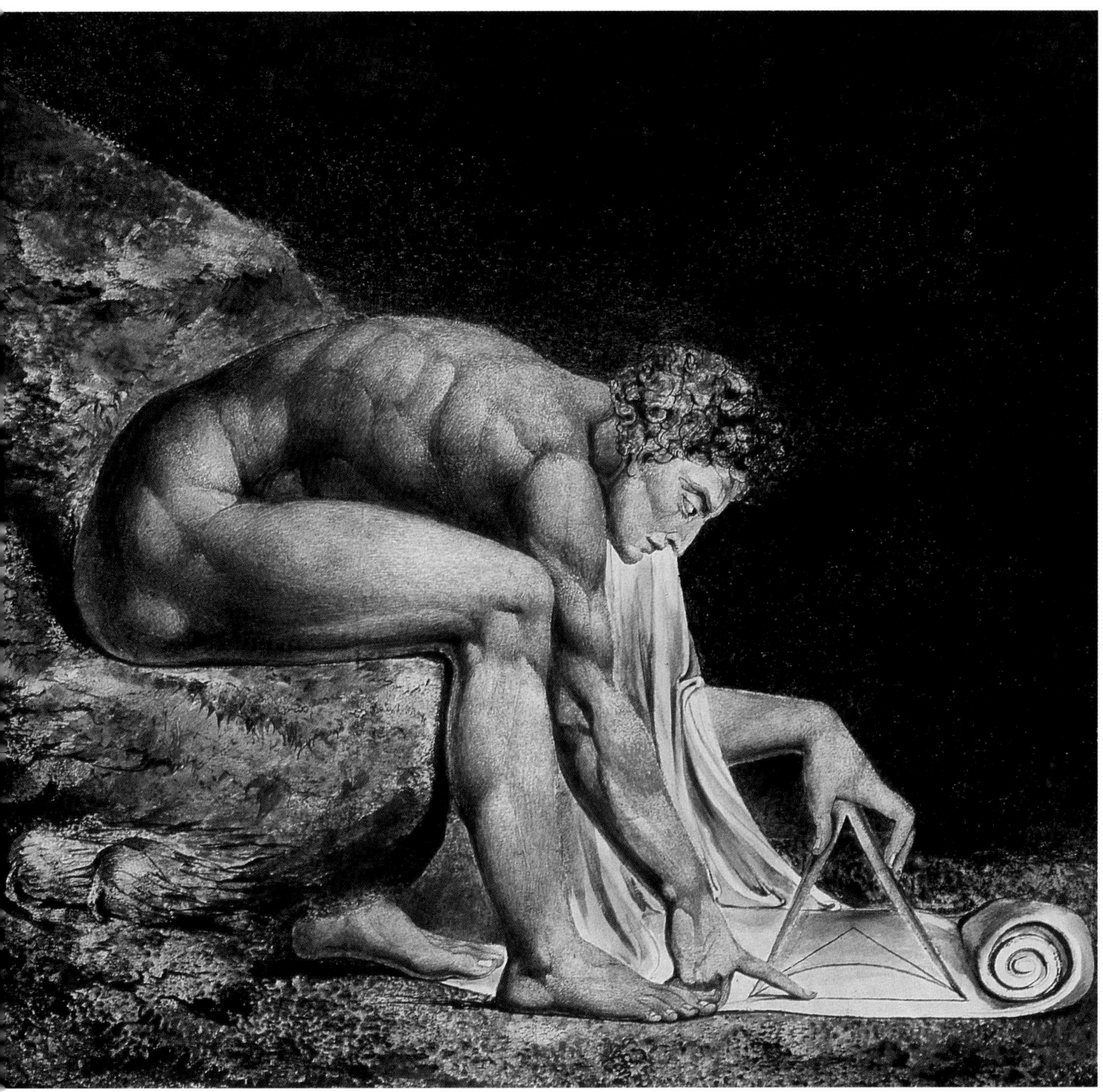

had important things to say about everyday events. The rainbow provides an excellent example. Newton's study of the colors which resulted when white light was passed through a prism could be applied immediately to the colors of the rainbow. His explanation of how rainbows are formed was seized on not only by writers, but also by artists inspired, as was Thomson in "The Seasons", by "the various twine of light by thee disclosed".

A "tyranny of science"?

One unexpected result of Newton's work was to emphasize the beginnings of a division between science and the arts, between scientific and literary language. This divide has continued to grow since, leading some years ago to the proposal by the English author and physicist C.P.Snow that they actually formed two different cultures. Voltaire, whose interests spanned both, recognized there was a problem. He commented unhappily in 1735:

"Fine literature is hardly any longer the fashion in Paris. Everyone works at geometry and physics. Everyone has a hand at an argument. Sentiment, imagination and the finer arts are banished. Not that I am angry that science is being cultivated, but I don't want it to become a tyrant that excludes all else."

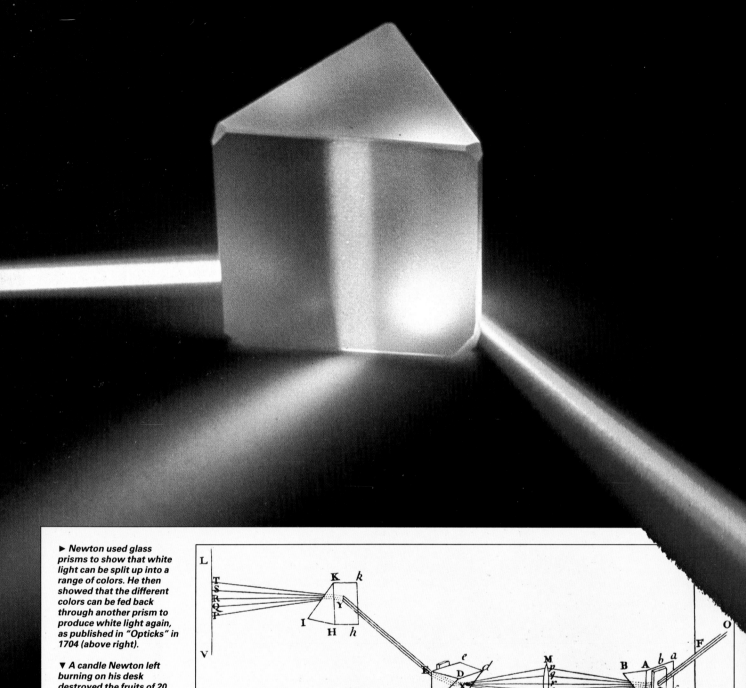

► Newton used glass prisms to show that white light can be split up into a range of colors. He then showed that the different colors can be fed back through another prism to produce white light again, as published in "Opticks" in 1704 (above right).

▼ A candle Newton left burning on his desk destroyed the fruits of 20 years' work on optics, and he had to start again from scratch. This sketch must postdate the disaster.

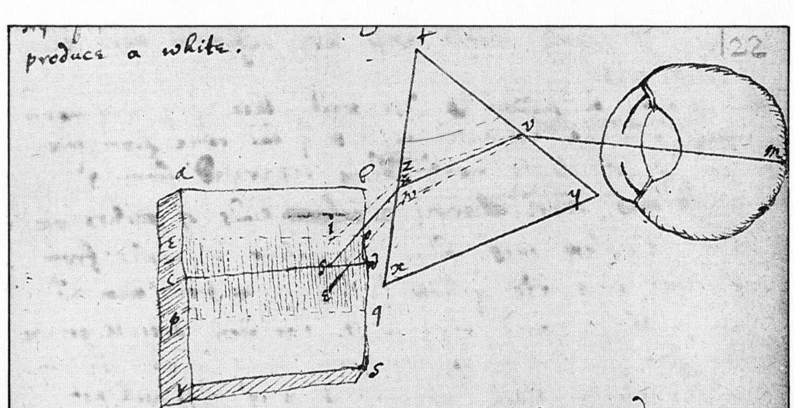

▼ Theodoric of Freibourg's "On the Rainbow" (early 14th century) was the standard work on the subject for over three centuries. This copy of a diagram by him comes from the account published by Jodocus Trutfetter in 1514.

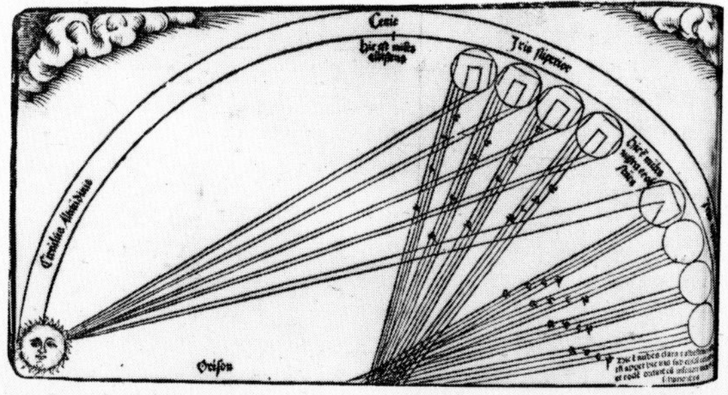

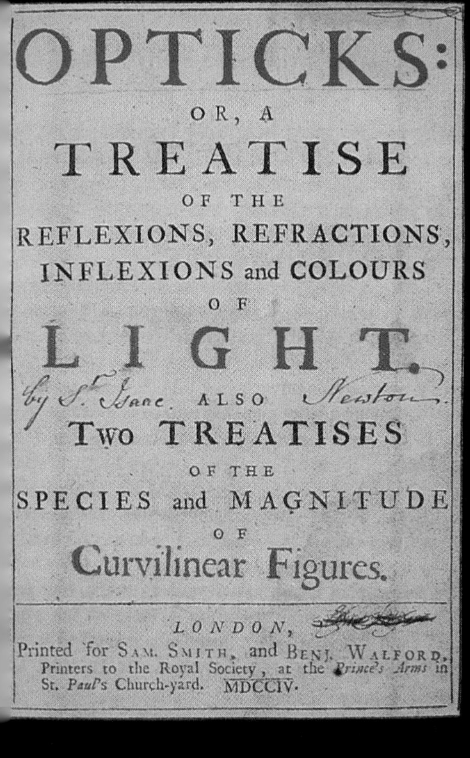
OPTICKS:
OR, A
TREATISE
OF THE
REFLEXIONS, REFRACTIONS,
INFLEXIONS and COLOURS
OF
LIGHT.
ALSO
Two TREATISES
OF THE
SPECIES and MAGNITUDE
OF
Curvilinear Figures.

LONDON,
Printed for SAM. SMITH, and BENJ. WALFORD,
Printers to the Royal Society, at the *Prince's Arms* in
St. *Paul's* Church-yard. MDCCIV.

The nature of light

The *Principia* was Newton's masterpiece; but, even in its absence, his other contributions would have made him a major figure of 17th-century science. The most important of the other topics on which he worked was light; his second important book, *Opticks*, was published in 1704. Since his days as an undergraduate student at Cambridge, Newton had been interested in the effect of a glass prism on sunlight. He carried out a series of experiments which by the early 1670s had led him to a new view of the nature of light. It had always been supposed that light was basically white, though colors could be added to it. Newton now showed that the colors of the rainbow, from blue to red, were actually the basic components: white light was simply a mixture of these colors. His work led him to two further topics concerning light – one practical and the other theoretical.

Newton's practical advance concerned the telescope. Early telescopes suffered from a number of defects. One which particularly affected the images seen was chromatic aberration, whereby the image was blurred by a surrounding rainbow of colors. Newton showed that such colored images were the inevitable result of white light passing through the lenses of the telescope; he claimed that this defect would always be present in a refracting telescope (i.e. one employing only lenses). Later, in the mid-18th century, it was shown that his claim was for once wrong: a better knowledge of the properties of glass allowed the design of refracting telescopes with little chromatic aberration. But his belief led Newton on to his important invention, in 1668, of the reflecting telescope, which uses a mirror rather than a lens to collect the light. Since color defects are caused by light passing though the lens, a mirror which reflects light from its surface introduces no color effects, so chromatic aberration can be eliminated. Over the years since Newton's invention, reflecting telescopes have been found to offer a number of advantages, and all large astronomical telescopes are now of this type.

Newton's theoretical advance concerned the ultimate nature of light. He proposed that it consists of small particles ("corpuscles") which are shot out from a source of light in much the same way that pellets are ejected from a shotgun. Although Newton was able to explain many of the known properties of light his ideas were in conflict with those of his contemporary, Christiaan Huygens, who proposed that light could be better understood as consisting of waves. At the time, Newton's corpuscular particle or emission theory of light won, wrongly in the view of 19th-century scientists, who found that new experiments on light could best be explained in terms of waves. However, light has proved to be remarkably difficult to understand: in the 20th century, we regard it as behaving sometimes like particles and sometimes like waves (➧ page 235). In Newton's day, at least, his picture of light helped, rather than hindered progress.

The Age of Reason

Two Newtonian approaches to nature

The 18th century, commonly referred to as the "Age of Reason", has also often been called the "Age of Newton and Locke", in the sense that it was profoundly influenced by the ideas of both Englishmen. Newton's own major advances fall into two compartments, one labeled "gravitation and motion" and the other "light". There was a fundamental difference in the way he approached these two subjects. The "Principia" was a highly theoretical treatise presented along the lines of Euclid's classical text on geometry. It showed the Universe as having a logical structure which could be properly interpreted by the application of mathematical methods. Newton's "Opticks" was different. It emphasized the importance of experimentation carried out without prior theorizing. Yet Newton also ended "Opticks" with a list of fascinating speculations regarding some of the major scientific questions of the day. Readers of his book would gain the impression that nature must be investigated piecemeal by laborious experiment, lightened by occasional speculation.

John Locke and reason

These two approaches, despite their differences, both influenced 18th-century thought: they were welded together by John Locke (1632-1704), one of the most influential of all philosophers. Locke was a friend and admirer of Newton, and had a wide acquaintance with leading scientists of the late 17th century. He concerned himself with the implications of the new scientific developments for understanding how human beings acquire a knowledge of the world around them. His main conclusion was that we can only obtain such knowledge via our senses – by experience, as he called it.

Data provided by experience of the world are subject to rational analysis by the mind so providing us with a true knowledge of what is happening. To Locke's contemporaries, this raised an immediate query – how did religious experience fit into his picture? Locke's answer was that faith must ultimately justify itself in terms of the same rational analysis as other experience:

"Reason therefore here, as contradistinguished to faith, I take to be the discovery of a certainty or probability of such propositions or truths, which the mind arrives at by deductions made from such ideas which it has got by the use of its natural facilities, viz., by sensation or reflection.

Faith, on the other side, is assent to any proposition, not thus made out by the deduction of reason, but upon the credit of the proposer, as coming from God in some extraordinary way of communication. This way of discovering truths to men we call revelation.

No proposition can be received for divine revelation, or obtain the assent due to all such, if it be contradictory to our clear intuitive knowledge, because this would be to subvert the principles and foundations of all knowledge."

Here Locke defines reason in a way that the 18th century came to accept as natural. But he pushed his argument a good deal further. He believed that a properly run state was one in which people lived together according to the light of reason. In such a society, each individual would have certain natural rights which could not be taken away from them. These ideas of Locke spread widely during the 18th century and were absorbed into both the United States Constitution and the French Revolution.

Locke's activities were aimed, in part, at applying to individuals and groups Newton's method of studying the physical world. He expected that a proper experimental approach would lead to generalizations about human behavior along the lines of Newton's laws.

◄ The French scientist Madame du Châtelet (1706-1749) translated Newton's "Principia" and helped disseminate Newtonian philosophy. Her companion, the writer Voltaire, had visited England in the 1720s and became a proponent of Newtonian ideas, once unpopular in France.

► Benjamin Franklin (1706-1790), was a leading scientist in the Newtonian tradition. He was also one of the five people involved in drafting the Declaration of Independence on 4 July, 1776 depicted in John Trumbull's painting. The Newtonian philosophy, as transmitted by Locke, is apparent in the 18th-century debate concerning the need for a system of checks and balances in government (as opposed to the absolute power wielded by monarchs). This was part of the background to the US Constitution.

◄ The English philosopher John Locke was a qualified doctor and a fellow of the Royal Society. Newton's influence is particularly evident in his examination of how the human mind gathers information. Newton's published writings were mainly concerned with exploring specific scientific problems. But a new way of looking at the world lay behind them. This Newtonian world view became widespread during the 18th century and influenced opinion over a wide range of topics. Locke was a prominent figure in Whig politics and an advocate of toleration. His ideas had a major impact on 18th-century philosophical and political thought.

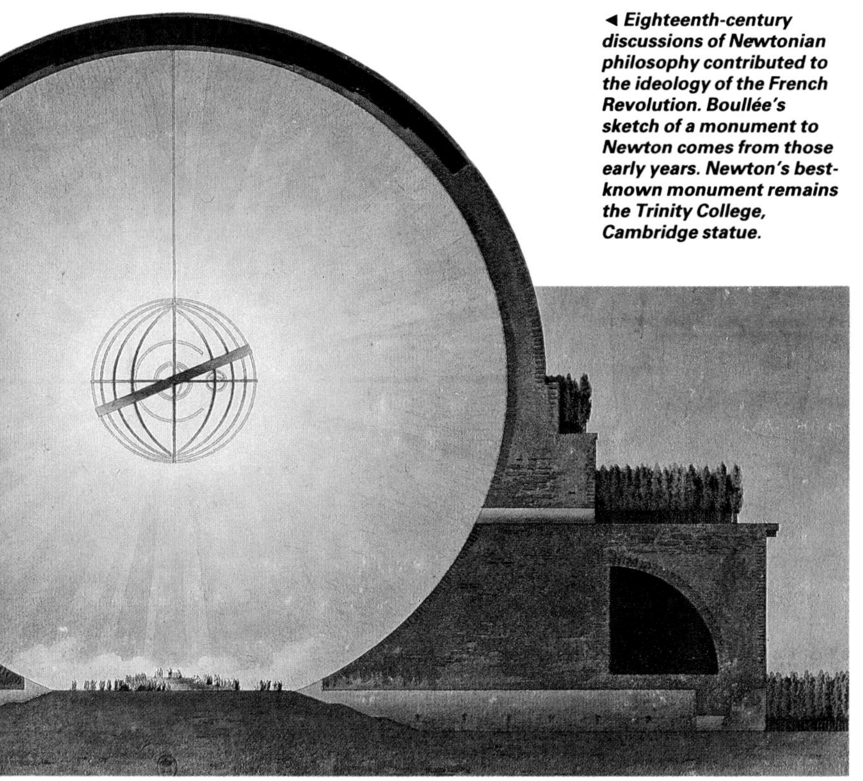

◄ Eighteenth-century discussions of Newtonian philosophy contributed to the ideology of the French Revolution. Boullée's sketch of a monument to Newton comes from those early years. Newton's best-known monument remains the Trinity College, Cambridge statue.

Followers and critics

This extension of the scientific approach to topics other than natural science was adopted by Locke's successors. The first major book on economics, "The Wealth of Nations" (1776) by the Scottish economist Adam Smith (1723-1790), was partly inspired by an admiration for Newton. It supposes that social activities are governed by their own laws, which can be discovered by careful study.

From the beginning, the Age of Reason had its critics. By the early 19th century, one strong strand of criticism had led to the flowering of Romanticism, but one of the most influential critics lived in the mid-18th century. He was David Hume (1711-1776), a Scottish philosopher who believed that there were limits to the role that reason could play in ordering human life. It could not, he asserted, act as a guide to morality. Rather reason, as defined by Locke, "is and ought to be the slave of the passions". The belief that investigating the Universe along the lines laid down by Newton would lead to a complete description of its nature on this reckoning was mistaken. Hume observed: "While Newton seemed to draw off the veil from some of the mysteries of nature, he showed at the same time the imperfections of the mechanical philosophy; and thereby restored her ultimate secrets to that obscurity in which they ever did and ever will remain."

"To myself I seem to have been only like a boy playing on the seashore" – Newton

◄ ▼ A coin press of the type in use when Newton was in charge as master of the Royal Mint in London from 1699. This could be used to produce medals, as well as coins. The medallion of Newton was struck at the Mint in the year of his death.

▼ Newton's portrait is held by the Greek goddess, Artemis. Halley's comet appears above, and the Big Dipper or Great Bear on the globe below. This sketch by William Stukeley, an 18th-century biographer of Newton, is a reminder that Newton, a founder of modern science, was also interested in astrology, alchemy, the occult and unconventional theology.

Newton in society

Newton's major work, the *Principia*, appeared at an important point in English political history. The unpopular King James II was attempting to restore some acceptance of the Roman Catholic church. Newton emerged as a leader of the opposition at Cambridge. When James was deposed in 1688 and William of Orange became king, Newton's importance in the world outside Cambridge grew. In 1696 he was appointed to the Royal Mint in London, where he played a significant role in revising the chaotic English coinage. By now his academic links with Cambridge were tenuous, and he finally resigned as professor of astronomy, on being elected (for a second time) member of parliament for the university in 1701. Not long after, he was appointed president of the Royal Society, a position he continued to hold until his death. But it was for his political activity that he was knighted in 1705, rather than for his abilities as a scientist.

When Newton was concentrating on a problem, other activities, including eating and drinking, were forgotten. But such periods of intense concentration only occupied a part of his long life. Despite his emergence as an important figure in society, however, Newton always remained a secretive man. Perhaps it was because, in the religious terms of the time, he held heretical views (he was an early Unitarian). This fact he kept so secret that no voice was raised against his burial in Westminster Abbey with the great of the land when he died in 1727. And yet his secretiveness may also seem appropriate; for he always saw nature as itself a secret which could only be interpreted with great effort and, even then, only partially: "I do not know what I may appear to the world, but to myself I seem to have been only like a boy playing on the seashore, and diverting myself in now and then finding a smoother pebble or a prettier shell than ordinary, whilst the great ocean of truth lay all undiscovered before me."

Antoine Lavoisier 1743-1794

5

The theory of phlogiston, a combustible "earth"... "Pneumatic" chemists discover "airs"...Lavoisier's experiments with water...and on decomposition of oxides...The oxygen theory of combustion...Lavoisier and followers advance the new chemistry... PERSPECTIVE...Vicissitudes of the Academy...The new chemists' equipment...Impact of the French Revolution...The developing language of science

"Chemistry is a French science; it was founded by Lavoisier of immortal fame." So wrote a French chemist in 1868. Needless to say, at a time of intense European nationalism and rivalry in science, as much as in politics, such a claim was received amid controversy. Nevertheless, even though Lavoisier could never have achieved what he did without the prior and contemporary investigation and intepretation of British, Scandinavian and German chemists, there is an essential grain of truth in the statement. For Lavoisier restructured chemistry from fundamental principles, provided it with a new language and fresh goals. To put it another way, a modern chemist on looking at a chemical treatise published before Lavoisier's time would find it largely incomprehensible; but anything written by Lavoisier himself, or composed a few years after his death, would present a modern reader with very little difficulty.

Lavoisier's upbringing

Antoine Laurent Lavoisier was born in Paris on 26 August 1743, the son of a lawyer who held the important position of solicitor to the Parisian Parlement, the chief Court of France. His wealthy mother, who also came from a lawyer's family, died when he was only five. Not surprisingly, Lavoisier's education was geared to his expected entry into the legal profession. This meant that he attended the best school in Paris, the Collège des Quatre Nations (known popularly as the Collège Mazarin). The building still survives and now houses the Institut de France, of which the French Academy of Sciences is part. Lavoisier spent nine years at the college, graduating with a baccalaureate in law in 1763, but his spare time was always to be devoted entirely to scientific pursuits.

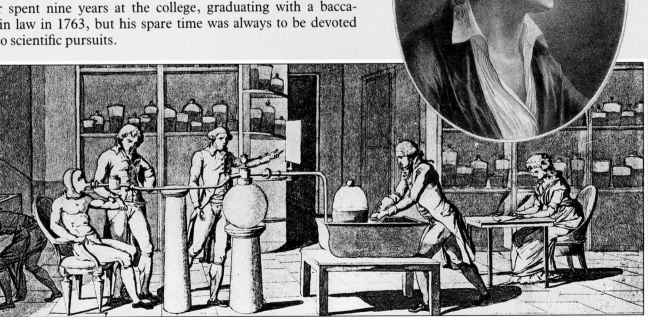

Author of the "Chemical Revolution"

Antoine Laurent Lavoisier was the author of the "Chemical Revolution" which took place in the 18th century. Compared with advance in other fields, these changes came late. The long postponement is explained by the sheer complexity of chemical phenomena, the failure to deal with the properties of pure substances, by the lack of any concept of the gaseous state, and by widespread confusion over the nature of chemical elements. For centuries, alchemical philosophers had supposed, as had Aristotle, that substances could be "transmuted" into each other.

Lavoisier restructured chemistry from fundamental principles, provided it with a new language and fresh goals. His system provided a firm foundation for Dalton's atomic theory at the beginning of the 19th century and allowed Berzelius, Dumas, Liebig and Pasteur to develop organic chemistry, from which the powerful ideas of valence and structure emerged in the 1860s.

Lavoisier's chemical elements were classified into a periodic table in 1869 by Mendeleev, but their arrangements and reactivities only received an explanation in the 20th century with the discovery by Curie of radioactivity and the theories of the electron and quantum mechanics.

▼ Lavoisier overturned a 100-year-old "phlogiston" theory of combustion and established the basis of modern chemistry. His wife's drawing depicts him conducting an experiment on respiration and herself taking notes.

Antoine Lavoisier – His Life, Work and Times

Chemistry before Lavoisier

Unlike astronomy and mechanics, whose subject matter had been transformed in the 16th and 17th centuries by Copernicus, Galileo, Newton and others, chemistry retained older patterns of thought well into the 18th century. For example, many chemists still did not think of chemical changes as due to interactions of a limited number of elements, or as involving rearrangements of groups of atoms. Chemical properties, too, were often explained in a not very enlightening way as due to the presence of "principles". Thus the fact that certain substances could burn, or were acidic in taste, was explained by supposing that they contained a "burning" or "acid" principle as one of their constituents.

These ideas, together with a good deal of observational and experimental evidence to support them, were codified at the beginning of the 18th century in an influential theory of matter published by the German chemist Georg Stahl (1660-1734). According to Stahl, all substances were ultimately composed from water and three varieties of earth, one of which was a combustible material called *phlogiston*, which was ejected into the atmosphere. In this theory, therefore, metals burned in air not because they absorbed something from the air, but because they lost phlogiston. This model received considerable support from the observation that metals could be recovered from what were called their "calces" (oxides) by combining them with a phlogiston-rich material such as charcoal.

Clearly phlogiston was a transferable agent (just as oxygen is in today's model of combustion) and it was this, rather than its supposed weightlessness, which made phlogiston of interest to contemporary chemists. To us, the principal snags of the theory seem clear: metals are elements, not compounds; metals, like any burning substance, gain in weight when they are burned, whereas if they lost phlogiston they would lose weight. In fact, it was not shown conclusively that metals did gain in weight until 1771, and this demonstration played a crucial role in making Lavoisier interested in combustion.

Ideas about respiration had scarcely advanced beyond those of Aristotle: breathing was thought to cool and ventilate the blood. However, in the 17th century, the Anglo-Irish natural philosopher Robert Boyle (1627-1691) had noticed analogies between respiration and combustion (for example, the way in which neither life nor flame persists in an enclosed volume of air) and speculated that both processes might be due to the presence of a "nitrous" substance in the air. To the phlogistonist, however, a flame ceased to burn, or a canary ceased to breathe, because the air had become saturated with phlogiston. Air was "purified" by plants absorbing phlogiston – thus completing a "phlogiston cycle" in nature.

Phlogiston theory (which was also used to explain acidity) seems like looking-glass chemistry from our modern viewpoint. Probably the principal reason why it appeared self-consistent to contemporaries is that until the end of the 18th century there was no concept of a gaseous state of matter. Chemical reactions only occurred between solids and liquids, and the air was regarded either solely as a waste bin for phlogiston fumes (which were reabsorbed by plants and animals), or as an element bearing physical properties, such as pressure.

It was not until 1727, when the English clergyman Stephen Hales (1677-1761) showed that large volumes of "air" could be released from the pores of solids and liquids, that chemistry became a three-dimensional subject.

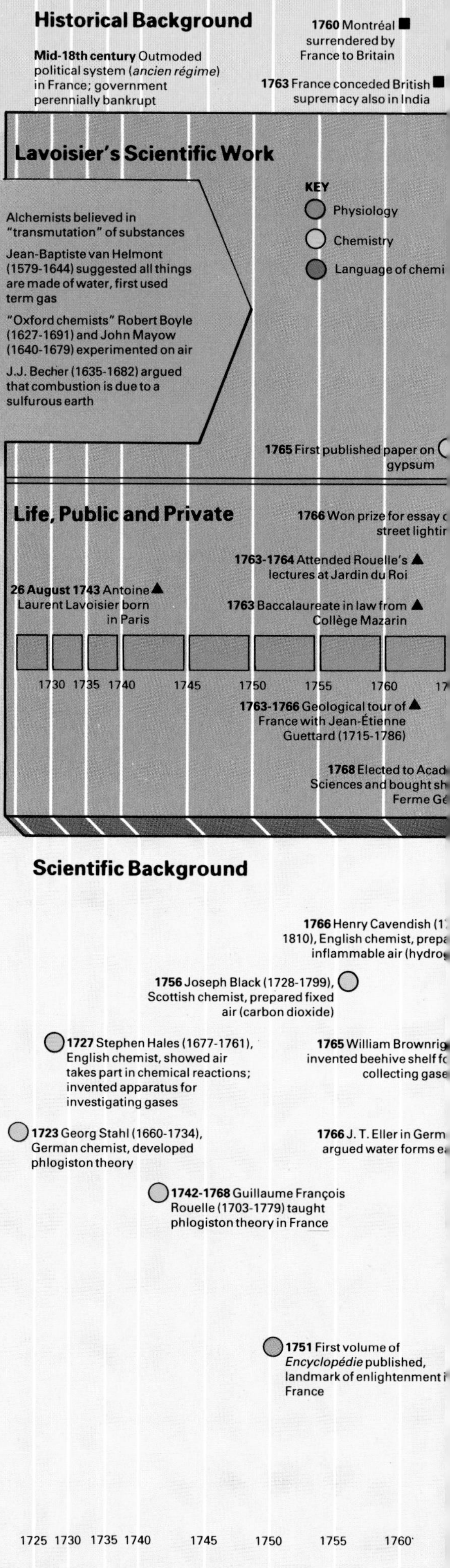

Historical Background

Mid-18th century Outmoded political system (*ancien régime*) in France; government perennially bankrupt

1760 Montréal surrendered by France to Britain

1763 France conceded British supremacy also in India

Lavoisier's Scientific Work

KEY
Physiology
Chemistry
Language of chemi

Alchemists believed in "transmutation" of substances

Jean-Baptiste van Helmont (1579-1644) suggested all things are made of water, first used term gas

"Oxford chemists" Robert Boyle (1627-1691) and John Mayow (1640-1679) experimented on air

J.J. Becher (1635-1682) argued that combustion is due to a sulfurous earth

1765 First published paper on gypsum

Life, Public and Private

26 August 1743 Antoine Laurent Lavoisier born in Paris

1763 Baccalaureate in law from Collège Mazarin

1763-1764 Attended Rouelle's lectures at Jardin du Roi

1766 Won prize for essay c street lightir

1763-1766 Geological tour of France with Jean-Étienne Guettard (1715-1786)

1768 Elected to Acad Sciences and bought sh Ferme Gé

1730 1735 1740 1745 1750 1755 1760 17

Scientific Background

1723 Georg Stahl (1660-1734), German chemist, developed phlogiston theory

1727 Stephen Hales (1677-1761), English chemist, showed air takes part in chemical reactions; invented apparatus for investigating gases

1742-1768 Guillaume François Rouelle (1703-1779) taught phlogiston theory in France

1751 First volume of *Encyclopédie* published, landmark of enlightenment i France

1756 Joseph Black (1728-1799), Scottish chemist, prepared fixed air (carbon dioxide)

1765 William Brownrig invented beehive shelf fc collecting gase

1766 Henry Cavendish (1 1810), English chemist, prepa inflammable air (hydro

1766 J. T. Eller in Germ argued water forms e

1725 1730 1735 1740 1745 1750 1755 1760

■ **1780s** Start of Industrial Revolution in England marked by expansion of cotton industry

1793 Louis XVI tried and executed; "Reign of terror", war dictatorship in France ■

■ **1795, 1807** France took large parts of Prussia

-**1771** First Pacific voyage of es Cook, who claimed ralia for Britain

■ **1778** France intervened on the side of the 13 colonies in the American War of Independence

■ **1789** French Revolution began with storming of the Bastille prison

■ **1804** Napoleon enthroned as Emperor of France

1785 Attacked theory of phlogiston

1787 Defined the chemical element and published *Method of Chemical Nomenclature* with Guyton, Claude-Louis Berthollet (1748-1822) and Antoine Fourcroy (1755-1809)

1805 Lavoisier's *Mémoires de chimie* published by his widow and collaborator (now married to Count Rumford)

1774 Published *Opuscules physiques et chimiques*

1772-1775 Showed metals burn by absorbing vital principle from air

1783 Showed that water is a compound of hydrogen and oxygen

1789 Published textbook *Elementary Treatise of Chemistry*

70 Proved that water cannot transmuted into earth xperiment started in 1768)

1777 Developed theory of gaseous state which involves heat as a principle of expansion

1789 Worked on respiration with Armand Séguin (c. 1765-1835)

1784 Worked with Pierre Simon Laplace (1749-1827) on calorific theory and animal heat using guinea pig

'68 Read Hales' *Vegetable aticks*

1776 Decided that all acids contain oxygen

1792-1794 Prepared a collected edition of his papers

▲ **1789** Founded new journal, *Annales de chimie*, to oppose De la Methérie's periodical

▲ **1775** Became hereditary nobleman and bought estates in countryside

▲ **1793** Arrested

▲ **1771** Married Marie-Anne Pierrette Paulze

▲ **1781** Elected to represent Third Estate of Orléanais

▲ **1788** Elected foreign member of Royal Society

▲ **5 May 1794** Executed in Paris

1775 1780 1785 1790 1795 1800 1805

▲ **1775** Moved into Paris Arsenal as a Commissioner of Gunpowder

1790 Joined Commission on metric system ▲

▲ **1791** Became Treasurer of Academy of Sciences and was sacked from gunpowder administration

1771-1772 C. W. Scheele (1742-1786), Swedish chemist, prepared "fire air" (oxygen)

1781 Scheele published book on gases

1798 Count Rumford (Benjamin Thompson) (1753-1814) concluded that heat is a kind of motion; he married Mme Lavoisier in 1804

1791 The Irish phlogistonist Richard Kirwan (1733-1812) adopted the new chemistry

1775 Priestley noticed "dew" is formed when hydrogen explodes with oxygen

1783 Cavendish "synthesized" water

1794 Priestley emigrated to America after Birmingham mob attacked his house; remained a convinced phlogistonist

1774 Joseph Priestley (1733-1804), English chemist, prepared dephlogisticated air (oxygen)

1802-1804 John Dalton (1766-1844), English chemist, developed atomic theory based upon Lavoisier's elements

1772 Daniel Rutherford (1749-1819), physician, in Scotland identified nitrogen in air

1787 Berthollet concluded that not all acids contain oxygen

1771-1772 Louis-Bernard Guyton de Morveau (1737-1816), French lawyer and chemist, showed that metals increase in weight when burned

1785-1795 J. C. de la Methérie attacked Lavoisier's theories in his periodical

1807-1822 Informal research Society of Arceuil trained scientists including Joseph Gay-Lussac (1770-1850) and François Arago (1786-1853)

1774 Pierre Bayen (1725-1798), French pharmacist, investigated the properties of calx of mercury (mercuric oxide)

1782 Guyton de Morveau urged reform of chemical language

1793 Academy of Sciences suppressed (in 1795 became 1st division of new National Institute)

1794 Foundation of the École Polytechnique

1787 Process for producing soda developed by Nicolas Leblanc (1742-1806)

On Lavoisier's foundations, organic chemistry developed by
J. J. Berzelius (1779-1848)
J. B. Dumas (1800-1884)
J. von Liebig (1803-1873)
L. Pasteur (1822-1895)

1869 Dmitri Mendeleev (1834-1907), Russian chemist, published periodic table of chemical elements

0 1775 1780 1785

Lavoisier – alone of the "pneumatic" chemists – was to reject the phlogiston theory of combustion

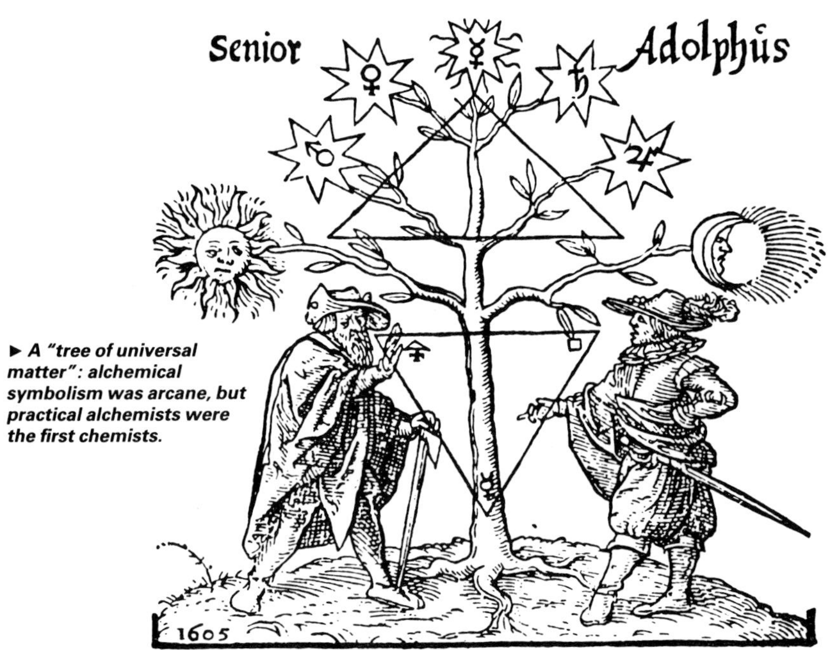

► **A "tree of universal matter": alchemical symbolism was arcane, but practical alchemists were the first chemists.**

"Pneumatic" chemists (those who studied the chemistry of airs released from solids and liquids), such as Joseph Black (1728-1799) in Scotland, Joseph Priestley (1733-1804) and Henry Cavendish (1731-1810) in England, and Carl Scheele (1742-1786) in Sweden, found in the latter part of the 18th century that there were many different varieties of air. All these chemists, however, thought of these "airs" in terms of the phlogiston theory. It was only Lavoisier who saw that airs (or gases as he preferred to call them) could also exist in solid or liquid form, and that the crucial factor which decided the physical state of a particular substance was the amount of heat that substance contained. It followed that, if gases were simply a different form of matter, like solids and liquids, their presence would have to be taken into account quantatively whenever the chemical balance was used. These new ideas – that air participated in chemical reactions, so that its gases would have to be included in chemical book-keeping, and that metals increased in weight when they burned – led Lavoisier to an original theory of chemical composition in which phlogiston had no place.

First steps in science: geology and lighting

One of the friends of the Lavoisier family was a cantankerous bachelor geologist named Jean-Étienne Guettard. Aware of Lavoisier's scientific bent, Guettard advised him, while still at the Collège Mazarin, to join a popular chemistry course being given by Guillaume-François Rouelle in the lecture rooms of the Jardin du Roi. Lavoisier's knowledge of contemporary ideas concerning the elements, acidity and combustion were probably derived from these lectures, as was his technical knowledge concerning the quantitative analysis of minerals. Guettard persuaded Lavoisier to accompany him as his assistant on his geological survey, which was carried out between 1763 and 1766. In their travels through the French countryside Lavoisier paid particular attention to water supplies and to their chemical contents. One mineral which particularly intrigued him was gypsum, popularly known as "plaster of Paris" because it was used for plastering the walls of Parisian houses.

Although Guettard's geological map was never published, Lavoisier's work on gypsum was presented to the Academy of Sciences in

▲ **Lavoisier's popular chemistry teacher in the lecture rooms at the Jardin du Roi in Paris, Guillaume-Francois Rouelle, dispensed in turn with hat, coat, wig and waistcoat as he warmed to his lectures.**

February 1765. With a clear, ambitious eye on being elected to the Academy, he had entered, the year before, the Academy's competition for an economical method of lighting Parisian streets – this was some 40 years before coal gas began to be used for the purpose. Although his study of candles, oil and apparatus did not win him first prize when the adjudication was made in 1766, it was judged the best theoretical treatment and King Louis XV ordered that the young man should be given a special medal.

Thus, by 1766, this ambitious man had succeeded in bringing his name before the small world of Parisian intellectuals. In the same year, two years before he reached his legal majority of 25, Lavoisier's father made a large inheritance over to him. To further his complete financial independence, in 1768 Lavoisier bought a share in the Ferme Générale, a private finance company which the government employed to collect taxes on tobacco, salt and imported goods in exchange for paying the state a fixed sum of money each year. Such a tax system was clearly open to abuse and consequently the *fermiers* were universally disliked and were to reap the dire consequences of their membership of the company during the French Revolution. Unfortunately, Lavoisier's suggestion that the *fermiers* should beat the smugglers by building a wall around Paris for customs surveillance was to lead to hostility toward him and to the circulation of the punning aphorism "Le mur murant Paris fait Paris murmurant" ("The wall enclosing Paris makes Paris mutter").

▲ The chemist's workroom was in transition between the alchemist's kitchen and the modern laboratory. A 1752 illustration from Diderot and d'Alembert's "Encyclopédie".

▼ Georg Stahl, the German inventor of the phlogiston theory of combustion which Lavoisier demolished. Stahl's work bridged the eras of alchemy and chemistry.

The French Academy of Sciences

The word "academy" is derived from the garden in Athens where Plato (♦ page 14) taught. During the Middle Ages the term applied to schools, and in the 15th and 16th centuries was adopted by the new associations promoting learning in science, literature and technology.

National Academies

The formation of national academies in the 17th and 18th centuries was an essential accompaniment to the intellectual transformation which historians describe as "the Scientific Revolution". The first truly successful scientific academy or society, which served as an inspiration and model for other countries, was the Royal Society of London incorporated in 1662. The French Royal Academy of Sciences was established in Paris four years later by King Louis XIV and his Chief Minister, Jean-Baptiste Colbert. Over the next 150 years, the German states, and Scandinavian and most other European countries and the United States followed suit.

These national academies, most of which survive today, had some common features – firstly, the desire to share and communicate information and ideas between members; secondly, a willingness to communicate the same findings to outsiders and to other societies by means of publication; and thirdly, a commitment to the experimental investigation of nature, often encouraged by the posing of problems, for the solutions of which prizes or premiums were awarded.

The French Academy of Sciences

Unlike the English Royal Society, whose fellows have never been salaried, the French Academy of Sciences was composed of 18 working academicians or "pensionnaires", paid by the government (until 1793 the Crown) to advise the state and to report on any official questions put to them as a body. There were also a dozen honorary members drawn from the nobility and clergy, a dozen working, but unpaid, associates ("associés") and, to complete the pecking order, a further dozen unpaid assistants ("élèves" or "adjoints"). The Academy also made room for its retired academicians and for foreign honorary associates.

Because of the tight restriction on the number of members, election to the Academy was a prestigious event in the career of a French scientist. This was in contrast to Britain's Royal Society, which for many years made it so easy for those with wealth or social status to become fellows that its fellowship lacked great prestige.

The three working grades of the French Academy, together with its aristocratic honorary membership, clearly reflected the rigid hierarchical structure of 18th-century French society. In practice, the academicians were allocated between the six sciences of mathematics, astronomy, mechanics, chemistry, botany and anatomy (or medicine). Biology and physics were added under Lavoisier's directorship in 1785. Such a distribution frequently prevented the election of a deserving candidate because the appropriate science section was "full". There was also a tendency to elect or to promote on grounds of seniority rather than merit.

Suppression and reconstitution

Because membership was restricted, vacancies often led to intense lobbying for position, factionalism, ill-feeling and sometimes (as with Lavoisier's own election in 1768) to the bending of rules. The repeated failure of the revolutionary Jean-Paul Marat, who fancied himself an expert chemist, to gain admission in the 1780s led him and others to oppose the Academy. Indeed, the Academy's close association with royal patronage and its reflection of the "corrupt" hierarchical structure of the "ancien régime", made inevitable its suppression in August 1793.

However, the revolutionary government could not manage without the expert advice of those who strove to advance science and the arts. In 1795, therefore, the leading learned societies in Paris were reconstituted as the National Institute, its first division or class being the former Academy of Sciences now divided into 10 sections containing six Parisian and six provincial members. Honorary membership was abolished, all distinctions of grade and salary were abolished and election to a vacancy was by the whole institute and not just by members of the appropriate scientific section as before. The age of the professional, salaried scientist had begun.

Bonaparte, Egypt and Arcueil

It is interesting to note that the artillery officer Napoleon Bonaparte was elected a member of the Institute's first class in 1797. He was to be a significant patron of French science during his period of power between 1799 and 1814. During his Egyptian campaign in 1798, Napoleon formed an Institute of Egypt, whose members included Lavoisier's disciple Claude Berthollet, who in turn formed an informal research group at his estate at Arcueil on the outskirts of Paris between 1807 and 1822. This Society of Arcueil trained and encouraged the young men – such as the chemist Joseph Gay-Lussac, (1778-1850), and the physicist François Arago (1786-1853) – who formed the core of France's scientific community following Napoleon's defeat in 1815.

▲ The Swiss-born revolutionary Jean-Paul Marat. Repeated rejection of his candidature fueled opposition to the rigidly hierarchical Academy which was temporarily suppressed in 1793.

► A visit by Louis XIV to the Academy of Sciences in its early years. Through the window can be seen the Paris Observatory under construction (1667-71).

▼ Lavoisier studied at Paris's best school the Collège des Quatre Nations; today the same building (below) houses the French Institute, which includes the Academy of Sciences and the Academy of Fine Arts. The engraving dates from about 1820.

Lavoisier's work on Academy reports ranged from hypnotism to gunpowder, from scientific farming, to glass-making, bleaching and dyes

Husband, academician and myth destroyer

In 1771, at the age of 28, Lavoisier further cemented his membership of the Ferme Générale by marrying the fourteen-year old daughter of a fellow member of the company. Despite their difference of age and their childlessness, their marriage was an extremely happy one. As a rich and talented man, Lavoisier was an obvious candidate for election to the prestigious Academy of Sciences. Although he had failed on the first attempt in 1766, by a modest bending of the rules to create an extra vacancy for him he was successfully admitted to the lowest rank of assistant chemist in 1768.

Much of Lavoisier's fortune was to be spent on the best scientific apparatus that money could buy – some of it so unique and complex that his followers had to simplify his experimental demonstrations in order to verify them. Lavoisier became a loyal servant to the Academy, helping to prepare its official reports on a whole range of subjects including – to select from one biographer's page-long list: the water supply of Paris, prisons, hypnotism, food adulteration, the Montgolfier balloon, bleaching, ceramics, the manufacture of gunpowder, the storage of fresh water on ships, dyeing, inks, the rusting of iron, the manufacture of glass and the respiration of insects.

The problem of the Parisian water supply came to Lavoisier's attention during the year of his election to the Academy, when the purity of water brought to Paris by an open canal was questioned. The test to decide whether water was drinkable involved evaporating it to dryness in order to determine its solid content. The use of this technique reminded academicians, including Lavoisier, of the long tradition in the history of chemistry that water could be transmuted into earth. Obviously, if this were the case, the determination of the solid content "dissolved" in water would tell nothing about its purity.

Although the transmutation of water into earth had been a basic principle of Aristotle's theory of the four elements water, earth, air and fire, the crucial experiment (published in 1648) had been that made with a willow tree sapling by the Fleming Jean-Baptiste van Helmont (1579-1644). Van Helmont had planted a weighed sapling in a weighed amount of earth and watered it daily for five years. Since, as we have seen, the role of air in chemical (and physiological) changes was ignored at this time, van Helmont naturally concluded that since the earth weighed the same, the considerable gain in weight of the tree after five years must be due entirely to the water. (Unknown to van Helmont, of course, carbon was being absorbed by photosynthesis, though a tree may well consist of 90 percent water.) Impressed by this experiment, van Helmont's British contemporary Robert Boyle had distilled rainwater in glass vessels and found that solid matter was precipitated. Although by the 1760s many chemists could no longer credit that such an apparently simple pure substance as water could be transmuted into an incredibly large number of complicated solid materials, it was seriously argued by a German named Johann Eller in 1766 that water could be changed into both earth and air by the action of fire or phlogiston.

It seems clear from the design of Lavoisier's experiment on water, which he began in October 1768, that he suspected that the earth described in Eller's experiment was really derived from the glass of the apparatus by a leaching effect. By weighing the apparatus before and afterward, and weighing the water before and after heating continuously for three months, Lavoisier showed that the weight of "earth" formed was more or less equal to the weight loss of the apparatus.

◄ *Marie-Anne Pierrette Paulze (1758-1836) married Lavoisier at 14, became his personal assistant, translated vital papers by Priestley and Cavendish, and illustrated chemical apparatus in Lavoisier's textbook (she was trained by David, painter of this 1788 portrait). She continued to host weekly scientific "salons" after Lavoisier's death, and in 1805 married American physicist Benjamin Thompson (Count Rumford); they soon parted.*

◄▼ *Dyers at Les Gobelins, Paris, in the 18th century. Lavoisier and Berthollet conducted researches into dyes. A hundred years earlier the industry was regulated by J.B. Colbert, the initiator of the Academy of Sciences.*

▲ *The British natural philosopher Robert Boyle found solid matter precipitated when he distilled rainwater. Lavoisier's proof that this derived from the apparatus itself required precise gravimetric experiments.*

▼ *The great burning lens made for the Academy of Sciences under the supervision of Lavoisier and other academicians in 1774. The principal lens was 132cm in diameter and comprised two sheets of glass with the lens-shaped space between them filled with wine vinegar. It was by using another large burning lens belonging to the Academy that Lavoisier was able in 1772 to show that when a metal "calx" (oxide) was roasted with charcoal, an enormous amount of "fixed air" was released.*

The experiment with water was enough to convince Lavoisier that Eller's contention that water could be transmuted into earth was completely fallacious. He also surmised that there was a much more plausible explanation of water's apparent change into vapor or air when heated – namely that heat, when combined with water and other fluids, might expand them. Conversely, when air was stripped of its heat it lost its extra volume and collapsed, or was "fixed", into either a solid or liquid condition. This was exactly the effect that Stephen Hales had found when he analyzed the air content of minerals and vegetables.

Lavoisier recorded these ideas in an unpublished essay in 1772. He had now established the basis for a theory of gases – though at this juncture Lavoisier knew nothing of the work of Priestley and others on pneumatic chemistry.

The stimulus of Guyton

In the spring of 1772 Lavoisier read an essay on phlogiston by a Dijon lawyer and part-time chemist, Louis-Bernard Guyton de Morveau (1737-1816). In a brilliantly-designed experimental investigation, Guyton showed that all his tested metals *increased* in weight when they were roasted in air; and since he still believed that their combustibility was caused by the loss of phlogiston, he supposed that phlogiston was so light that it "buoyed up" the bodies which contained it. Most members of the French Academy, including Lavoisier, thought that Guyton's explanation was absurd and Lavoisier immediately saw that a more likely explanation was that, somehow, air was being "fixed" during the process of combustion and that this air caused the increase in weight. It followed that the "fixed air" should be released when the "calces" of metals were decomposed – just as had been suggested by Hales's earlier experiments.

Lavoisier was able to verify this in October 1772 by using a large burning lens belonging to the Academy. When litharge (an oxide of lead) was roasted with charcoal an enormous volume of "air" was, indeed, liberated.

Chemical Apparatus

"Workers by fire"

The action of heat is the easiest way of making two or more substances combine together, or of decomposing a substance. Furnaces, ovens, lenses, mirrors and sand and water baths (for indirect controlled heating) were thus major tools of alchemists – those early chemists who strove to enrich their lives both spiritually and financially by transforming worthless dross into silver and gold. For centuries chemists were known as "workers by fire" and rejected socially as "sooty empirics". Although a pair of weighing scales, or chemical balance (which had reached a sophisticated form by the 16th century) played a significant role in their manipulations of matter, the alchemists' workshops inevitably tended to resemble kitchens, or blacksmiths' workshops, rather than today's sophisticated teaching and research laboratories (which only emerged in the 19th century).

Water baths and stills

Hindsight has shown that alchemical objectives were illusory (at least at the level of chemical manipulation). Nevertheless, as a result of their investigations the alchemists developed many pieces of apparatus which are still in use, e.g. the water bath – reputedly invented by the semilegendary Greek alchemist, Mary the Jewess, in the early 3rd century AD. Their most important contribution was probably to develop the "art of distillation". Prehistoric earthenware vessels with an exaggerated rim forming a channel for the collection of perfume – and volatile oil distillates – have been frequently discovered. By the 1st century AD Greek alchemists had succeeded in adding vertical tubes to the still head for the air-cooling of the distillate. Some time during the 11th and 12th centuries such apparatus had led to the discovery of alcohol. Water cooling of volatile distillates was practiced by the 16th century as the large numbers of practical handbooks on distillation published during the Renaissance make clear.

The collection of gases

Apparatus for the preparation, collection and study of gases was a necessary factor in the chemical revolution of the 18th century. (The Greek word "gas" means chaos, a 16th-century expression for the mysterious, invisible and sometimes pungent products of certain reactions which seemed to resist containment within closed vessels; if the attempt was made they escaped with explosive chaotic force!) Experiments with the newly-invented air-pumps in the 1660s allowed Robert Boyle and John Mayow (1640-1679) to deduce that combustion and respiration were related phenomena. Mayow, an Oxford doctor, succeeded in collecting various "airs" over acid solutions by using an inverted cupping glass – a device used by doctors for expressing mother's milk, or blood from veins. However, such techniques did not allow "airs" to be differentiated. In 1727 Stephen Hales hit upon a way to isolate the "air" produced from a heated solid. In order to estimate accurately the amount of "air" produced, and to remove any impurities, Hales "washed" his "airs" by passing them through water before collecting them.

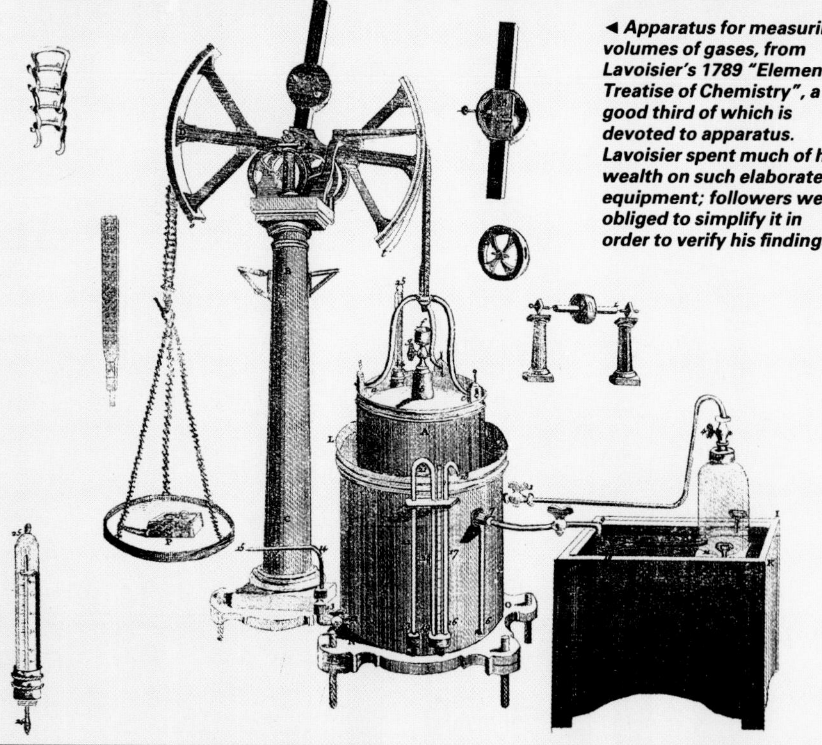

◀ Apparatus for measuring volumes of gases, from Lavoisier's 1789 "Elementary Treatise of Chemistry", a good third of which is devoted to apparatus. Lavoisier spent much of his wealth on such elaborate equipment; followers were obliged to simplify it in order to verify his findings.

◀ *"Alchemist in Search of the Philosopher's Stone Discovers Phosphorus" (1771-1795, by Joseph Wright). Hamburg alchemist Hennig Brand discovered phosphorus by distilling urine, about 1669.*

▶ *John Mayow proved that respiration reduces the volume of air. Mayow used the cupping glass also to collect gases over solutions.*

▼ *Distilling apparatus, from a German publication of 1607.*

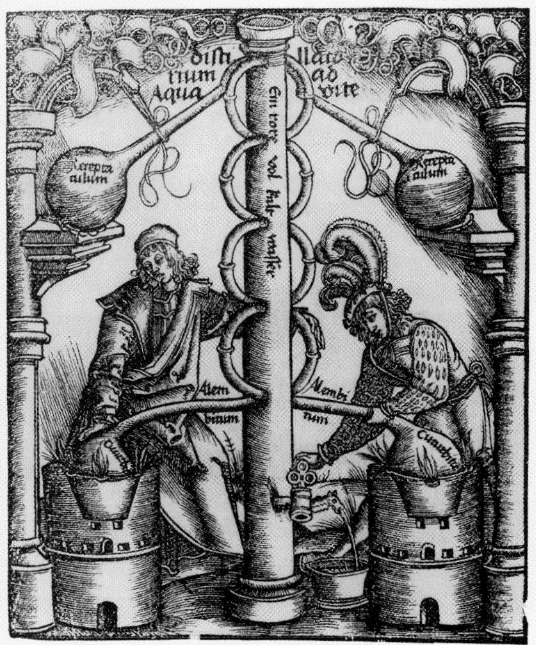

Black's demonstration that "fixed air" (our carbon dioxide) was different from ordinary air, encouraged Henry Cavendish, Joseph Priestley and others to develop Hales's apparatus to study different varieties of air. An incentive here was the invention of soda water by Priestley which encouraged interest in the potentially health-giving properties of mineral waters generally. In 1765, while investigating spa waters, the English doctor, William Brownrigg, invented a simple shelf with a central hole to support a receiving flask or gas holder. This "pneumatic trough" enabled gas samples to be transferred from one container to another and for gases (as Lavoisier called them) to join solids and liquids on the chemical balance sheet. It was Cavendish who began the collection of water-soluble gases over mercury, but Priestley who brought their study and manipulation to perfection. Lavoisier only rarely used the simple pneumatic trough; instead he invented an expensive and sophisticated gasometer. A good third of "Elementary Treatise of Chemistry" (1789) was devoted to chemical apparatus. This was replaced in 1827 by Faraday's book, "Chemical Manipulation", which remained the bible of chemical instrumentation and techniques for most of the 19th century.

In committing himself to the hypothesis that ordinary air was responsible for combustion and for the increased weight of burning bodies, Lavoisier actually demonstrated that he was quite ignorant of contemporary chemical work on the many different kinds of airs which can be produced in chemical reactions. In Scotland, a decade earlier, Joseph Black had succeeded in demonstrating that what we call "carbonates" (e.g. magnesium carbonate) contained a fixed air (carbon dioxide) which was fundamentally different in its properties from ordinary atmospheric air. Unlike the latter, it turned lime water milky and it would not support combustion. A few years later, Henry Cavendish studied the properties of a light inflammable air (hydrogen) which he prepared by adding dilute sulfuric acid to iron. These experiments were to stimulate the astonishing industry of the Unitarian clergyman, Joseph Priestley, who, between 1770 and 1800 prepared and differentiated some 20 new "airs". These included (in our terminology) the oxides of sulfur and nitrogen, carbon monoxide, hydrogen chloride and oxygen.

Hence, although largely unknown to Lavoisier in 1772, there was already considerable evidence that atmospheric air was a complex body and that it would be by no means sufficient to claim that air alone was responsible for combusion. Lavoisier seems to have been aware of his chemical ignorance. So, with the firm intention of bringing about, in his own words, "a revolution in physics and chemistry", he spent the whole of 1773 studying the history of chemistry – reading everything that chemists had ever said about air, or airs, since the mid-17th century and repeating their experiments.

Ironically, far from clarifying his ideas, his new familiarity with the work of pneumatic chemists now led him to suppose that carbon dioxide in the atmosphere was responsible for the burning of metals and the increase of their weight.

Discovery of "dephlogisticated air" or oxygen

Two things led Lavoisier to change his mind. Firstly his attention was drawn by Pierre Bayen (1725-1798), a French pharmacist, to the fact that, when heated, the calx (oxide) of mercury decomposes directly to the metal mercury without the addition of charcoal. No fixed air was evolved. This observation also made it difficult to see how the phlogiston theory could be right. Here was a calx regenerating the metal without the aid of phlogiston!

The mercury calx had also come to the attention of Joseph Priestley. In August 1774 Priestley heated the calx in an enclosed vessel and collected a new "dephlogisticated air" which he eventually found supported combustion far better than ordinary air did. These experiments were reported directly to Lavoisier by Priestley when he was on a visit to Paris during the autumn of 1774.

Bayen's and Priestley's observations, together with his own experiments on the oxide of mercury, caused Lavoisier to revise his hypothesis of 1774. In March 1775 Lavoisier read a paper to the Academy of Sciences "on the principle which combines with metals during calcination and increases their weight" in which, still more confused, he identified the principle of combustion with "pure air" and *not* any particular constituent of the air. This suggestion, which was published in May, was seen by Priestley, who gently put Lavoisier right in a book published at the end of 1775. This, perhaps with further experiments of his own, finally led Lavoisier to the oxygen theory of combustion.

Of his oxygen theory of combustion, Lavoisier wrote, in 1778: "The principle which unites with metals during calcination, which increases their weight and which is a constituent of the calx is: nothing else than the healthiest and purest *part* of air, which after entering into combination with a metal, (can be) set free again; and emerge in an eminently respirable condition, more suited than atmospheric air to support ignition and combustion."

Because this "eminently respirable air" burned carbon to form the weak acid, carbon dioxide, Lavoisier called the new substance "oxygen", meaning "acid-former".

Thus, by 1777, half of Lavoisier's chemical revolution was over. Oxygen gas was an element containing heat (or caloric, as Lavoisier called it) which kept it in the gaseous state. On reacting with metals and nonmetals the heat was released and the oxygen element joined to the substance, causing it to increase in weight. In respiration, oxygen burned the carbon in foodstuffs to form the carbon dioxide exhaled in breath, while the heat released was the source of animal warmth. Lavoisier and his fellow countryman, the mathematical physicist Pierre Simon Laplace (1749-1827), demonstrated this quantitatively with a guinea pig in 1783 – the origin of the expression "to be a guinea pig". Respiration was a slow form of combustion. The non-respirable part of air (later called nitrogen) was exhaled unaltered.

At first glance, in this new theory, the properties of phlogiston

◄ Henry Cavendish, eccentric millionaire experimenter and convinced believer in the phlogiston theory of combustion. Cavendish demonstrated that water comprised oxygen and "inflammable" air, named hydrogen or "water-former" by Lavoisier.

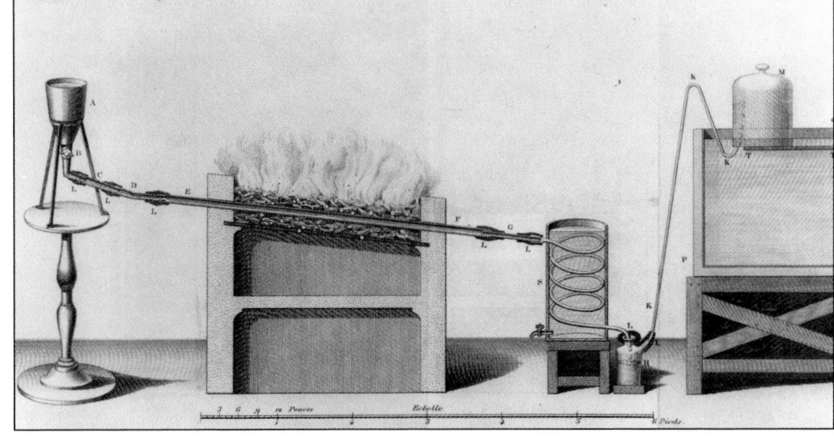

► Apparatus used by Lavoisier to decompose water by passing it over heated iron.

◄ Napoleon as victim of a pneumatic experiment by "the great chemist Lavoisier in Paris". In this German plagiarism of an English cartoon the demonstrator's assistant (at the Royal Institution in the original) is Humphry Davy.

◄▲ Joseph Priestley pioneered teaching of practical chemistry and modern history at the new Nonconformist academy at Warrington, England. He wrote (1767) a textbook on electricity, and between 1770 and 1800, discovered some 20 "airs" including oxygen, ammonia, hydrogen sulfide, carbon monoxide, and oxides of sulfur and nitrogen. He refined chemical apparatus in his laboratory (left) and was first to collect water-soluble gases over mercury. A political and religious radical, Priestley emigrated to the United States in 1794. In science Priestley remained attached to the phlogiston theory discredited by Lavoisier.

seem to be transferred to oxygen gas. In reality, although Lavoisier waited some years before spelling out his new theory in detail, there were major differences. Heat was absorbed or emitted during most chemical reactions, and was present in all substances, whereas phlogiston was usually supposed absent from substances that did not burn. When added to a substance, heat caused expansion, or a change of state; above all, heat could be measured with a thermometer whereas phlogiston could not.

The problem of water

In 1777 Lavoisier did not suggest that chemists should abandon the theory of phlogiston. The main reason for this was that the phlogiston theory could explain why an inflammable "air" (hydrogen) was evolved if the oxide was used. Lavoisier's gas theory gave no hint why these two reactions behaved differently. The problem was that moisture (i.e. a compound of hydrogen and oxygen) is so ubiquitous in chemical reactions that it was easy to overlook its presence.

It was Joseph Priestley who first noticed the presence of water when an electrical spark was passed through a mixture of air and "inflammable air". He described this observation in 1781 to Cavendish, who repeated the experiment and reported in 1784:

"...it appeared that when inflammable air and common air are exploded in a proper proportion, almost all the inflammable air, and near one-fifth of the common air, lose their elasticity and are condensed into dew. It appears that this dew is plain water."

Incidentally, it was this same experiment which led Cavendish to record that a small bubble of uncondensed air remained. Long afterward, the Scots chemist William Ramsay (1852-1916) was to recall Cavendish's observation when he showed in 1894 that the bubble contained the hitherto unknown family of inert gases, argon, neon, etc. Cavendish did not interpret his results as demonstrating the synthesis of water, for he was a convinced believer in phlogiston. For Lavoisier, however, Cavendish's work was evidence that water was not an element. Assisted by Laplace, he quickly showed that water could be synthesized by burning inflammable air and oxygen together in a closed vessel.

Whether or not Lavoisier was really the first to grasp that water was a compound of hydrogen (meaning "water-former") and oxygen, he could now explain why metals dissolved in acids to produce hydrogen. This, he asserted, came not from the metal (as the phlogistonists claimed) but from the water in which the acid oxide was dissolved.

Science and the Revolution

▲► "All year the peasant toils, just to be able to pay the tax collector". As a tax collector, as well as a brilliant chemist, Lavoisier's own fate was inevitably tied to revolutionary developments. Right: the storming of the Bastille, 14 July 1789, painted by "Cholet", a participant.

The necessity for reform

When Lavoisier was born, France was still a monarchy with power lying firmly in the hands of the Crown and aristocracy, together with the hierarchy of the Roman Catholic Church. These autocratic, and sometimes corrupt, minority groups, which were virtually exempt from taxation, were the landlords of the majority of the people – the peasants, merchants, teachers and bankers from whom France's wealth was derived. Agricultural depression, a rise in population, and a succession of expensive wars (including French intervention in the American War of Independence in 1778) led the nation toward bankruptcy in the 1780s. The necessary reformation of political and economic structure was to precipitate the Revolution in 1789.

The mobilization of scientists

The head of the tribunal which tried Lavoisier for treason in 1794 is reputed to have declared publicly "The Republic has no need for experts." In practice, they were indispensable and Lavoisier and his fellow academicians played very important roles in supporting the French Revolution and sustaining the country during the European Wars it created. Their activities formed a precedent for the mobilization and politicization of scientists in later warfare: science was to be exploited and scientists were to be applied.

Introduction of the metric system

During the Revolution, and before its brief suppression in 1793, the French Academy of Sciences standardized weights and measures by

▲ In the drive for self-sufficiency following the blockading of French ports, scientists played an important role. Among industrial-scientific developments promoted by the revolution was the "Leblanc process" of producing soda, invented in 1787 and introduced in 1818 at the St Rollex Chemical Works, Glasgow, Scotland (above).

introducing the metric system based upon meters (one ten-millionth of the distance between the North Pole and the Equator) and grams (the mass of one cubic centimeter of water at 0°C). Although not legally binding in France until 1840, the metric system soon became adopted internationally by the scientific community. Metrication also embraced the calendar, which was "rationalized" into weeks lasting ten (renamed) days and twelve (renamed) months of exactly 30 days each. Extra holidays or festivals were added to make the secular and astronomical years agree. No other country followed this revolutionary precedent, which Napoleon scrapped at the end of 1805. Currency was also decimalized; it was because of its associations with revolution in France and America that British governments avoided decimalization until 1970.

The Leblanc process

During the blockading of French ports by other countries, scientists were directed toward useful military and economic work, where they showed considerable resourcefulness. For example, since metals for munitions could not be imported, techniques were devised for extracting metals, such as copper, from scrapped church bells. Saltpeter (potassium nitrate), an essential ingredient of gunpowder, had to be manufactured instead of imported from India. A technique was developed whereby crude calcium nitrate formed in the organic decay of manure heaps in barnyards was converted into saltpeter by heating it with soda (sodium carbonate). Since the latter was in short supply (and also needed for making soap and glass) the Academy offered a monetary prize in 1782 for a method of making it in large quantities. By 1788 Nicholas Leblanc had developed the complicated process whereby salt was added to sulfuric acid to produce sodium sulfate and the latter was then converted to soda with limestone and coke. This "Leblanc process", despite its inefficiency and lethal waste products, was to form the basis of the "alkali", or soap-making, heavy chemical industry that developed in Great Britain after 1820.

The École Polytechnique

Perhaps the most significant effect of revolution was the realization that France needed experts in large numbers – especially engineers and doctors. This need led in 1794 to the establishment of the École Polytechnique, whose students were given advanced courses in mathematics, engineering and chemistry. Most of the next generation of France's research scientists were graduates of this school, which continues to this day to produce France's most talented intellectuals. With the enthronement of Napoleon as Emperor in 1804, France made great strides scientifically. French scientists were encouraged to make the country self-sufficient in tobacco, dyestuffs, cotton and sugar (from sugar beet) and prizes were awarded for scientific work and industrial development. Scientific recruitment was encouraged by the establishment of a system of school secondary education (the "lycée" system) which still forms the basis of French schooling.

Through soirées, chemical demonstrations and publications, Lavoisier and his young associates set out to dispel the "vague principle" of phlogiston

▲ **Lavoisier with his collaborator Claude Berthollet.**

The new chemistry

Lavoisier was now in a position to bring about a revolution in chemistry by ridding it of phlogiston and by introducing a new theory of composition. His first move in this direction was made in 1785 in an essay attacking the concept of phlogiston. Since all chemical phenomena were explicable without its aid, it seemed highly improbable that the substance existed. He concluded:

"All these reflections confirm what I have advanced, what I set out to prove [in 1773] and what I am going to repeat again. Chemists have made phlogiston a vague principle, which is not strictly defined and which consequently fits all the explanations demanded of it. Sometimes it has weight, sometimes it has not; sometimes it is free fire, sometimes it is fire combined with an earth; sometimes it passes through the pores of vessels, sometimes they are impenetrable to it. It explains at once causticity and non-causticity, transparency and opacity, color and the absence of colors. It is a veritable Proteus that changes its form every instant!"

By collaborating with younger assistants, whom he gradually converted to his way of interpreting combustion, acidity, respiration and other chemical phenomena, and by holding twice-weekly soirees at his home for visiting scientists where demonstrations and discussions could be held, Lavoisier gradually won over a devoted group opposed to the idea of phlogiston. Since the *Journal de physique* was controlled by people who believed in phlogiston, Lavoisier and his disciples founded their own journal, the *Annales de chimie* in 1788. This is still a leading chemical periodical.

Three particularly important converts to the new chemistry were Guyton (whose work had earlier catalyzed Lavoisier's interest in combustion), Claude-Louis Berthollet (1748-1822) and Antoine Fourcroy (1755-1809). Guyton, in particular, was very much exercised by the

◄ *Antoine Fourcroy was co-author, with Lavoisier, Guyton and Berthollet, of the "Method of Chemical Nomenclature" (1787).*

► *The title page of Lavoisier's "Elementary Treatise of Chemistry" (1789), with a book-reading cherub and a retort, lists Lavoisier's membership of learned societies in Paris, Orléans, London, Bologna, Basle and Padua.*

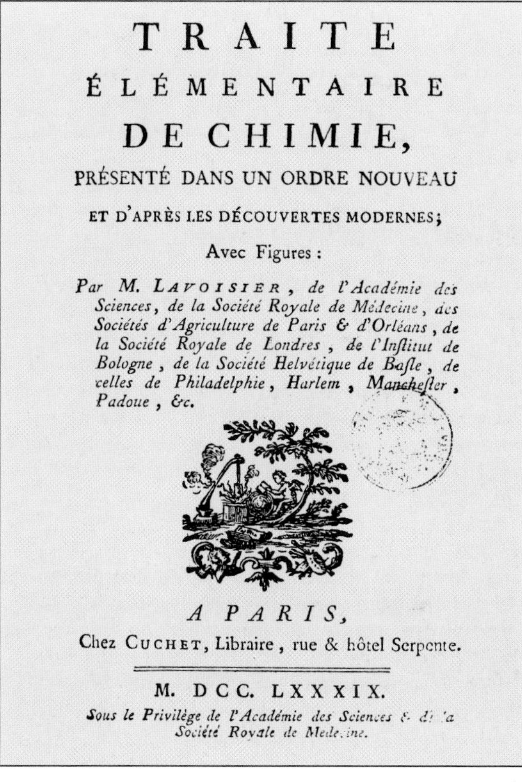

TRAITE
ÉLÉMENTAIRE
DE CHIMIE,
PRÉSENTÉ DANS UN ORDRE NOUVEAU
ET D'APRÈS LES DÉCOUVERTES MODERNES;

Avec Figures :

Par M. Lavoisier, de l'Académie des Sciences, de la Société Royale de Médecine, des Sociétés d'Agriculture de Paris & d'Orléans, de la Société Royale de Londres, de l'Inſtitut de Bologne, de la Société Helvétique de Baſle, de celles de Philadelphie, Harlem, Mancheſter, Padoue, &c.

A PARIS,
Chez Cuchet, Libraire, rue & hôtel Serpente.

M. DCC. LXXXIX.

Sous le Privilège de l'Académie des Sciences & de la Société Royale de Médecine.

inconsistent use of terms by chemists and pharmacists. It was while the four men were collaborating on a new language of chemistry that Guyton was converted to Lavoisier's new chemistry. Because this new language was also the vehicle of anti-phlogiston chemistry, it aroused much opposition. Nevertheless, through translation, it rapidly became, and still remains, the international language of chemists.

Lavoisier's final piece of propaganda for the new chemistry was a textbook published in 1789 called *An Elementary Treatise of Chemistry*. Together with Fourcroy's larger text (published in 1801), this became a model for chemical instruction for several decades. In it Lavoisier defined the chemical element as any substance which could not be analyzed further by chemical means. Such a definition enabled him to identify some 33 basic substances (including some which later proved to be compound bodies).

By the mid-1790s, opposition to phlogiston had triumphed, and only a few prominent chemists, such as Joseph Priestley (who emigrated to the United States in 1794) continued to believe in it. Unfortunately by then the French Revolution had put paid to the possibility that Lavoisier would apply his insights to fresh fields of chemistry.

▼ *Lavoisier was arrested along with other former shareholders in the tax-collecting Ferme Générale, including his father-in-law. It was Lavoisier's own suggestion that a wall be built round Paris to control smugglers that earned him particular hostility from, among others, Marat. On 5 May ("16 Floréal, year II") he and 27 others were tried, sentenced, and guillotined.*

The Language of Science

1 🜂	2 🜁	3 🜄	4 🜃	5 f.🜁	6	7	8	9 X	10 M.🜃	11	12 ∴	13
27 ⊗	28	29 ⊖	30 ⦶	31	32	33	34	35	36	37	38	39

The debt to classical languages

Newton wrote his "Mathematical Principles of Natural Philosophy" (1687) in Latin, but published his other great work "Opticks" (1704) in English. This change of language reflects the 17th century opinion, advanced strongly by Galileo, that science ought to be communicated in the language of ordinary discourse. Although there are exceptions, Latin had largely ceased to be used by scientists by the middle of the 18th century. Nevertheless, its effects have been long-lasting.

Not only does science contain more than an average number of Latin (and Greek) words, but scientists have continued to derive new technical expressions from these languages. Michael Faraday consulted a classics scholar and scientist, William Whewell (1794-1866), when he wanted the invented terms anode, cathode and ion to describe phenomena in electrolysis, and the same scholar supplied Charles Lyell with the words Eocene and Miocene for geological time periods. (Whewell also coined the word "scientist", itself, in 1833 – meaning someone who spent most of their time in scientific investigation; oddly, the word did not catch on until the 1930s.)

Precision and humor

One of the ways a comedian can make us laugh is to use the fact that words in most west European languages can have several meanings, depending upon context. It is because language is potentially ambiguous, imprecise and punnable that scientists over the centuries have had to develop rigorous definitions and an agreed and precise vocabulary. Here Latin and Greek are still useful, though words may also be borrowed from other languages or even from everyday speech ("parasite", meaning an organism which obtains its food from another organism, was borrowed in 1727 from the everyday meaning of a social sponger).

Half the task of scientific education is learning to speak and write the very precise, and therefore rather dull and humorless, language of science. As one critic has put it, "A scientific paragraph says precisely what it means, and no more; it reads as if it had been composed by a robot, with oil for blood and cogs for corpuscles."

It was partly as an attempt to inject some humor into the recondite theories of particle physics that, during the 1950s, the American physicist Murray Gell-Mann introduced terms such as "strangeness" and "quarks" (a nonsense word borrowed from James Joyce's "Finnegans Wake", a novel based upon the ambiguities of languages). Particle physicists have followed his lead and physics currently abounds with oddities such as "boojum" (from Lewis Carroll), "color", "charm" and "truth". Nevertheless, such words have very precise meanings in their new contexts.

The language of the old chemistry

Alchemical and chemical texts written before the end of the 18th century can be difficult to read because of the absence of any common chemical language. Greek, Hebrew, Arabic and Latin words were used, there was widespread use of analogy in naming chemicals or in referring to chemical processes, and the same substance might receive a different name according to the place from which it had been derived. Examples of such language are "Aquila coelestis" for ammonia, "father and mother" for sulfur and mercury, "gestation" as a metaphor for reaction, "butter of antimony" for deliquescent antimony chloride, and "Spanish green" for copper acetate. Names might also be based upon properties – smell, taste, consistency, crystalline form, color – or uses. Although several of these names have lingered on as "trivial" names, Lavoisier and his colleagues decided to systematize nomenclature by basing it solely upon what was known of a substance's composition. Since the theory of composition chosen was the new oxygen system, Lavoisier's suggestions were intitially resisted by those who believed in the existence of the material phlogiston. Clearly adoption of the new nomenclature involved a commitment to the new chemistry.

The French system

Following the inspiration of the Swedish naturalist Carolus Linnaeus, who had, in 1737, begun to systematize the naming of plants, Louis-Bernard Guyton de Morveau, a lawyer-chemist from Dijon, suggested in 1782 that chemical language should be based upon three principles: substances should have one fixed name, names should reflect composition when known (if unknown they should

▲ Symbols used in alchemy, astrology and old chemistry. Similar notation was used in Denis Diderot's "Encyclopédie" (1751-1772). (1) Fire (2) Air (3) Water (4) Earth (5) Fixable air (6) Clay (7) Siliceous earth (8) Fusible earths (9) Talc (10) Magnesia (11) Earth of alum (12) Sand (13) Gold (14) Silver (15) Copper (16) Tin (17) Lead (18) Mercury (19) Iron (20) Zinc (21) Antimony (22) Regulus of antimony (23) Arsenic (24) Cobalt (25) Orpiment (26) Cinnabar (27) Tutty (28) Vitriol (29) Sea salt (30) Niter (31) Borax (32) Alum (33) Tartar (34) Fixed alkali (35) Potash (36) Vinegar (37) Wine (38) Urine (39) Essential oil (40) Fixed oils (41) Sulfur (42) Phosphorus (43) Phlogiston (44) Soap (45) Glass (46) Caput mortuum (47) An hour (48) A day (49) A night (50) To distill (51) To precipitate (52) A retort.

▼ Scientific and commercial exchange were greatly assisted by the standardization, simplification and brevity introduced with the metric system. Lavoisier served on the commission which recommended its introduction. The liter (1), gram (2) and meter (3) replaced the pint, pound and ell respectively.

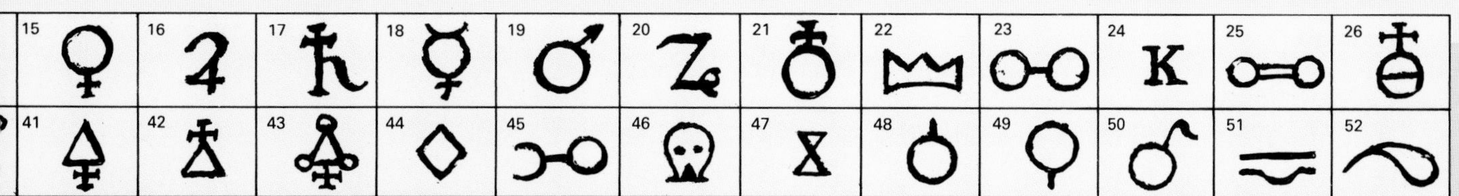

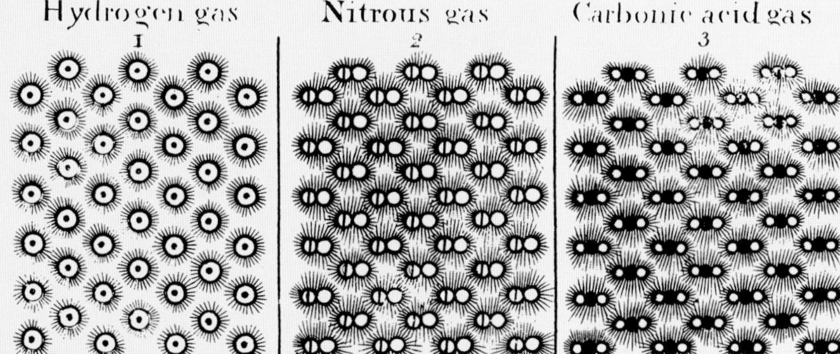

◄▲▼ **John Dalton, who originated the modern chemical atomic theory. The Englishman turned to chemistry after publishing his "Meteorological Observations" in 1793. In "A New System of Chemical Philosophy" (1808) he published a consistent atomic theory, proposing that the identical atoms of an element differ from those of all other elements, in every way, including weight. His was the first table of atomic weights (1803). Above: Dalton's notation for equal numbers of particles of hydrogen, nitric oxide (nitrous gas) and carbon dioxide (carbonic acid gas). Below: his formula for alum.**

◄ **In 1813 the Swedish chemist J.J. Berzelius introduced the modern method of chemical notation: the formula for alum – $KAl(SO_4)_2.12H_2O$ – is simple compared with Dalton's (above). Berzelius determined atomic weights for many elements and discovered three new ones.**

be noncommittal), and names should generally be chosen from Greek or Latin roots.

Lavoisier believed that language should express clear and distinct ideas. Much struck by these suggestions for systematizing the crude and inconsistent nomenclature of chemists and pharmacists, Lavoisier called de Guyton to Paris. In 1787, together with two of Lavoisier's anti-phlogistonist disciples, Berthollet and Fourcroy, they published the 300-page "Méthode de nomenclature chimique", which appeared in English and German translations a year later.

One-third of this book consisted of a dictionary which enabled a reader to identify the new name of a substance from its older ones. For example, "oil of vitriol" became "sulfuric acid" and its salts "sulfates" instead of "vitriols". "Flowers of zinc" became "zinc oxide". Perhaps the most significant assumption in the nomenclature was that substances which could not be decomposed were simple (elements), and that their names should form the basis of the entire nomenclature. Thus the elements oxygen and sulfur would combine to form either sulfurous or sulfuric acids depending upon the quantity of oxygen combined. These acids when combined with metallic oxides would form the two groups of salts, sulfites and sulfates.

Chemical symbols

The French system also included suggestions for ways in which chemicals could be symbolized by geometrical patterns: elements were straight lines at various inclinations, metals were circles, alkalis were triangles. However, although they represented an attempt to produce a simpler and more systematic notation than the alchemists', such symbols were inconvenient for printed use and never became widely established. Typographical difficulties were also a factor in preventing the adoption of the atomic symbols introduced by the English chemist John Dalton (1766-1844) in 1803. Dalton represented elementary atoms by circles, distinguished by shading or by an alphabetical symbol. The system was fine for simple compounds, but became unmanageable with complicated organic compounds.

In 1813 the Swedish chemist J. J. Berzelius (1779-1848) introduced the modern notation whereby an atom of an element is symbolized by the initial letter of its name, using two letters where elements might otherwise be confused: H, hydrogen; K, potassium (kalium, from the Arabic); C, carbon; but Cu, copper (from cuprum). Berzelius's symbols, which could be easily arranged algebraically to represent compounds, became generally adopted from the mid-1830s onward, when chemists also began to represent reactions by means of equations.

Lavoisier was condemned as a shareholder in the nation's tax-collecting company, despite his outstanding scientific achievements and services to France

The aftermath

Although Lavoisier's continued research after 1789 – for example, beginning some promising work on the analysis of organic substances – he found his official activities as an academician and *fermier* taking up more and more of his time, in particular with both technical and administrative problems.

On 14 July 1789, Revolution broke out with the storming of the Bastille prison in Paris. In fear of their lives, Crown and aristocracy renounced their privileges while the Declaration of the Rights of Man was drawn up by a National Assembly composed of the third estate – peasant farmers, merchants, teachers, and bankers. National unity was short-lived, however, as the more radical Jacobins maneuvered for political power and the downfall of monarchy. War with Austria and Prussia was to prove the excuse for the king's execution on 21 January 1793.

In the period of terror and anarchy which followed, Lavoisier was to lose his life. Undoubtedly, Lavoisier supported the initial phase of the Revolution. He had worked hard within the Academy in improving the quality of gunpowder and in devizing the metric system in 1790 – he was both secretary and treasurer of the commission appointed to come up with a uniform system of weights and measures. As commissioner of the Treasury in 1791 he set up a system of accounts of unprecedented punctuality. But Lavoisier's services to France and his international reputation were, in the words of one historian, "as dust in the balance when weighed against his profession as a *fermier-général*". On 24 November 1793, Lavoisier and his fellow shareholders (including his father-in-law) were arrested and charged, ludicrously, with having mixed water and other "harmful" ingredients in tobacco, charging excessive rates of interest and witholding money owing to the Treasury.

Although later investigations by historians have revealed the worthlessness of these charges, they were more than sufficient in the aptly-named "Age of Terror" to ensure the death penalty. Even so, there is some evidence that Lavoisier, alone of the *fermiers*, might have escaped but for the evidence that he corresponded with France's political enemies abroad. The fact that his correspondence was scientific did not, in the eyes of his enemies, rule out the possibility that Lavoisier was engaged in counter-revolutionary activities with overseas friends.

▼ *Pointing the way: the monument to Lavoisier was erected beside La Madeleine church in Paris a hundred years after his death, but was scrapped in World War II.*

▲ *France's new generation of scientists: J. L. Gay-Lussac, Humboldt's collaborator, and J.-B. Biot take measurements on their pioneering balloon ascent in 1804.*

A monument to Lavoisier

Lavoisier was guillotined on 8 May 1794. The mathematician Joseph Lagrange (1736-1813) commented: "It required only a moment to sever his head, and probably one hundred years will not suffice to produce another like it." The words were prophetic. Following the centenary of the French Revolution, a statue was erected by international subscription in the 1890s to commemorate Lavoisier. Some years later it was discovered that the sculptor had copied the face of the phlosopher Marie Jean Condorcet, the Secretary of the Academy of Sciences, whose dates were the same as Lavoisier's but who died in prison. Lack of money prevented alterations being made and, in any case, the French argued pragmatically that all men in wigs looked alike anyway. The statue was melted down during World War II and has never been replaced.

Lavoisier's real memorial is the science of chemistry itself. For even in the late 20th century, chemistry is still based upon the foundations established 200 years ago by Antoine Lavoisier.

Alexander von Humboldt 1769-1859

6

Humboldt a "great scientist"?...Upbringing in Frederick the Great's Prussia..."Neptunist" versus "Plutonist" geology...South American travels...Humboldt develops the new science of biogeography...Wallace's and Weber's lines...Wegener and continental drift... Magnetism studies...PERSPECTIVE...Exploration... German universities take the lead...Associations for science...Scientists' joint international ventures

The name of Friedrich Heinrich Alexander von Humboldt hardly trips off the tongue as does that of Isaac Newton, say, or Marie Curie. Nor does it spring as readily to the modern mind for inclusion as one of the world's great scientists. Yet he was an important catalyst.

The scientific revolution of the 17th century was much concerned with attempts to interpret the world about us in terms of mechanical models. During the 18th century this approach proved increasingly useful in "natural philosophy" – such sciences as physics and chemistry. But it was of much less value for studying "natural history" – environmental or biological problems. The scientific revolution that culminated in Newton therefore gave greater aid to natural philosophy than to natural history. Humboldt's importance is that he helped focus attention on the significance of scientific explanation in natural history. He linked the world view of Newton in the 17th century with that of Darwin in the 19th century.

Whole Earth studies

Exploration of the Earth dates back to antiquity, but long voyages with specific scientific aims only began in the 17th century. These voyages, as by the Englishman, Halley, and the Frenchman, Jean Richer (1630-1696), were stimulated by the desire to observe objects outside the Earth, but soon included studies of the Earth, such as Halley's work on the Earth's magnetism. In the 18th century, the importance of "scientific" voyages increased: Sir Joseph Banks, probably the most influential English scientist of Humboldt's day, was a botanist on Captain Cook's first voyage round the world.

The principal sciences studied on these voyages were biology and geology, which became more organized toward the end of the 18th century. Linnaeus' scheme of classification transformed biology, while the arguments of Werner and Hutton led to modern geology. Humboldt contributed to, and stimulated, work across this broad front. Above all, he demonstrated the value of international cooperation in solving global scientific problems.

His advocacy led to major 19th-century developments in Earth sciences – geomagnetic studies under Gauss, mobile continents as proposed by Wegener, and oceanography. His interest in biogeography reappeared in Wallace, and his South American travels stimulated Darwin – the two cofounders of evolutionary theory.

▼ ***Humboldt and his traveling companion Aimé Bonpland.***

Alexander von Humboldt – His Life, Work and Times

The young Humboldt

Alexander von Humboldt was born in Berlin in 1769. He was one of two brothers, the elder, Wilhelm, a philologist and diplomat, was also to play a significant role in German history. Germany as it is known today did not exist then; instead there was a conglomerate of independent states. Berlin was the capital of the most important of these states, Prussia, but was still a relatively small provincial town. Nevertheless, by the time of Humboldt's birth its ruler, Frederick the Great, had become one of the leading figures in European politics. He had also developed a reputation in intellectual circles: Voltaire was a visitor to his court, and J. S. Bach wrote music for him.

Frederick paid particular attention to the needs both of his army and of the civilian administration. Humboldt's father was an army officer who came from a family long involved in the administration. Humboldt's mother was wealthy in her own right. In consequence, the two Humboldt boys were both well off and immediately involved in the political and intellectual life of the capital.

The Humboldt brothers were taught by private tutors in their youth, and then went on together to university. Alexander's studies at the University of Göttingen, though brief, were of major importance for his future activities. There he met the German Georg Forster (1754-1794), who had been the naturalist on one of Captain Cook's voyages round the world. Humboldt was captivated by Forster's descriptions of distant lands, and the two were soon (1790) off together on a grand tour of Europe.

Their trip took in Britain, which gave Humboldt his first taste of a sea voyage – he thoroughly enjoyed it – and also introduced him to the naturalist Sir Joseph Banks (1743-1820). Banks had also traveled as a naturalist with Cook, but unlike Forster he was a wealthy man, by now President of the Royal Society and one of the major figures in British science of the time.

By the end of his grand tour, Humboldt had decided to travel widely and, in particular, to visit the Southern Hemisphere . He realized, however, that he needed a better grounding in the relevant sciences, if his travels were to produce results of scientific value. So he entered (1791) the Academy of Mines at Freiberg in Saxony, then one of the leading institutions for teaching geology in Europe. The head of the school was Abraham Werner (1749-1817), an important figure in the history of geology.

◀ ***A "statesman of science", Sir Joseph Banks had already served half his 40 years as president of the Royal Society by the time the young Humboldt met him in London. Banks earlier accompanied James Cook on his first (1768-1771) voyage to the Pacific. He played an important role in developing the Botanic Gardens at Kew and helped promote science in general.***

Historical Background

1740-1786 Rule of Frederick II saw Prussia become Europe's leading military power

■ **1775-1783** American Revolution

1789 French Revolution b

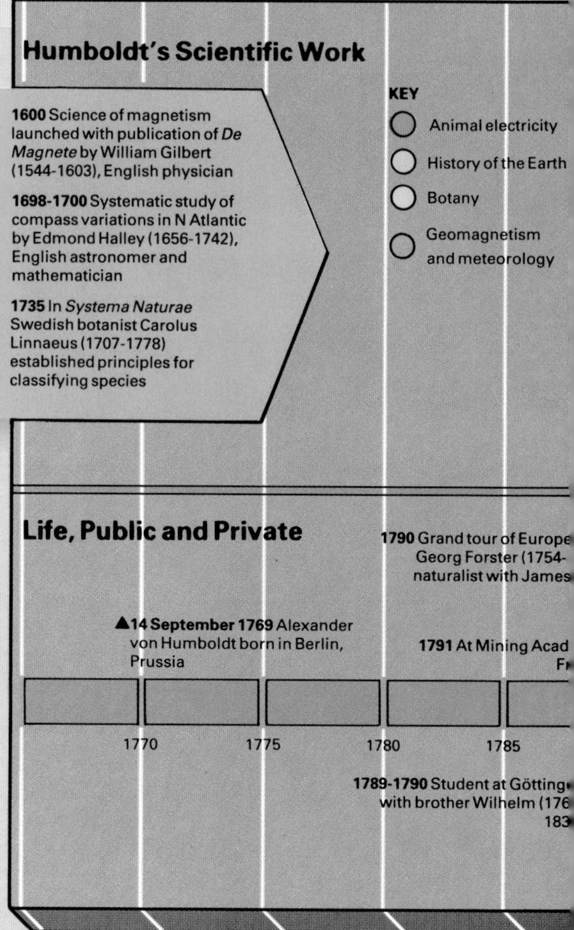

Humboldt's Scientific Work
1600 Science of magnetism launched with publication of *De Magnete* by William Gilbert (1544-1603), English physician
1698-1700 Systematic study of compass variations in N Atlantic by Edmond Halley (1656-1742), English astronomer and mathematician
1735 In *Systema Naturae* Swedish botanist Carolus Linnaeus (1707-1778) established principles for classifying species
KEY
Animal electricity
History of the Earth
Botany
Geomagnetism and meteorology
Life, Public and Private
▲**14 September 1769** Alexander von Humboldt born in Berlin, Prussia
1790 Grand tour of Europe Georg Forster (1754- naturalist with James
1791 At Mining Acad F
1770
1775
1780
1785
1789-1790 Student at Götting with brother Wilhelm (176 183

Scientific Background

1788 Linnean Society, Londo founded, first of specialize scientific societie

1768-1771 James Cook (1728-1779), British navigator, made first voyage to the Pacific with Joseph Banks (1743-1820) as chief naturalist

1766-1769 Louis de Bougainville (1729-1811), French navigator, visited South America and investigated Pacific islands

1775 Abraham Werner (1749-1817), German geologist, began to teach "Neptunist" theory of origin of rocks

1765 1770 1775 1780 1785

1803 Louisiana purchase by USA from France

1810-1824 S American wars of liberation; Simón Bolívar prominent

1814-1815 German map redrawn, expanding Prussia

1807 Prussia lost half of territory to France

1795 By treaty with France, now Europe's first power, Prussia ceded the left bank of the Rhine

1804-1815 Period of reform in Prussia

1815 Final defeat of Napoleon at Waterloo

1848 Revolutions shook Germany, France, Italy, Hungary

1793 Published *Florae fribergensis*, including results of studying plant physiology

1819 Developed idea of isotherms

1797-1798 In Tyrol, systematic, detailed meteorological observations

1806 Observed "magnetic storm" (and invented term)

1836 Encouraged establishment of magnetic stations round the world

1799 Pioneer work measuring elevation of plateau/*meseta* of central Spain

1808-1827 In Paris worked with French scientists, especially Gay-Lussac on composition of air

1797 Results of 4,000 experiments on animal electricity published in *Experiments on the excited muscle and nerve fiber*

1805-1834 Published over 30 volumes based on his S American experiences, including (1814-1829) *Personal Narrative*

1799-1804 Carried out botanical, zoological and geological studies in America

1808 *Aspects of Nature*, his most popular book in Germany, an "aesthetic treatment of natural history"

1829 Excursion to Russia; yielded data for isothermal map; resulting in discovery of diamonds in Urals

1845-1862 Publication of *Cosmos*, a 5-volume description of the physical universe

1794 First meeting with Johann von Goethe (1749-1832), poet and scientist

1809 Friendship began with François Arago (1786-1853), French physicist and mathematician

1828 Organized meeting of German Association for Natural Science

1829 At invitation of Tsar, led expedition to Urals and Siberia

1797 Inspector of Mines; set up free mining school

1804 Returned to Europe, staying in Berlin

6 May 1859 Died in Berlin

1795 1800 1805 1810 1815 1820 1825 1830 1835 1840 1845 1850 1855

1798 Visited Paris

1808 Settled in Paris as member of a delegation to negotiate with Napoleon, staying until 1827

1827 Returned to Berlin

1848 Revolution Humboldt hailed as leading German liberal

1799-1804 Traveled to South, Central and North America with Aimé Bonpland (1773-1858), French botanist

1849-1859 Retained favor of king, but lived under police surveillance

1826-1852 At Giessen University J. von Liebig set up first practical chemistry teaching laboratory

1840 Liebig noted value of phosphates in agriculture, spurring trade in S American guano

1810 University of Berlin founded, by Wilhelm Humboldt

1843 Importance of large-scale ocean currents recognized, especially by Matthew Maury (1806-1873), American oceanographer

1799 Royal Institution founded in London

1822 First meeting of German Association for Natural Science, in Leipzig; followed by British Association for Advancement of Science (1831)

1828 onward Heinrich Dove (1803-1879), German meteorologist, developed new science of climatology

1853 First International Meteorological Conference, in Brussels

Luigi Galvani (1737-1798), an anatomist, announced his ...overy of "animal electricity"

1800 Alessandro Volta (1745-1827), Italian physicist, reported invention of electric battery

1833 Göttingen Magnetic Union of observation stations set up by Karl Gauss (1777-1855), German mathematician and student of magnetism

1843 Heinrich Schwabe (1789-1875), German astronomer, discovered 11-year periodicity of sunspots

About 1800 In the "heroic age" of geology, Werner's Neptunist views clash with the "Plutonist" ones of Scot James Hutton (1726-1797)

1852 Sir Edward Sabine (1788-1883), British geophysicist, discovered solar-terrestrial interaction

1804 Humboldt's altitude record, set on Chimborazo, broken by Joseph Gay-Lussac (1778-1850), French chemist, in hot-air balloon ascent

1808 Gay-Lussac discovered that gases combine in simple proportions by volume

Scientists influenced by Humboldt's writings and who visited South America included:

Louis Agassiz (1807-1873), Swiss American geologist and zoologist ...

Charles Darwin (1809-1882), British geologist and biologist ...

and Alfred Wallace (1823-1913), British biologist

1915 In *Origin of Continents and Oceans* Alfred Wegener (1880-1930) presented continental drift theory

1795 1800 1805 1810 1815 1820 1825

Geology's heroic age

Werner provided the first widely accepted scheme for the classification and interpretation of rocks and landscapes. By the latter part of the 18th century, adequate classification had become essential for the development of natural history. Increasingly frequent voyages from Europe to distant parts of the world had led to the recognition of plants, animals, rocks and landscapes which differed strikingly from those common in Europe. Existing schemes of classification were seen to be too limited and lacking in a scientific basis. For plants and animals, the system of nomenclature introduced at this time – by the Swedish botanist Carolus Linnaeus (1707-1778) – is still in use. It designates related groups by two Latin names, the generic and the specific. For example, both the relatedness and the distinctiveness of a horse and a zebra are indicated by the binomial names *Equus caballus* and *Equus zebra*. The human species is referred to as *Homo sapiens*. In this case, no other closely related group now exists, though the bones of extinct groups – *Homo neanderthalis*, for example – have been found.

Linnaeus extended his classification beyond plants and animals to rocks, but there it proved less successful. The approach which lies behind modern methods was pioneered by Werner at Freiberg. Werner related his classification scheme to a comprehensive theory of how rocks formed. He believed that the early Earth was totally covered by a vast ocean. This gradually disappeared, leaving behind various types of rocks deposited as layers (or strata) in sequence one above the other. Werner's theory, which thus supposed that all rocks were sedimentary – that is, deposited from water – was referred to as Neptunist geology (after Neptune, the god of the oceans in classical times). It

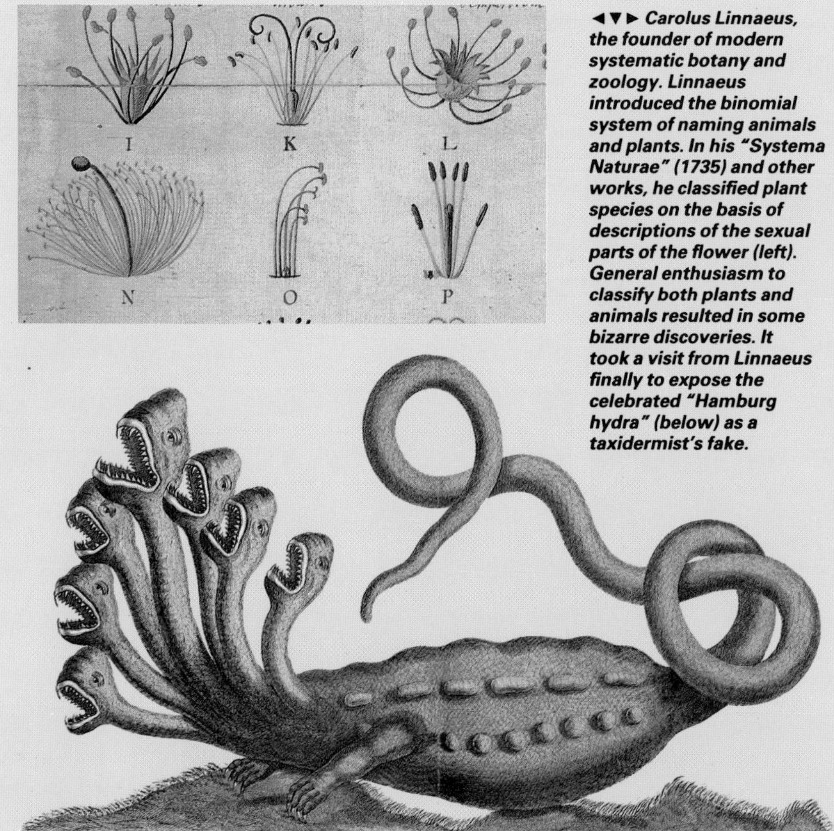

◀▼▶ *Carolus Linnaeus, the founder of modern systematic botany and zoology. Linnaeus introduced the binomial system of naming animals and plants. In his "Systema Naturae" (1735) and other works, he classified plant species on the basis of descriptions of the sexual parts of the flower (left). General enthusiasm to classify both plants and animals resulted in some bizarre discoveries. It took a visit from Linnaeus finally to expose the celebrated "Hamburg hydra" (below) as a taxidermist's fake.*

◄ The salt mines of Cracow were visited by Humboldt, who was appointed assessor of mines in 1792.

►▼ The "Neptunist" Abraham Werner (right), Humboldt's teacher at Freiberg's school of mines, taught that all rocks had a sedimentary origin. Rival "Plutonists", headed by James Hutton (below), believed the chief mechanism to be heat, a force observed in eruptions of Vesuvius at the time.

soon came to be opposed by a totally different interpretation of the history of rocks. James Hutton (1726-1797) a Scotsman educated at the University of Edinburgh, believed that all changes apparent in the rocks should be explicable in terms of forces still at work today. He correspondingly discounted the idea of an early ocean, and gave pride of place to heat as the mechanism for molding rocks. Such heat is most evident in volcanic activity, associated in classical times with Pluto, the god of the underworld. Consequently, Hutton's views were referred to as Plutonist geology. It is often claimed that modern geology emerged from the conflict between Neptunist and Plutonist views, and the period just before and after 1800 when the conflict was at its height has been called the heroic age of geology.

This was not simply a scientific debate. Neptunist geology envisaged a once-and-for-all, fairly rapid process for the formation of rocks, whereas Plutonist geology favored an extended, continuing process. Religious views at the end of the 18th century, as in Newton's day, favored a relatively recent date for the formation of the Earth. Neptunist views fitted in better with this belief, and had the additional bonus that the supposed worldwide ocean could be related to Noah's Flood. Neptunist geology was therefore regarded as more respectable, while Plutonist geology was associated with rationalism and radicalism.

As a student of Werner, Humboldt naturally began his scientific career as a Neptunist. After several years of activity in mining, the opportunity arose for his long-awaited travels – to South America. There the evidence of the landscape converted him to Plutonist views. Nevertheless, he was left with a strong sense of the complexity of geological interpretation.

Economic Advantage

Travel has often been associated with economic gain. Indeed, some famous voyages have been almost entirely motivated by the expected financial return. For example, Marco Polo's trip from Venice to China in 1271-1295 formed part of a more general Venetian attempt to open up new trade routes. But such pioneering exploration also brought back information of interest to science.

Humboldt finds guano and diamonds

Humboldt, on his 1799-1804 visit to South America, collected mainly items of interest to contemporary scientists, but one of the specimens he brought back – a piece of guano – proved to be of major economic significance. Guano is a type of deposit found on a number of islands off the coast of Peru, as well as at places on the mainland. It has no single source, but a major component is the excreta of seabirds. This dries into solid layers which can build up to considerable depths over the years, because there is little rain in the region to wash it away. Humboldt handed his specimen to the leading French chemists for analysis. They found that it contained large quantities of phosphate minerals. There the matter rested until 1840, when the German chemist Justus von Liebig (1803-1873) published an influential book which called attention to the importance of phosphates as fertilizers of the land in agriculture. Within a few years a lively trade in guano had started. Over the following decades millions of tons were imported from South America to increase the production of food in Europe, and guano became a significant factor in the South American economy.

In 1829, some 30 years after his South American expedition, Humboldt was invited to travel extensively throughout Russia and Siberia. In Brazil, he had noticed that diamonds were often to be found in the same localities as gold and platinum. The Ural mountains in Russia contained both metals, so when Humboldt visited them he instigated a (successful) search for diamonds.

Prospecting

Hit-or-miss methods were much commoner in 19th-century prospecting for minerals, as was obvious in the gold rushes, or in the search for oil. What was needed was a fuller understanding of geology, which Humboldt and his contemporaries tried to obtain. For example, not long after Humboldt's Russian trip, the geologist William Clarke (1798-1878) was dispatched from England to Australia. He carried out a detailed study of where coal occurred in New South Wales using a newly developed method of looking for the characteristic fossils to be found in the rocks above and below known seams of coal. Thus it became easier to pinpoint where new coal seams might be found.

Economic plants

It was in botany and the cultivation of plants that science and the economy came together most obviously. Nearly all the natural rubber in use comes from a Brazilian tree species. Originally, rubber was only used for waterproofing, and it was of little economic consequence. However, in the latter part of the century, the introduction of rubber tires led to a rapid increase in demand. Brazil naturally tried to retain control over the product, but in 1876 seedlings were smuggled out, cultivated at Kew Gardens in London and introduced into Malaya: nowadays Southeast Asia is the main source of natural rubber.

Some of the consequences of the trade in plants were less predictable. In the period round 1870, many European vineyards were ruined and others almost put out of business by the phylloxera bug, which had been accidentally imported from America. When it was recognized that native American vines were immune to the bug, roots were hastily imported from America and susceptible European vines grafted to them so averting an economic crisis. Equally, once cultivation of the American potato had started in Europe, it soon became a staple crop.

► Rubber latex tapped on a plantation in Malaysia. The tree Hevea brasiliensis is native to South America and rubber is known to have been used by the Aztecs in the 5th century AD. Cultivation of rubber was introduced to Southeast Asia and West Africa via Kew Gardens in the 19th century.

▼ Explorers and naturalists combined their skills with entrepreneurs and their financial backers, both to export an enormous range of natural products, and to spread their cultivation to other parts of the world. Economic plants from South and Central America include – apart from rubber – the fruits tomato, papaw, pineapple, guava and avocado. Among the vegetables are beans of the genus Phaseolus; cucurbits including the squashes, marrow, pumpkin and gourd; paprika, chili and sweet peppers; maize; potato; cassava and sweet potato. Nuts from the continent include Brazil, cashew and pecan nuts. Tobacco, cocoa and mahogany are other major crops from tropical and subtropical America.

◄ In addition to thousands of botanical specimens, Humboldt brought back a piece of guano which, as fertilizer, became a major export to Europe.

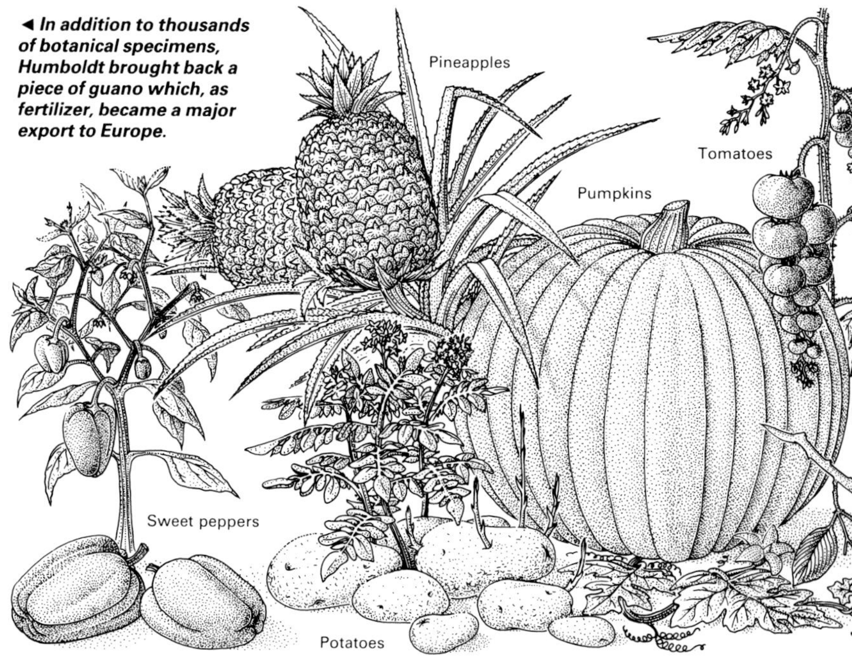

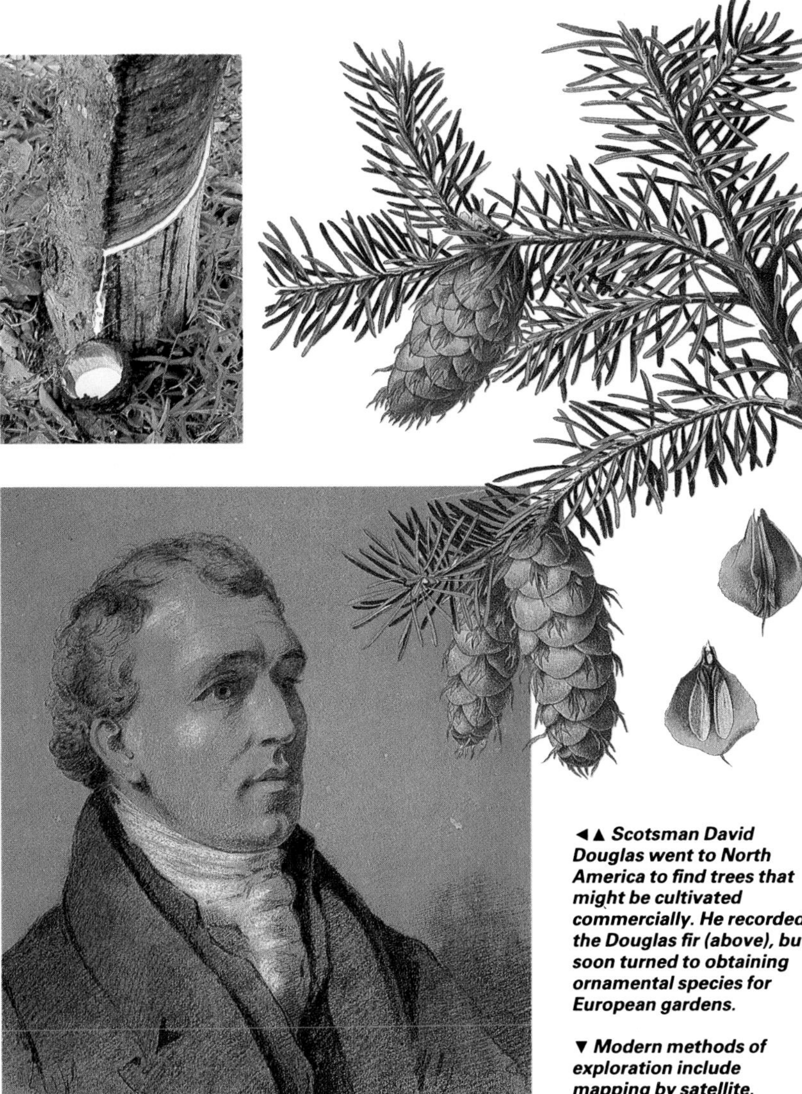

◄▲ **Scotsman David Douglas went to North America to find trees that might be cultivated commercially. He recorded the Douglas fir (above), but soon turned to obtaining ornamental species for European gardens.**

▼ **Modern methods of exploration include mapping by satellite. Improved geological theory helps locate likely deposits of important minerals.**

Ornamentals

The cultivation of the potato in the latter half of the 18th century coincided with more wide-ranging attempts to introduce plants into Europe from other continents. The English naturalist Joseph Banks (1743-1820) brought back new plants from his 1768-1771 voyage with Captain Cook. As a result of his efforts, the Royal Botanic Gardens at Kew became the first institution to decide on a policy of systematic overseas exploration for new types of plants. This approach soon became more general. In the 1820s the Horticultural Society of London dispatched a young Scotsman, David Douglas (1798-1834), to North America for just such a purpose. The motivation for his trip was economic: he was instructed to find fruit trees that might be cultivated commercially in Britain. But he soon concentrated more on looking for plants that could be cultivated in European gardens. During the 19th century, this desire for a wider range of garden plants – of no great direct economic importance – was a basic motivation behind a number of botanical expeditions.

Power politics and mineral exploration

In the 20th century, geology has provided the best examples of financial return from scientific exploration. Most of the minerals of economic importance in the 19th century occurred widely throughout the world. This is not true of some of the industrially important minerals today. The political consequences of such limited distribution were apparent in the oil crisis of the 1970s, when the oil-producing countries tried to exploit their monopoly. A different sort of problem can be found in current moves to exploit the mineral resources of the sea floor. In this case, the developing countries fear that they will lose out in the race to retrieve these minerals, because they lack the necessary technology. So they want to impose restrictions on the exploitation of what has hitherto been considered international territory.

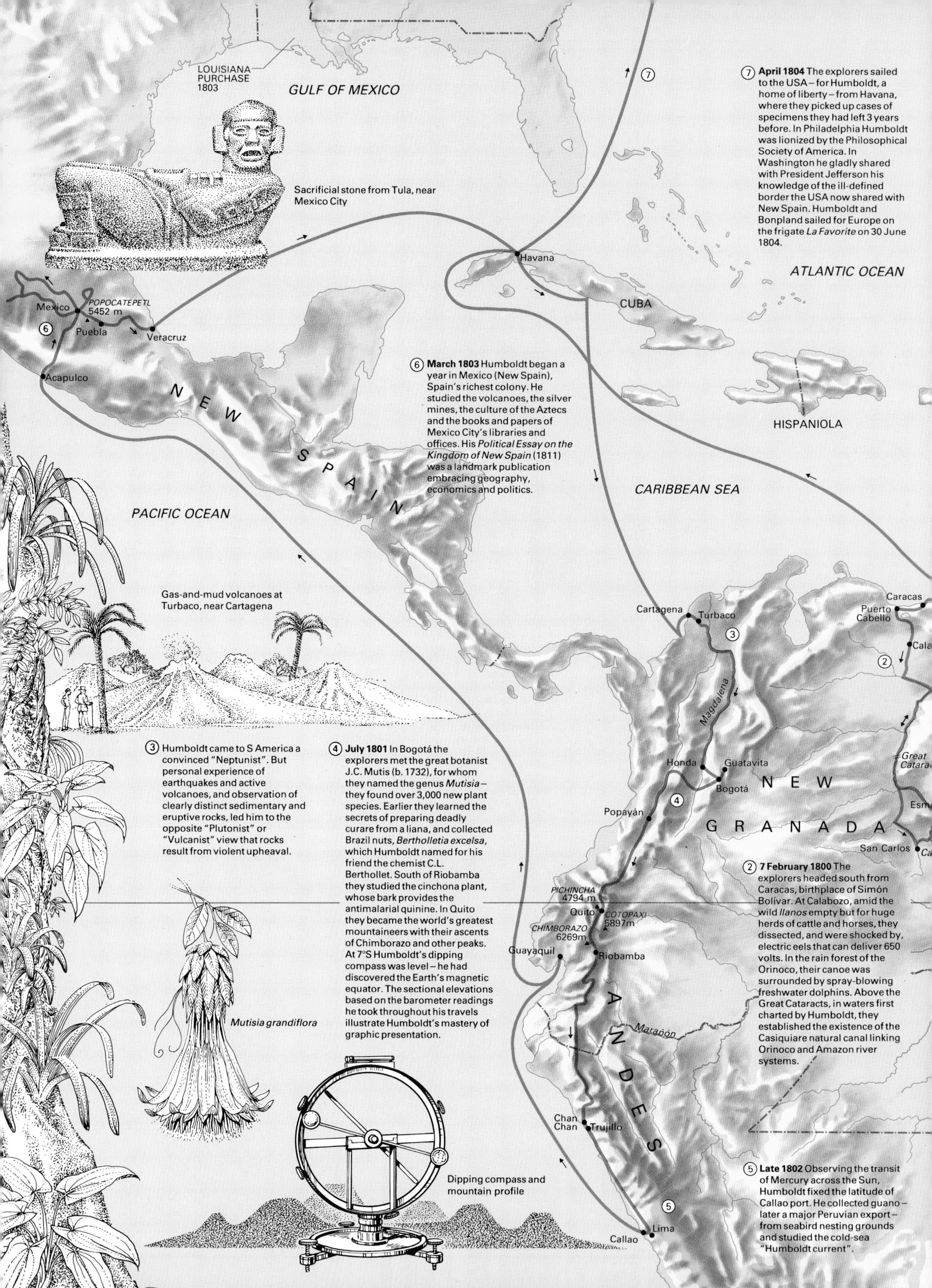

Sacrificial stone from Tula, near Mexico City

Gas-and-mud volcanoes at Turbaco, near Cartagena

Mutisia grandiflora

Dipping compass and mountain profile

(7) **April 1804** The explorers sailed to the USA – for Humboldt, a home of liberty – from Havana, where they picked up cases of specimens they had left 3 years before. In Philadelphia Humboldt was lionized by the Philosophical Society of America. In Washington he gladly shared with President Jefferson his knowledge of the ill-defined border the USA now shared with New Spain. Humboldt and Bonpland sailed for Europe on the frigate *La Favorite* on 30 June 1804.

(6) **March 1803** Humboldt began a year in Mexico (New Spain), Spain's richest colony. He studied the volcanoes, the silver mines, the culture of the Aztecs and the books and papers of Mexico City's libraries and offices. His *Political Essay on the Kingdom of New Spain* (1811) was a landmark publication embracing geography, economics and politics.

(3) Humboldt came to S America a convinced "Neptunist". But personal experience of earthquakes and active volcanoes, and observation of clearly distinct sedimentary and eruptive rocks, led him to the opposite "Plutonist" or "Vulcanist" view that rocks result from violent upheaval.

(4) **July 1801** In Bogotá the explorers met the great botanist J.C. Mutis (b. 1732), for whom they named the genus *Mutisia* – they found over 3,000 new plant species. Earlier they learned the secrets of preparing deadly curare from a liana, and collected Brazil nuts, *Bertholletia excelsa*, which Humboldt named for his friend the chemist C.L. Berthollet. South of Riobamba they studied the cinchona plant, whose bark provides the antimalarial quinine. In Quito they became the world's greatest mountaineers with their ascents of Chimborazo and other peaks. At 7°S Humboldt's dipping compass was level – he had discovered the Earth's magnetic equator. The sectional elevations based on the barometer readings he took throughout his travels illustrate Humboldt's mastery of graphic presentation.

(2) **7 February 1800** The explorers headed south from Caracas, birthplace of Simón Bolívar. At Calabozo, amid the wild *llanos* empty but for huge herds of cattle and horses, they dissected, and were shocked by, electric eels that can deliver 650 volts. In the rain forest of the Orinoco, their canoe was surrounded by spray-blowing freshwater dolphins. Above the Great Cataracts, in waters first charted by Humboldt, they established the existence of the Casiquiare natural canal linking Orinoco and Amazon river systems.

(5) **Late 1802** Observing the transit of Mercury across the Sun, Humboldt fixed the latitude of Callao port. He collected guano – later a major Peruvian export – from seabird nesting grounds and studied the cold-sea "Humboldt current".

◀▼ ***Humboldt's travels with the botanist Bonpland in 1799-1804 are one of the greatest ever "voyages of discovery". Equipped with his scientific training, phenomenal energy and appetite for knowledge, Humboldt pursued a range of researches that were to earn him a reputation as the last great polymath.***

Soon after landing, he set about compiling a list of Southern Hemisphere stars by magnitude; in Lima he observed a transit of Mercury. He correlated plant distribution with altitude and habitats, made notes on fauna, surveyed topographical features and achieved feats of mountaineering, studied the source and properties of curare and quinine, and became a convinced "Plutonist".

Collared inca hummingbird, *Coeligena torquata*, of Venezuela

(1) **16 July 1799** The travelers left the typhoid-ridden corvette *Pizarro* at Cumaná, New Andalusia – "a fabulous and extravagant place". Bowled over, they explored the rain forest, collecting plants. Meticulously, Humboldt recorded the regular daily "atmospheric tide" of the tropics; an extraordinary shower of meteors ("urine of the stars" to the Indians); and his first, major, earthquake.

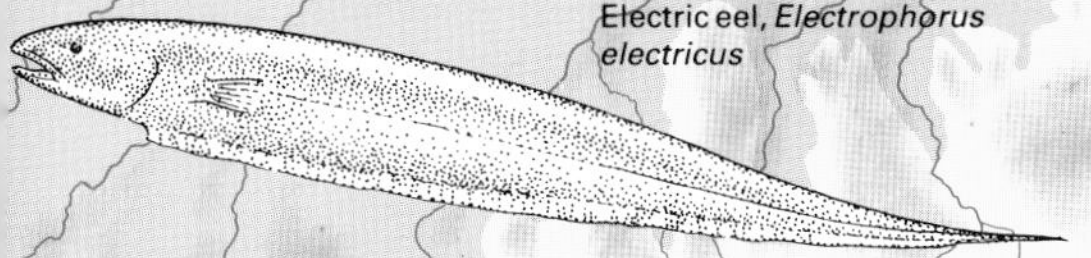

Electric eel, *Electrophorus electricus*

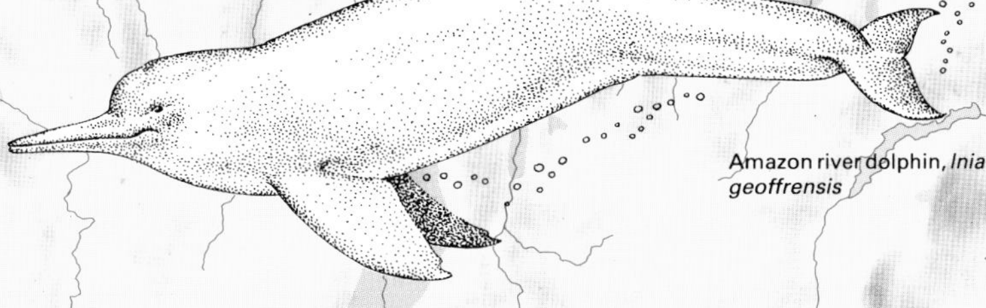

Amazon river dolphin, *Inia geoffrensis*

South American travels

In Humboldt's time, travel to South America was much more difficult than it became a few decades later. Centuries before, the Pope had divided the newly discovered continent between Spain and Portugal, and these two countries still discouraged visitors to their domains. (The situation changed in the early decades of the 19th century as the countries in South America fought to achieve independence.) It was partly because of these difficulties that South America attracted Humboldt. The last scientist to carry out work there had visited the continent over half a century before: the area seemed ripe for an expedition such as those mounted by Captain Cook's naturalists.

In 1799 Humboldt received permission to visit the Spanish colonies in South America. He soon left Europe with one companion, the French naturalist Aimé Bonpland (1773-1858). They spent the next five years in exploration of the continent. They amassed a vast amount of data: for example, some 60,000 plant specimens were collected, of which more than 10 percent were previously unknown to European scientists. Along with the specimens, Humboldt acquired a mass of related knowledge. It was already known, for example, that some South American Indians prepared a poison, used in hunting and warfare, which paralyzed their victims. Humboldt learnt how to prepare this substance – curare – from plants and brought some back to Europe, where its properties attracted much medical attention.

Humboldt planned originally that he and his companion Bonpland would spend some time in Cuba before exploring South America. However, an outbreak of typhoid on board led to their ship docking instead at Cumaná in Venezuela. This was used as a base for some weeks before their first major excursion in September 1799, when they explored the highlands south of the town. In November they departed by sea for Caracas, the capital of Venezuela. Humboldt's intention was to establish whether a link existed between the two great river systems in the north of the continent – the Orinoco and the Amazon – as rumors suggested. Humboldt and Bonpland set out for the headwaters of the Orinoco in February 1800. They first crossed the great plains by foot, then embarked on a long canoe journey up the Orinoco. In May, local Indians carried their canoe overland to a tributary of the Rio Negro (itself a major tributary of the Amazon). Near San Carlos, on the frontier between Brazil and Venezuela, they found the Casiquiare canal, the reputed waterway linking the Rio Negro to the Orinoco. Following this, they discovered that it did, indeed join the headwaters of the Amazon to those of the Orinoco. They traveled back down the Orinoco, reaching Angostura (now called Ciudad Bolívar) in June, and finally arriving back at Cumaná in August, 1800. In November the two companions sailed to Cuba, whence Humboldt thought they might explore the west coast of North America. However, he changed his plans, and the two sailed instead for Peru.

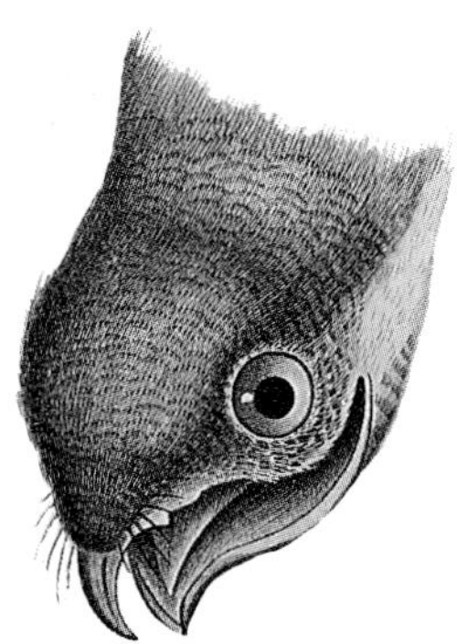

▲◄ *In a cave near Caripe, the expedition's landfall, Humboldt discovered the oilbird. Adults feed their young on oily fruit until they reach 100-150 percent of adult weight. Humboldt observed that the Indians boiled the young birds to give oil for heat and lighting. The "fat bird of Caripe" is the only nocturnal fruiteating bird in the world. Humboldt and Bonpland's account, the "Zoological Observations", also illustrated a local species of nightjar (left).*

▲ *End of a forest. A pioneer ecologist, Humboldt noted extensive destruction by European colonists of the forests of South America, and forecast possible long-term effects on climate.*

In March 1801, Humboldt and Bonpland landed at Cartagena, in Colombia, and decided to reach Lima via a route through the Andes, a journey that was to take them into Ecuador as well as Peru. Their first destination was Bogotá, where Bonpland fell ill. It was not until September, therefore, that they set out on the next stage to Quito, a journey which proved hardest of all their travels in the Andes. The two had by this time become excellent mountaineers, and, from Quito, they carried out the climb which brought them most fame in Europe. In June 1802, they came within 500m of the top of Chimborazo (6,269m), then thought to be the highest mountain in the world (Mount Everest is 8,848m high).

After six months in Quito, they traveled on to Lima, a city which Humboldt greatly disliked. He and Bonpland soon went down to the coast, where, in February 1803, they set sail for Mexico. They spent the next twelve months exploring the country (Humboldt left relatively few notes on this stage of his travels). In March 1804, they sailed to Havana to pick up the scientific collections Humboldt had left there, and then they took passage to the United States. Their exploration of South America had ended.

Early biogeography

The significance of South America for Humboldt was its impact on his thinking. Seeing the animals and plants in this new environment, and comparing his observations with what he knew of Europe, led Humboldt to develop the new science of biogeography. Determining the distribution of living things on the Earth's surface – which is the concern of biogeography – may seem a rather obvious activity. Yet it played a key role in the development of 19th- and 20th-century

▼ *End of a civilization – an Aztec pictogram recorded by Humboldt. During 12 months exploring Mexico, Humboldt produced maps of Mexico and the Louisiana Territory.*

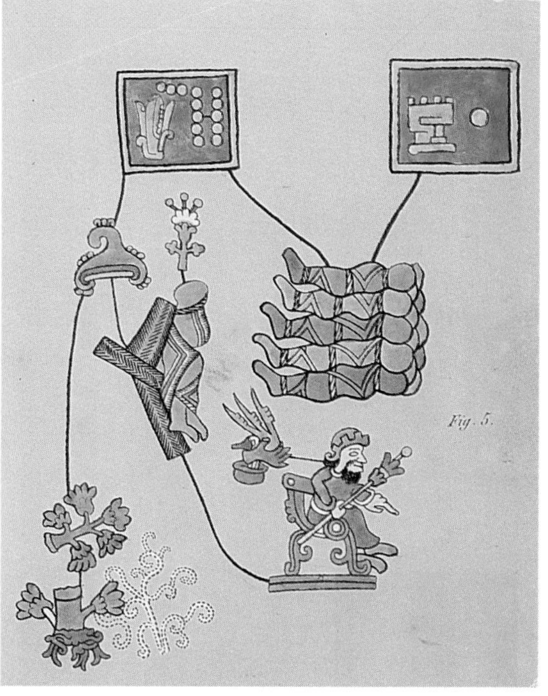

► *Fifty years after Humboldt's pioneering work in biogeography, A.R. Wallace, the "father of zoogeography" (and co-originator of a new theory of evolution), began his researches in Malaysia. Wallace discovered a line of sudden change in the balance between the very distinct Oriental and Australasian faunas. As modified by T.H. Huxley, "Wallace's Line" roughly follows the continental shelf of Southeast Asia. Along "Weber's Line" Australasian and Oriental species of animal occur in roughly equal numbers.*

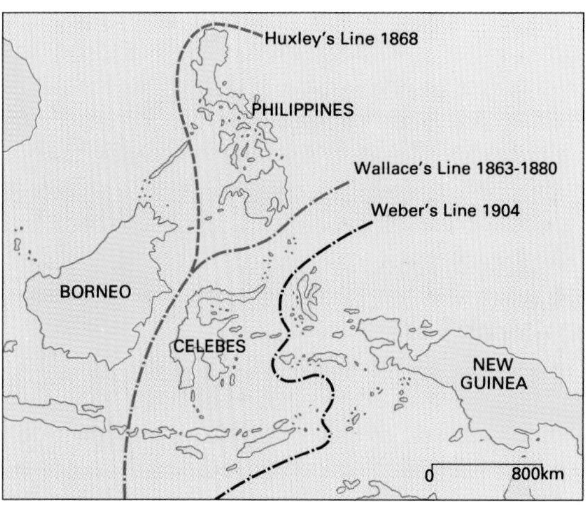

biology and geology. As the 19th century progressed, it became apparent that there could be surprisingly large biological differences over quite small geographical areas. The English naturalist A.R. Wallace (1823-1913), who can be claimed to be, with Charles Darwin (1809-1882), a cofounder of modern evolutionary concepts, came to his idea of evolution during a biogeographical study of the islands near Malaya. There he discovered that there was a remarkably clearcut difference in the animals and plants on either side of a line that could be drawn between these geographically close islands.

This line – still known today as the Wallace line – actually finds an explanation not in biology, but in geology. It has become apparent in the 20th century that the surface of the Earth is mobile, not stationary as our ancestors believed. The most famous example of this mobility was actually studied by Humboldt. It appears in the complementary outlines of the east coast of South America and the west coast of Africa. Humboldt suggested that the South Atlantic was really the remnant of a vast former river which had carved out a channel between South America and Africa. The two sides curved in the same way because they were the opposite banks of this single river. His idea fitted in with Neptunist theory. It also provided an explanation for a story that has intrigued Western man for centuries – the disappearance of Atlantis. Nowadays the event, if accepted at all, is not supposed to have happened in the Atlantic. In Humboldt's day, however, it was a piece of information that needed explaining.

Though Humboldt's explanation of the Atlantic was discarded long ago, his biogeographical approach has provided some of the fundamental data which have led to our modern understanding. In this case, biogeography has been applied not to living organisms, but to fossils. These show that in times past the living organisms on the east coast of South America were closely related to those on the west coast of Africa. But why should there be this historical identity between two regions separated by many hundreds of miles?

The presently accepted answer was proposed in 1912 by another Berliner, the geologist Alfred Wegener (1880-1930). He suggested that continents can move across the surface of the Earth, so that sometimes they congregate together, and at others they separate. On this view, South America and Africa originally formed a single continent which split down the middle, and the fragments have since drifted apart. The original plant and animal populations were the same on either side of the divide, but, as time has passed, they have evolved in somewhat different directions.

▼ *The explorer and surveyor's essential companion – an early 19th-century theodolite in the final form evolved largely by the great English instrument maker, Jesse Ramsden (1735-1800). Easily the most important surveying instrument in the 19th century, and much used by Humboldt on his travels, the theodolite can measure simultaneously both horizontal angles between points, and their elevations.*

Early development of an integrated science of the Earth, and of modern science in universities, owes much to Germany

The fact that Wegener was a Berliner is important. Humboldt believed that the study of the Earth should bring togeher all the different sciences involved – geology, meteorology, oceanography, biology and so on. This vision of an integrated science of the Earth was maintained in Germany, but did not really take root in other countries. It was in this tradition that Wegener, a meteorologist, was prepared to speculate on the geological problem of continents. Since World War II, the concept of integrated study of the Earth has become more popular and has been applied, in particular, to the problems raised by continental drift.

Humboldt not only preached the need to see our environment as a whole: he practiced such study himself. Two illustrations, still of importance today, can be taken from his South American tour. Humboldt habitually recorded temperatures wherever he traveled. One set of temperatures he obtained was of the Pacific Ocean off the west coast of South America. He was able to show from his observations that there was an extended cold current of water flowing past the coast of Peru. This Humboldt current – as it is known today – is on the same sort of scale as the Gulf Stream, and is of considerable significance for the economy of the region. Practical importance also characterizes the second example of Humboldt's observations. Noting the extensive destruction of the South American forests by European immigrants, he deduced that this might have long-term climatological consequences – a point that is being made with increasing frequency today. In this kind of study he was pioneering the science that we now label "ecology".

Humboldt's Paris years

His activity in South America brought Humboldt great acclaim. Within a short time of his return in 1804 to Europe, he was recruited to aid the Prussian state. In 1806 Napoleon beat the Prussian army at the battle of Jena and occupied Berlin. Humboldt was sent as a member of the Prussian delegation to Paris in the following year. It was hoped that his scientific prestige and his influence with numerous acquaintances in Paris might help reduce French demands. In the event, his presence seems to have made little difference to the negotiations, but it allowed Humboldt to settle in Paris. He remained there for much of the following 20 years.

◄ Simón Bolívar, the Liberator of South America. In Paris in 1804, Humboldt told the young Bolívar that he believed the Spanish American colonies to be ripe for independence. The two men stayed in touch until Bolívar died in 1830, having led victorious revolutionary wars of independence in Venezuela, Colombia, Peru and Bolivia (named for him).

Humboldt's abiding interest in the economic and political issues facing Latin America is recorded in books he wrote on two other Spanish colonies: the world's richest mining country – Mexico or "New Spain" (1811), and the island of Cuba, with its slave-based plantation economy.

New approaches in education

Many European universities were backward in the 18th century, their teaching inefficient and their research nonexistent. They were content to cater for the traditional professions, law, medicine and the Church, and their main concern was with handing on traditional material based on Latin or Greek sources. One German official in the 18th century remarked that the purpose of medical school training was to ensure that people were buried methodically! Not all universities were the same. Colleges in the United States such as Harvard, although founded (1636) on the Oxford/Cambridge model, had teaching that was more liberal than at the English universities.

One of the first universities to take a more modern approach was the one attended by the brothers Humboldt. The Hanoverian royal family, who ruled in both England and Hanover, saw themselves as equal in status to the rulers of Prussia, the most powerful German state. As one way of asserting this point, the new University of Göttingen was established in 1737, followed by a Göttingen Academy of Sciences five years later. The two bodies interacted from the beginning, and the city soon became one of the main academic centers in Europe. This is what attracted the Humboldt brothers to the university.

Wilhelm Humboldt and university reform

Not long afterward, the remaining, mainly moribund, German universities were galvanized by the Napoleonic wars. Many were closed and an agonized debate followed on the future of the university system. Within a short time a reformed system, based on the Göttingen model, came into existence. Influential in pushing this reform forward was Humboldt's elder brother, Wilhelm (1767-1835). He pressed the view that the primary task of universities should be the development of individual character. Professors should explore topics with their students, rather than providing little but rote learning. The emphasis should be on the search for new knowledge. As chief of the state education department in Prussia (1809-1810), Wilhelm had an excellent opportunity for implementing his ideas. One result was the establishment in 1810 of the University of Berlin, which became one of the great universities of the world during the 19th century.

Science in the Universities

▲◄ **The famous chemistry laboratory of Justus von Liebig at the Giessen University in 1840. Many young scientists from all over Europe began their researches here. Germany saw the appearance of the research group of students attached to the university professor, and of the research laboratory. Strong links between research and industry resulted in total dominance of the German dye-making industry by the late 19th century (left: German artificially produced ultramarine dye).**

▲▼ **The Humboldts attended Göttingen University, the model for reforms promoted by Wilhelm (below) as education minister.**

Research, laboratories and PhDs

The reform of German universities led to an increase in science teaching and established their growing reputation for scientific research. It was from Germany that the modern concept of a university spread.

Intending applicants for academic jobs in Germany soon began to attach themselves to university professors who already had a proven record as successful researchers. Nineteenth-century Germany saw the appearance of the research group and the research laboratory, both of which play such important roles in modern science. As the German research community grew, the custom arose of awarding a doctorate in philosophy (PhD) to scientists who had carried out original research work. Researchers came from far afield for their German PhD. A typical example was the British physicist John Tyndall (1820-1893) who studied at Marburg under Professor Robert Bunsen (1811-1899).

University funding and industry

As scientists trained in Germany began to obtain positions in their own countries, the idea of the PhD as an entry certificate to the scientific profession grew. Before the end of the 19th century, PhD training had been introduced into the United States. Britain followed suit in the present century. By this time the German concept of a university professor as someone who both taught and carried out research was becoming widespread. In Germany, professors were paid by the state and were therefore seen as civil servants. In English-speaking countries, though universities might be supported by the state, the funding was channeled to them indirectly and academic staff were not regarded as employees of the state. This could lead to differences in the way research developed. For example, central control, as in Germany, may give more prestige to scientific research, but may also be less flexible in trying out new ideas.

Before the latter part of the 19th century, important technology – the railways, for example – had been developed by essentially rule-of-thumb methods with little scientific basis. Britain continued to follow this approach through much of the 19th century, but Germany and other countries began to rely increasingly on scientific input. This paid off in the growth of new, science-based industries, such as those involving large-scale preparation of chemicals. As overseas industries began to pull ahead of those based in Britain, the previous industrial world leader, so British scientists increasingly clamored for more attention to, and funding for, scientific research. But British industrialists were slow to respond. Their suspicion extended in some cases until after World War II. The role that science played in this war was seen as a vindication of its industrial importance.

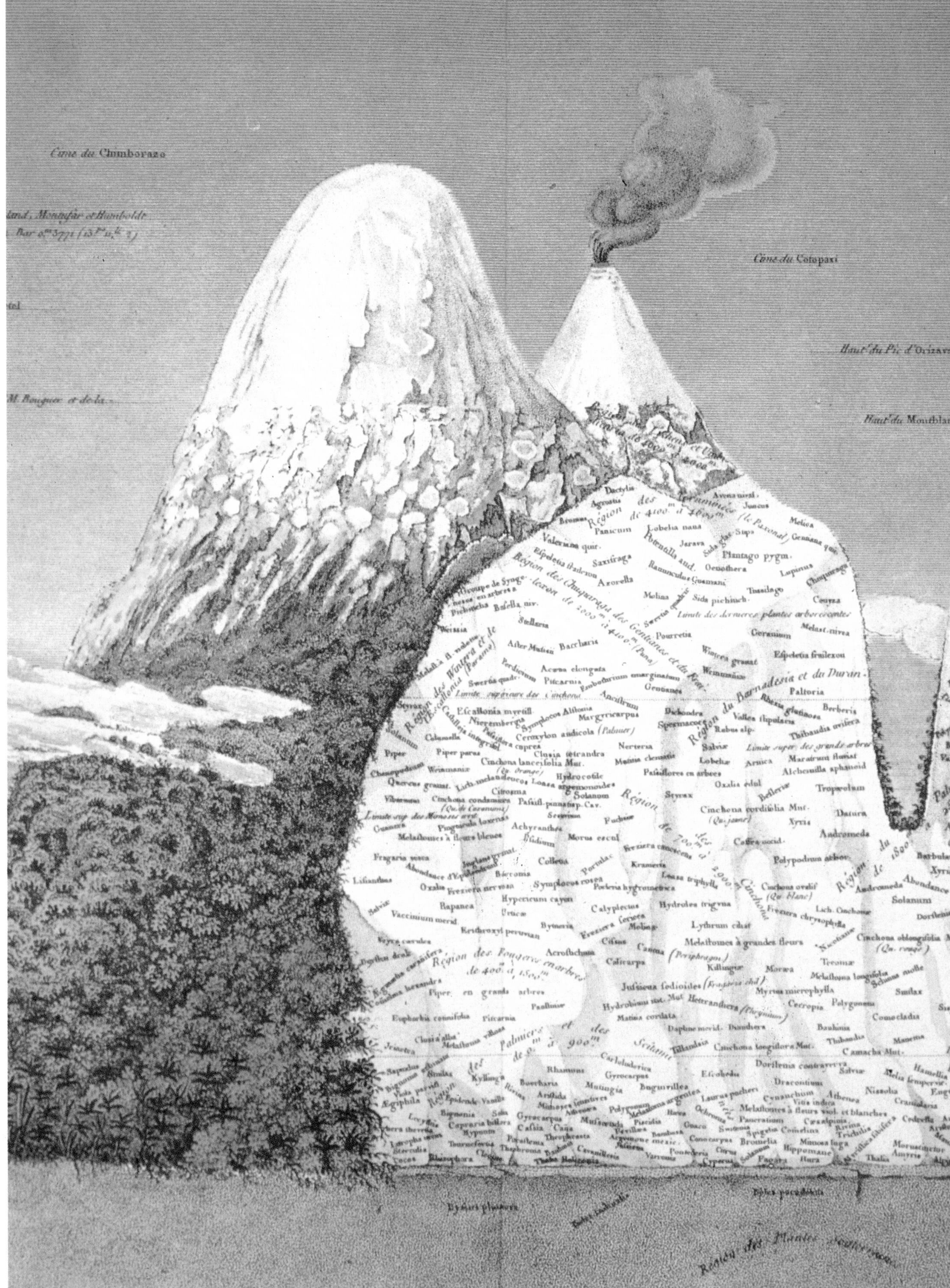

Cime du Chimborazo
Cime du Cotopaxi
Haut. du Pic d'Orizava
Haut. du Montblanc
Région des Graminées de 4100.m à 4600.m (le Paxonal)
Région des Chuquiraga des Gentianes et du Frailexon de 2000 à 4100.m (Puna)
Région des Wintera et de Escallonia (Paramo)
Région du Barnadesia et du Duranta
Région des Fougeres en arbres de 400.m à 1500.m
Région de 700.m à 2900.m des Cinchona
Région des Palmiers et des Scitaminées de 0.m à 900.m
Limite supérieure des Cinchona
Limite super. des grands arbres
Limite des dernieres plantes arborescentes
Limite sup. des Mimoses arbr.
Région des Plantes Souterraines
Dactylis
Agrostis
Avena
Juncus
Bromus
Panicum
Lobelia nana
Melica
Valeriana quir.
Jarava
Stipa
Plantago pygm.
Espeletia frailexon
Saxifraga
Potentilla and.
Oenothera
Lupinus
Chuquiraga
Azorella
Ranunculus Guzmani
Tussilago
Molina
Sida pichinch.
Couyza
Bafella niv.
Pourretia
Geranium
Melast. nivea
Aster Mutisii
Baccharis
Espeletia frailexon
Acæna elongata
Pitcarnia
Gentianes
Paltoria
Escallonia myrtill.
Margyricarpus
Dichondra
Berberis
Nierembergia
Symplocos Alstonia
Vallea stipularis
Ceroxylon andicola (Palmier)
Thibaudia myrifera
Columella
Piper
Clusia tetrandra
Nerteria
Salviæ
Arnica
Cinchona lancifolia Mut.
Lobeliæ
Hydrocotile
Passiflores en arbres
Oxalis edul
Loasa argemonoides
Solanum
Styrax
Tropæolum
Cinchona condaminea
Cinchona cordifolia Mut.
Datura
Achyranthes
Morus escul
Fuchsia
Andromeda
Gunnera
Mélastomes à fleurs bleues
Colletia
Polypodium arbor.
Fragaria vesca
Symplocos rosea
Kramerie
Loasa triphylla
Barbula
Oxalis
Hypericum cayen.
Calyplectus
Hydrolea trigyna
Cinchona ovalif (Qu. blanc)
Solanum
Rapanea
Vaccinium merid.
Lythrum cihat
Erythroxyl peruvian
Mélastomes à grandes fleurs
Cinchona oblongifolia Mut. (Qu. rouge)
Acrostichum
Piper en grands arbres
Jussieua fedioides
Myrtus microphylla
Cecropia
Polygonum
Smilax
Euphorbia cotinifolia
Pitcarnia
Daphne merid.
Bauhinia
Comocladia
Clusia alba
Cinchona longiflora Mut.
Thibaudia
Camacha Mut.
Rhamnus
Gyrocarpus
Dorstenia contrayerva
Draconitum
Kyllinga
Boerhavia
Bugainvillea
Escobedia
Salviæ
Ægiphila
Bignonia
Capraria biflora
Cassia
Cnnna
Mimoses fensitives
Melastomes argentea
Laurus pucheri
Cynanchum
Athenea
Nissolia
Mélastomes à fleurs viol. et blanches
Pancratium
Cæsalpinia
Commelina
Bromelia
Hippomane
Mimosa fuga
Theobroma
Conocarpus

Humboldt preferred living in Paris because it was scientifically lively and he found the political atmosphere more congenial than in Berlin. He supported Lavoisier's radical new chemistry (◆ page 104) and proclaimed his opposition to slavery.

Humboldt used his time in Paris to write up a full account of his travels. Though he produced more than 30 volumes between 1805 and 1834, bankrupting himself in the process, he never completed all he had to say on the topic. He renewed his interest in experimentation (though hindered now by an almost useless right arm – another legacy of his South American trip). He worked initially with the French chemist, Joseph Gay-Lussac (1778-1850). Their main concern was to see whether samples of air always had the same composition, regardless of where and when they were collected. They determined the amount of oxygen in the air by burning it with hydrogen, so producing water. Gay-Lussac realized that the volume of hydrogen consumed was always simply related to the amount of oxygen present, and saw that this must have implications for the physical nature of these substances. In fact, the observation was subsequently regarded as a strong clue to the atomic nature of matter.

Humboldt's lack of training in mathematics did not prevent him from emphasizing the importance of quantitative measurements in science. In particular, he looked for measurements that indicated something important about the Earth as a whole. He introduced the idea of mapping average seasonal temperatures over the globe, and such maps have formed the basis for discussions of climate ever since. He also stressed the importance of allowing for temperature differences with height when discussing the distribution of plant life. The vegetation on a high mountain near the Equator could have much in common with that found at sea level in northern Europe. Such a global view informed his attempts to look at probems of immediate practical importance – such as which parts of the Earth might be suitable for the growth of vines and the production of good-quality wine.

◄ The ecology of plants is recorded in this cross-section of mountains in Ecuador entitled "Geography of Equatorial Plants", part of Humboldt's mammoth publication of his findings during his 19 years in Paris. Latin and common names record species' distribution as noted by Humboldt. The cones of Chimborazo and Cotopaxi (and the height of Vesuvius) are shown.

▼ The suspension bridge over the river Chambo near Penipe in Peru at the southern limit of Humboldt's travels, drawn from a sketch done by Humboldt himself.

Scientific Associations

► *Title of a report on the 15th meeting of the British Association for the Advancement of Science.*

The German Association for Natural Science

One result of the growing German interest in science after the Napoleonic wars was a demand for "a great yearly meeting of the cultivators of natural science and medicine from all parts of the German fatherland". The main advocate of this, Professor Lorenz Oken (1779-1851) at Jena University, organized the first such meeting in 1822. It was held in a beer cellar in Leipzig. The scientists who came together there were mainly liberal thinkers. In the prevailing reaction to preceding revolutionary decades, the Deutscher Naturforscher Versammlung which they formed had to struggle for existence in its early years. This changed in 1828, when Humboldt agreed to organize a meeting of the society in Berlin. Humboldt's wide range of acquaintances allowed him to prepare a wide-ranging program of social events and to obtain royal favor for the meeting.

The spread of scientific associations

The German association had a predecessor, set up a few years before, in Switzerland; but it was the German initiative that was copied in other countries. The British followed suit in 1831, forming the British Association for the Advancement of Science, and their example was followed in turn by American scientists, who established the American Association for the Advancement of Science in 1848. By the middle of the 19th century, similar bodies existed in France, Hungary, Italy and Scandinavia, and, by the end of the century, they had spread as far afield as India and Australia. The aims and organization of these associations were generally similar. Meetings were held annually, but not at a fixed place. These meetings brought together amateur and professional scientists over a wide range of scientific disciplines. Some lectures were aimed at the entire audience, but most of the discussions were held in sections, each devoted to a particular subject area (physics, chemistry, etc). This arrangement – which still continues today – dates back in part to Humboldt's arrangements for the Berlin meeting.

Local scientific societies

The latter part of the 18th century and the early 19th century saw the creation of local scientific societies in a number of countries. Their appearance reflected, on the one hand, the growing attention being paid to science and to its practical applications and, on the other, the difficulties of travel and interchange of information. Local societies could sometimes play a major role in the advancement of science. In Britain, for example, the Lunar Society of Birmingham included among its members the English chemist Joseph Priestley (1733-1804), the Scottish engineer James Watt (1736-1819), who developed the steam engine, and Erasmus Darwin (1731-1802), the extremely ingenious grandfather of Charles Darwin.

▼ ***In 1828 Humboldt organized an historic meeting in Berlin of the German Association for Natural Science. The medal (right) commemorates the event. Left: a meeting of the Geological Section of the British Association, Birmingham, 1849.***

Amateurs and professionals
One of Humboldt's guests at the Berlin meeting was the English scientist, Charles Babbage (1792-1871), the "father of the computer". Babbage, full of enthusiasm for the idea of an association, believed that science was in decline in Britain. British science was dominated by amateurs, and this was leading, Babbage and others believed, to a lower level of support for scientific research than in other countries. Babbage could back these arguments by pointing to the fiercely contested election for the position of President of the Royal Society in 1830. The Duke of Sussex, a son of George III and a former contemporary of Humboldt at Göttingen University, was preferred to the eminent scientist John Herschel (1792-1871). The formation in 1831 of the British Association was therefore welcomed as an alternative venue where scientists from Britain and overseas could meet and discuss research.

Scientific associations today
Ironically, in view of the debate at the time of its formation, the British Association became the focus for meetings of both amateur and professional scientists as the 19th century progressed, while the Royal Society gradually became the preserve of professional scientists. This divergence between national academies and the associations also occurred in other countries. The associations have also tended to be more concerned with the interaction of science and society. For example, they have hosted recurring debates from the 19th century onward concerning the ways in which scientific education and research should be organized and funded. In more recent years, they have provided a forum for bringing together people with a variety of backgrounds to discuss such interdisciplinary topics as the environment. The increasing specialization and professionalization of science has made such debates more necessary.

The annual meetings continue to form the backbone of the associations' activities. Most associations have also been involved for many years in activities between meetings, from the production of widely circulated journals to various kinds of scientific education, especially among younger age groups. One of the original purposes of the associations – to provide a means of communication between professional scientists – is now generally less important.

Return to Berlin

Humboldt was very fond of Paris and prolonged his period of residence there for as long as he could. However, his main source of funding was now the retainer paid him by the king of Prussia. By 1826, William Frederick decided that Humboldt had been away from Berlin for long enough. He apparently wrote to Humboldt saying: "I cannot allow you to extend your stay any further in a country which every true Prussian should hate." Consequently, Humboldt left Paris in 1827 and after a short visit to London, settled again in Berlin. He found the change depressing: the social life was far below the level of Paris or London, and the political regime was oppressive. But his lectures at the University of Berlin and elsewhere drew huge crowds, which pleased him. Though his talks were devoted primarily to the scientific understanding of the Earth, they had the additional aim of countering the romantic, anti-scientific philosophy of Hegel and Schilling then popular in Berlin. He set to work to convert his lectures into a book, which eventually appeared in five volumes under the title *Cosmos* (1845-1862).

In 1827 the Russian minister of finance sought Humboldt's advice on mining. This soon led to an invitation for him to study the geology of the Ural mountains at the Tsar's expense. In April 1829 Humboldt's expedition set out for St Petersburg (now Leningrad), where they were given a rousing reception. In May they moved on to Moscow, and from there traveled eastwards – by land as far as the ancient town of Nizhni Novgorod (now Gorki) and then by barge along the Volga river. Humboldt stopped to study the Urals, as required by the Tsar, but then thrust on further through Siberia to the Chinese border. This was the farthest point eastward that his group reached. They returned along the Irtysh river to Omsk and back overland to the Urals again. After a long diversion southward to investigate the Caspian Sea, they finally reached Moscow in November. Humboldt was greeted enthusiastically by the Tsar, so that he was not able to get back to Berlin until after Christmas. This, the last of his major excursions (➧ page 128), provided, he remarked, a great contrast with his South American travels.

Mapping Earth's magnetic field

The type of global observation that most interested Humboldt concerned the Earth's magnetism. It had been known for many years that the needle of a magnetic compass does not usually point precisely toward the geographical North Pole. The energetic Englishman Edmond Halley (1656-1742) himself measured compass deviations from true North in the Atlantic. Unfortunately, it was soon found that the deviation of the compass needle at any point gradually changed, so his map went out of date as a navigator's aid. But Halley's work on magnetism started a whole new field of investigation – the physical properties of the whole Earth, or, as it is now called, geophysics.

Humboldt, of course, found Halley's earlier studies very congenial to his own view of science, so he proceeded to push them further. Halley had concentrated on the deviation of the compass needle from true North; Humboldt also looked at the way in which a compass needle could point down toward the Earth. He measured this "dip" along with the ordinary deviation of the compass wherever he went on his travels. Combining his own data with observations of other travelers, Humboldt began to map the Earth's magnetism over a good fraction of its surface.

It was Humboldt's study of the Earth's magnetic field that led him to promote international cooperation between scientists

Humboldt's researches into the Earth's magnetism soon attracted the attention of one of the greatest mathematicians of all time, Karl Friedrich Gauss (1777-1855), who spent most of his life at Humboldt's old university, Göttingen. Gauss was greatly interested in the Earth's magnetism, and could do what Humboldt could not – turn the measurements into a mathematical picture. For the first time the complexity of the Earth's magnetism began to emerge.

Humboldt was fascinated by an unexpected property of compass needles: the way in which they would sometimes start to oscillate backward and forward. He called these disturbances "magnetic storms" and wondered whether they occurred simultaneously all over the Earth. In 1836 he wrote to the Duke of Sussex, President of the Royal Society, suggesting that the Royal Society establish observatories. His successful efforts to secure the setting up of magnetic and meteorological observation posts, not only in British possessions but also in Russia, marked an important development in scientific cooperation

Observing the transits of Venus

Astronomers had long been uncertain about the size of the Solar System. Just how far away was the Earth from the Sun? The English astronomer Edmond Halley (1656-1742) pointed out that one way of resolving the problem was to measure the position of the planet Venus as it crossed the Sun's surface. Depending on the position of the observer on Earth, Venus would appear to cross differing parts of the Sun's surface. These differences could be used, along with the known positions of the various observers on Earth, to give the distance of Venus from the Earth. This, in turn, could be applied to give a value for the distance of the Earth from the Sun. Such passages of Venus across the Sun – called "transits of Venus" – only happen at rare intervals. (As seen from the Earth, Venus usually passes either above or below the Sun.) Two transits occurred in the 18th century: one in 1761, and the other in 1769. To observe these transits from a number of places round the Earth required more resources than any single nation could afford. Consequently, a number of countries agreed to mount a cooperative campaign. At some sites – for example, in North America – there were already resident astronomers. Others, especially in the Southern Hemisphere, had to be manned by astronomers sent from Europe. One of Captain Cook's voyages to the Pacific was planned, in part, to carry out measurements of Venus. This is the earliest example of truly international cooperation. Up until then research projects were confined within the same country.

Humboldt's magnetic observation stations

It was difficult enough to mount an international cooperative effort for a specific major event. Global measurements which required frequent repetition over an extended period presented a step up in difficulty. The only way was to have scientists in countries all round the world.

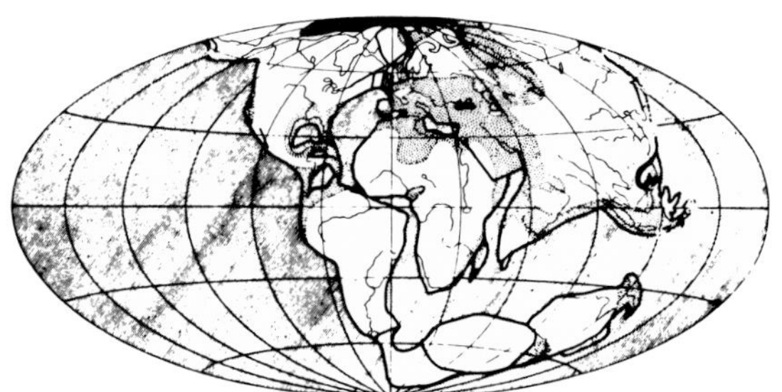

◀ A solar flare forms during an abrupt change in the Sun's magnetism. Humboldt postulated that "magnetic storms" caused the periodic inexplicable oscillations of compass needles on Earth. Sunspot counts and worldwide magnetic observations subsequently established this link with the Sun's activity.

▲ Alfred Wegener's 1915 map shows the supercontinent Pangaea splitting up in Eocene times, some 50 million years ago. His theory of continental drift remained unpopular with scientists until it was confirmed in the 1950s by paleomagnetism – a science started by Humboldt 160 years earlier.

Scientific Cooperation Worldwide

In the mid-1830s, the Göttingen Magnetic Union was set up by Karl Friedrich Gauss to coordinate magnetic measurements made at different stations – to begin with in Germany, then from Ireland across Europe. Humboldt, who had prompted this move, went further afield.

Following his visit in 1829, he successfully appealed to the government of Russia for observation stations to be set up there. In 1836 Humboldt wrote to the Duke of Sussex, a fellow student at Göttingen who by this time was President of the Royal Society in London, with the result that stations were established in the British colonies worldwide. These stations also typically collected weather data. International meteorological conferences were convened (at Brussels in 1853 and Vienna in 1873) and agreement was reached to pool weather data collected by nations in various parts of the world.

▼ Straddling a crevasse, a Snow-Cat of the Trans Antarctic Expedition, part of the International Geophysical Year, 1957-1958. Scientists from 66 countries cooperated in the IGY, probably still the largest ever international scientific effort. An early success in this field of international collaboration was the network of observation stations established in the 1830s on Humboldt's initiative.

International "Polar Years"

A proposal from the Germans, that a cooperative effort should be made to collect magnetic and meteorological data from the polar regions, led in 1882-1883 to a special campaign (mainly in the Arctic), since labeled the "First Polar Year". The initiative for the Second Polar Year in 1932-1933 came again from the Germans – more specifically from one of Alfred Wegener's closest colleagues. Much more attention was paid to the Antarctic this time, but the basic motivation remained magnetic and meteorological observation. Then, in 1957-1958, came the most important of these events – the International Geophysical Year. This was planned as a Third Polar Year, but it was decided instead to make observations over the whole of the Earth's surface. New measurement techniques were introduced, the most startling being the launch of the first artificial Earth satellites. The International Geophysical Year thus marked the beginning of the space race between the United States and the Soviet Union.

Nowadays, exploration is not the main reason for international scientific cooperation: the concern is primarily with sharing expensive research resources and valuable expertise. Where research facilities are very expensive – as in high-energy physics or astronomy – it has become commonplace for scientists to share instrumentation and other resources funded by more than one country. Sharing of expertise commonly takes place between developed and developing countries, when experts from the former visit the latter to give advice. These visits may be arranged directly between nations, but are more commonly associated with intergovernmental organizations, such as UNESCO. It is interesting that such visits to South America have recently included experts concerned with the destruction of rain forests and the impact on the environment – a topic which Humboldt first brought to public attention.

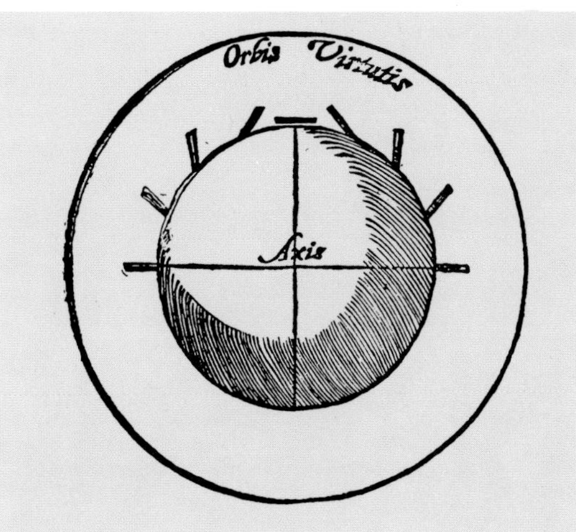

▲ Compass needles dipping in the Earth's magnetic field or "sphere of power" (orbis virtutis), from William Gilbert's "De Magnete" (1600). Humboldt measured "dip" wherever he traveled.

between nations. Edward Sabine (1788-1883), was the leading expert in charge of the British program. By the early 1850s, Sabine knew from these observations that the number of magnetic storms varied periodically, reaching a peak every eleven years. At this time, Sabine's wife was translating into English Humboldt's most important book, *Cosmos*. In this work Humboldt described some little-known observations by a German amateur astronomer, Heinrich Schwabe (1789-1875). These consisted simply of counting the number of spots visible on the Sun's face each day over a period of many years. Schwabe showed that this number varied regularly with a maximum every 11 years. Schwabe's results, when combined with Sabine's, suggested that the Earth's magnetism was being affected by activity on the Sun. This discovery started off a whole new area of investigation into solar-terrestrial relations. Today many artificial satellites carry instruments to examine such interactions between the Sun and the Earth.

Besides studying the magnetism of the whole Earth, Humboldt noticed that individual rocks on the Earth's surface sometimes had a peculiar magnetism of their own. His interest in this started after his mining course at Freiberg, when he was an inspector of mines.

While an inspector of mines for the Prussian government Humboldt chanced on a large rock which was appreciably magnetic but magnetized in the opposite direction to the Earth, so that the North pole of the compass pointed to the South end of the rock. Humboldt continued to look for such rocks throughout his life, but the pay-off only came long after his death. A major problem of Wegener's theory of continental drift (◀ page 119) lay in finding evidence to support it. Over the last few decades it has become apparent that rock magnetism can be used for this purpose. Rocks which formed by solidifying from a hot liquid have been found to retain an imprint of the Earth's magnetism at the time they solidified. Rocks which formed at the same time all over the world should obviously record the same picture of the Earth's magnetism. But it proved impossible at first to fit these records together: the various rocks seemed to indicate differing magnetic directions at different parts of the Earth's surface. The solution was only obtained when it was supposed that there had been relative motion between the continents from which the rocks were collected. In particular, the older the rocks studied, the closer South America and Africa had to be together. The splitting of the two continents had proceeded just as Wegener had proposed.

Humboldt would have been delighted with this example of interaction between different branches of science. He explained in his book, *Cosmos*: "I ever desired to discern physical phenomena in their widest mutual connection, and to comprehend nature as a whole...the separate branches of natural knowledge have a real and intimate connection." In the year he died, 1859, one of the great unifying concepts of science appeared: Darwin explained his ideas on evolution in the book *Origin of Species*. An enthusiastic reader of Humboldt's *Personal Narrative* (1814-1829) of his travels in South America, Darwin remarked: "I shall never forget that my whole course of life is due to having read and re-read as a youth his *Personal Narrative*." Darwin's comments reflect Humboldt's scientific importance: he persuaded others to look at the Earth in a different way.

▲ *Humboldt at 77 – an early photograph taken in 1847. In an age of increasing specialization in science, Humboldt made significant contributions in many different fields. Celebrated as an explorer as well as a scientist, he is said to have been second in fame only to Napoleon during his lifetime.*

▼ *The last great excursion of the liberal Alexander von Humboldt was in 1829 at the invitation of Tsar Nicholas I, ruler of perhaps the most reactionary power in Europe. In 25 weeks Humboldt traveled headlong from St Petersburg (Leningrad) via Moscow to the Urals, the Chinese border, the Altai mountains and the Caspian Sea. The ferry crossing of the Irtysh River at Tobolsk in Siberia (below) was one of dozens made by the expedition, which yielded data for his world isothermal map, and the discovery, in the Urals' gold mines, of the first Russian diamonds.*

Michael Faraday 1791-1867

7

A blacksmith's son...Davy at the Royal Institution... Faraday's contribution to chemistry...Committee man, lone researcher, and Sandemanian...Invention of the electric motor, transformer and dynamo...Optics: "Faraday rotation"...The underlying unity of the world...PERSPECTIVE...Sects and scientific education... Giving science to the people...From amateurs to paid scientists...Developments in the electrical industry

Many people feel that genius will make itself felt regardless of the circumstances in which it is placed. It is certainly true that some great scientists have had to surmount major obstacles in order to pursue their interests. Michael Faraday is an obvious example. His father was originally a blacksmith in the north of England, but trade worsened with the outbreak of the French Revolution, so the family moved south in the hope of finding more work. Michael Faraday was born shortly after the move, on 22 September 1791, in the village of Newington in Surrey (now a part of London). The family moved again to north of the Thames, but Michael's father suffered increasingly from ill-health which restricted the work he could do, and finally led to his death in 1810. Money was therefore very tight; indeed, when he was ten years old, Michael Faraday's basic diet was one loaf of bread a week.

It is hardly surprising that there was little to spare for schooling. As Faraday himself commented later: "My education was of the most ordinary description, consisting of little more than the rudiments of reading, writing and arithmetic." As soon as possible, Faraday went out to work. He began as an errand boy with a local bookseller and bookbinder, and then at the age of 14 was apprenticed to learn the trade. He soon became competent – some of the volumes he bound still survive.

A believer in the unity of nature

The new approach to chemistry established by the French scientist Antoine Laurent Lavoisier, and the electric battery devised by the Italian physicist Alessandro Volta, were two major scientific developments at the end of the 18th century. The British chemist Humphry Davy combined the two in the new process of electrolysis. Faraday, following him, established the basic laws of electrolysis. This was a fundamental contribution to chemistry, but Faraday also shaped the future course of physics. In 1820, Hans Oersted in Denmark discovered that an electric current can affect the needle of a magnetic compass. Faraday used this result to build the first electric motor. He subsequently showed that magnetism can equally be converted into electricity, and constructed a dynamo. He also made the first electrical transformer, so providing all the basic components needed to create an electrical industry. He was guided in this work by a belief in the fundamental unity of nature. It led him to results unsuspected by his contemporaries – for example, to the fact that light can be affected by magnetism.

Faraday's electromagnetic theory differed considerably from that of his contemporaries. Because he was no mathematician, it required the work of another British scientist, James Clerk Maxwell, to convert his ideas into quantitative results. Maxwell's consequent prediction that electromagnetic waves exist led to the discovery of radio waves by the German physicist, Heinrich Hertz. Faraday's ideas supposed that space is itself possessed of important properties. In this century Albert Einstein used this concept in gravitation; it has more recently been applied to quantum mechanics.

▼ Faraday in 1842 during his last period of fruitful research.

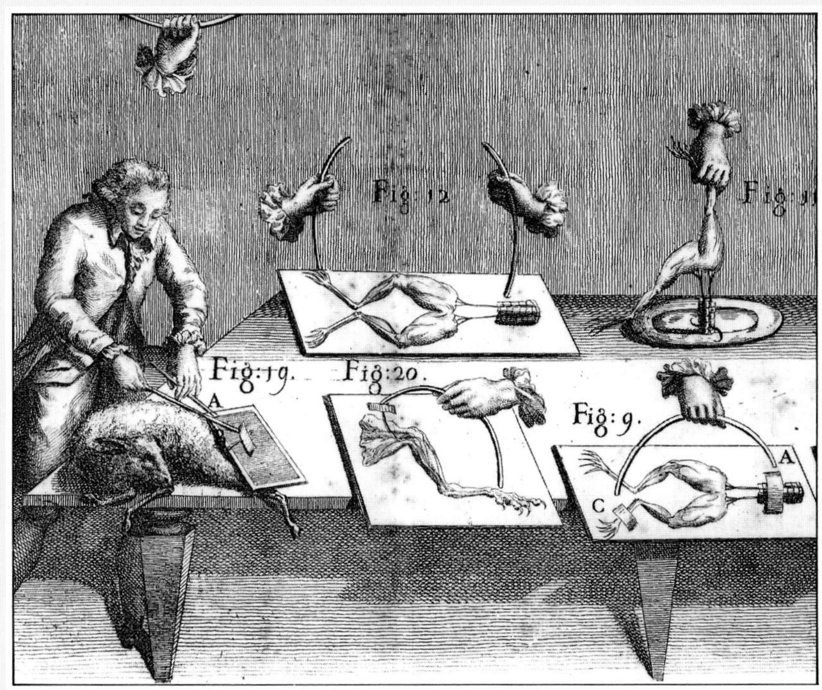

◄ Luigi Galvani demonstrates "animal electricity" (1791). Humboldt carried out 4,000 such experiments. In 1800, Alessandro Volta invented the battery, showing that animal matter was not needed in order to produce electricity.

Michael Faraday – His Life, Work and Times

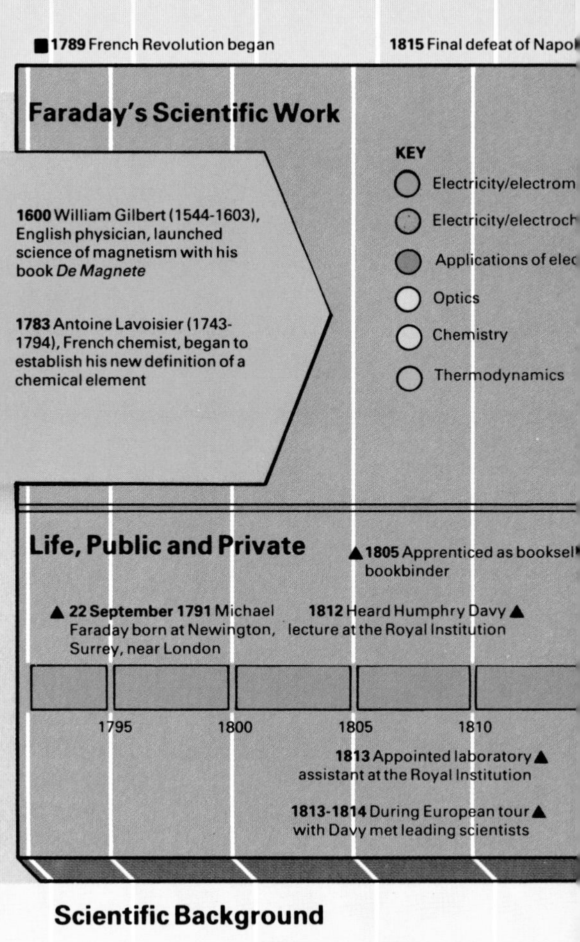
Historical Background
1789 French Revolution began
1815 Final defeat of Napo
Faraday's Scientific Work
KEY
Electricity/electrom
Electricity/electroch
Applications of elec
Optics
Chemistry
Thermodynamics
1600 William Gilbert (1544-1603), English physician, launched science of magnetism with his book De Magnete
1783 Antoine Lavoisier (1743-1794), French chemist, began to establish his new definition of a chemical element
Life, Public and Private
1805 Apprenticed as booksel bookbinder
22 September 1791 Michael Faraday born at Newington, Surrey, near London
1812 Heard Humphry Davy lecture at the Royal Institution
1795
1800
1805
1810
1813 Appointed laboratory assistant at the Royal Institution
1813-1814 During European tour with Davy met leading scientists

Davy and the Royal Institution

Now surrounded by books, Faraday became an avid reader. His interest soon centered on science, with a preference for descriptions of chemical and electrical experiments. He began to repeat these experiments himself, and joined a new scientific club, the City Philosophical Society, which put on lectures, discussions, or demonstrations each week. Here he began to think about the scientific problems that were to occupy the rest of his life: his first talk to the society was on the nature of electricity.

Faraday's involvement in science became known to customers of the bookshop, and one of them gave him tickets to hear the scientist Humphry Davy (1778-1829) lecture at the Royal Institution in central London. The Royal Institution had been established in **1799**, on the initiative of Benjamin Thompson (1753-1814), a Massachusetts-born American who spent the latter part of his life in Europe. After serving in the government of the elector of Bavaria, he became Count Rumford in 1791, then toward the end of that decade took up residence in London. Here he sought ways of applying scientific ideas to everyday life – the Rumford domestic stove was his invention. The original intention in setting up the Royal Institution was to show workers, by means of teaching and research, how science could be applied to their activities. However, the Royal Institution depended on subscriptions to keep it going, and this meant that more emphasis had to be placed on lectures aimed at wealthier people. Rumford soon quarreled with his co-organizers and returned to the Continent (where he subsequently married Lavoisier's widow, Marie Anne Paulze). However, before going, he brought Humphry Davy from the West Country to lecture at the Royal Institution.

Davy was an outstanding researcher, and he was also an excellent lecturer who quickly attracted large audiences. By the time Faraday attended his lectures in 1812, Davy had made the Royal Institution one of the major scientific centers in Britain. Davy's particular interests, like Faraday's, were chemistry and electricity.

▲ *The library of the Royal Institution, which was founded to promote applied science. It was intended both for teaching and research, and had a lecture theater and laboratories. The library was planned to service both activities.*

Scientific Background

1800 William Nicholson (1753-1815) and Anthony Carlisle (1768-1840), British scientists, used new Voltaic pile to decompose water by electrolysis

1807-1808 Humphry Davy (1778-1829), British chemist, developed applications of electrolysis; eventually identified 6 new metallic elements

1799 Royal Institution founded in London, on initiative of Rumford

1798 Count Rumford (Benjamin Thompson) (1753-1814), American-born scientist, reported relationship between mechanical work and heat

1791 Luigi Galvani (1737-1798), Italian anatomist, published work on animal electricity

1800 Alessandro Volta (1747-1827), Italian physicist, reported invention of his electric battery or "pile"

1800 Thomas Young (1773-1829), British scientist, revived wave theory of light; in 1807 published his physical concept of energy

1806 Josef van Fraunhofer (1787-1826), German instrument-maker, began to develop high-quality optics, notably in spectroscopy

1802-1804 John Dalton (1766-1844), British chemist, developed his atomic theory

1790 1795 1800 1805 1810

1845-1849 Potato famine in Ireland

1853-1856 Crimean War

1861-1865 Civil War in United States

1837 Accession of Queen Victoria

1839 First British Opium War with China

1853 David Livingstone's discovery of Victoria Falls heralded European expansion in Africa

1867 Reform Bill enfranchised male urban working class in Britain

1832 Reform Bill enfranchised middle class in Britain

1848 Revolutions shook Germany, France, Italy, Hungary

1825 Discovered benzene during work to assist gas industry in bottling gas under high pressure

1824-1831 Researched on production of good-quality glass

1823 Liquefied chlorine

1832 Deposited sealed envelope (opened 1937) at Royal Society proposing that electromagnetic waves exist analogous to waves on water's surface

1845 Observed "Faraday rotation" of polarized light in magnetic field and "diamagnetism"

1821 Devised simple electric motor

1831 Invented the electrical transformer, discovering electromagnetic induction; invented a dynamo; and introduced "lines of force"

1833 Determined 2 laws of electrolysis which remain the basis of electrochemistry

1861 His *Chemical History of a Candle* published, based on children's Christmas lectures

1821 Married Sarah Barnard

1840 Became elder of the Sandemanian Church

1824 Elected fellow of the Royal Society

1833 Appointed Fullerian Professor of Chemistry at the Royal Institution

1841 Fell ill and spent several months in Switzerland

25 August 1867 Died at Hampton

20 1825 1830 1835 1840 1845 1850 1855 1860 1865 1870

1825 Appointed director of the laboratory at the Royal Institution

1865 Resigned from Royal Institution

1858 Moved into house at Hampton Court, west of London, given by Queen Victoria

26 Initiated Friday night at the Royal Institution

1827 Initiated Christmas lectures for children at the Royal Institution

1837 Samuel Morse (1791-1872) in USA and Charles Wheatstone (1802-1875) in Britain patented electric telegraphs, developed with J. Henry's assistance

1846 Smithsonian Institution founded, Washington DC; Joseph Henry its secretary and director

1870s Thomas Edison (1847-1931) in US and Joseph Swan (1828-1914) in Britain produced carbon-filament lamps

Early 1830s Joseph Henry (1797-1878), US physicist, studied electromagnetic induction; made large electromagnets and an early electric motor; invented electromagnetic telegraph

1851 Great Exhibition, London

1876 Alexander G. Bell (1847-1922), Scottish-born, US scientist, invented the telephone

1831 British Association for Advancement of Science founded

1840s First industrial dynamos (for electroplating) installed in Birmingham, England

1848 American Association for Advancement of Science established

1823 Davy suggested liquefaction of chlorine (accomplished by Faraday)

1841 James Joule (1818-1889), British physicist, indicated equivalence of mechanical heat and work

1847 Hermann von Helmholtz (1821-1894), German physicist and physiologist, gave mathematical formulation to law of conservation of energy

1869 Dmitri Mendeleev (1834-1907), Russian chemist, published periodic table of elements

827 Georg Ohm (1789-1854), erman physicist, formulated Ohm's Law on resistance of electricity conductors

1824 Dominique Arago (1786-1853), French physicist, discovered that a rotating copper disk deflects a magnet

1845 William Thomson (Lord Kelvin), British mathematician and physicist, provided a quantitative treatment of "lines of force"; later formulated 2nd law of thermodynamics

1850 R. J. E. Clausius (1822-1888), German physicist, first stated 2nd law of thermodynamics

1877 James Dewar (1842-1923), British physicist, worked on liquefaction of gases

1820 André Marie Ampère (1775-1836), French mathematician and physicist, began study of forces set up between electric current-carrying conductors

1850 John Tyndall (1820-1893), British physicist and popularizer of science, studied magnetism of crystals; succeeded Faraday at Royal Institution in 1865

1855 James Clerk Maxwell (1831-1879), British physicist and mathematician, developed Faraday's ideas on "lines of force"; later predicted light is a form of electromagnetic wave

1820 Hans Oersted (1777-1851), Danish physicist, discovered the magnetic effect of an electric current

1858 Julius Plücker (1801-1868), German physicist and mathematician, studied electrical discharges through gases at low pressure

1876 William Crookes (1832-1919), British physicist, invented his evacuated glass "Crookes tube"

1833 Göttingen Magnetic Union of observation stations set up by Karl Gauss (1777-1855), German mathematician

1826-1852 First practical chemistry teaching laboratory set up at Giessen University by J. von Liebig (1803-1873), German chemist

1845 August Hofmann (1818-1892), German chemist, began investigation of benzene and coal-tar derivatives

1880s Buildings lit by electricity in cities from New York to Moscow

1883 Svante Arrhenius (1859-1922), Swedish chemist, worked out theory of electrolysis

1887 Heinrich Hertz (1857-1894), German physicist, discovered radio waves

1897 J. J. Thomson (1856-1940), British chemist, established existence of electron

1901 Guglielmo Marconi (1874-1937), Italian engineer, transmitted radio signals from England to N America

1820 1825 1830 1835 1840 1845

Lavoisier called oxygen the "acid carrier" but Davy disproved him, showing that muriatic acid contained not oxygen but the element chlorine

Faraday follows Davy's work on electrolysis

During his first years in London, Humphry Davy brilliantly combined chemistry and electricity in his work on electrolysis. The invention of the electric battery, reported by the Italian physicist Alessandro Volta (1745-1827) in 1800, sparked off a series of investigations into how it worked and what effects it could produce. Almost immediately, two British scientists – William Nicholson (1753-1815) and Anthony Carlisle (1768-1840) – showed that the battery could be used to split up water into its basic constituents of hydrogen and oxygen. If two pieces of metal, attached respectively to the positive and negative sides of the battery, were suspended in a bowlful of water, hydrogen began to appear at one of the metal pieces and oxygen at the other. This process of breaking down compound substances by passing electricity through them came to be called "electrolysis". Davy saw that it might provide a means of identifying new chemical elements. Within a few years of arriving at the Royal Institution he had used electrolysis to identify six new metals – potassium, sodium, calcium, strontium, barium and magnesium. His results were so highly esteemed that he was awarded a prize established by Napoleon, even though France and Britain were then at war.

Davy made a great impression on Faraday, who, as he came to the end of his apprenticeship, realized he wanted more than anything else to become a full-time scientist. In October 1812, Davy was temporarily blinded by an explosion in his laboratory. Faraday managed to become his secretary for a few days while his eyes recovered. Subsequently, Faraday begged Davy to be considered for any scientific post that came available. Not long afterward, the laboratory assistant at the Royal Institution was dismissed for fighting, and Faraday was appointed in his place. At last he could practice science, even if at a menial level.

As a laboratory assistant, Faraday's work was routine – much of it relating to chemical analysis, since Davy was officially professor of chemistry. However, the routine was soon broken. In 1812 Davy married a wealthy widow, and, the following year, he resigned from the Royal Institution in order to make an extended tour of Europe. He obtained permission to take Faraday with him as an assistant.

▲ Humphry Davy came to the Royal Institution from Bristol, where he had studied gases. His public lectures became very popular, and his researches founded the science of electrochemistry. Davy achieved fame for inventing the miner's safety lamp.

► From the day in 1813 when Faraday became laboratory assistant to Davy at the Royal Institution, he spent much of his time in the laboratory. He is seen here surrounded by bottles of chemicals, and with electrical apparatus on the right.

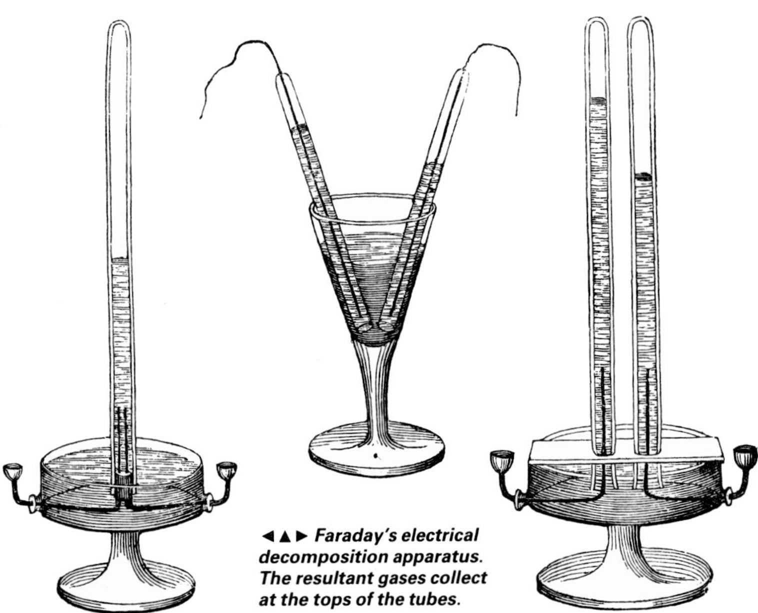

◄▲► Faraday's electrical decomposition apparatus. The resultant gases collect at the tops of the tubes.

► The large electric battery built at the Royal Institution in 1807. It was used in Davy's, and later Faraday's, electrochemical experiments.

▼ Volta's electric battery, or "pile". Silver and zinc disks are separated by pads soaked in water.

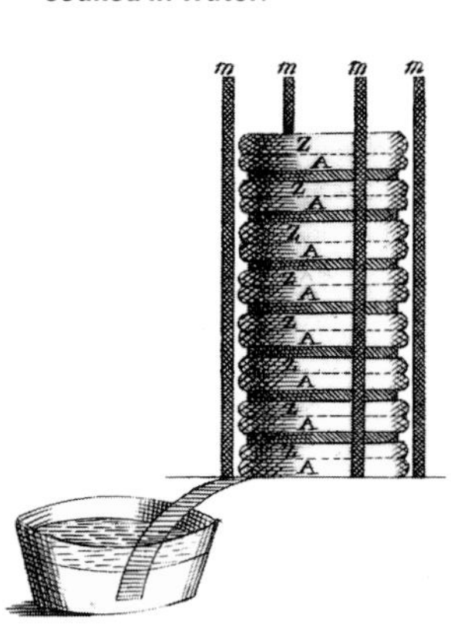

In 18 months, Faraday met many of the leading European scientists. In Paris, he observed Davy discover yet another new element. French scientists showed Davy a new substance that had been obtained from seaweed. They thought it was a compound substance, but Davy was sure it was an element. In a short time, he had proved his point and called the new element iodine. It had many properties in common with chlorine gas – an element that fascinated both Davy and Faraday. In fact, the first important piece of chemical research Faraday carried out was the liquefaction of chlorine in 1823.

Faraday also followed Davy in his enthusiasm for electrolysis. Realizing the need to standardize names used, he turned to the philosopher and mathematician William Whewell (1794-1866) at Cambridge for advice. It was Whewell, a scholar of classics, who called the pieces of metal attached to the positive and negative sides of a battery the anode and the cathode, respectively, and the two together he labeled electrodes. In the early 1830s Faraday proposed two laws of electrolysis which have ever since formed the basis of electrochemistry (the study of chemical reactions brought about by the passage of electricity).

With his study of electrolysis, Faraday provided the first clue to the existence of a unit of electricity – later called the electron

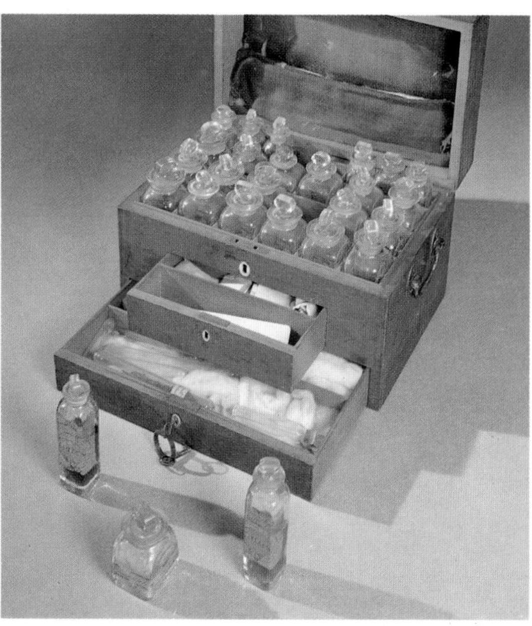

▲ Faraday's chemical cabinet containing his basic chemicals.

▼ Blackfriars glassworks in London, where Faraday began his work on optical glass.

▲ Faraday (above) became interested in water pollution, especially of "Father Thames".

The laws of electrolysis and their implications

Faraday's first law of electrolysis stated that the amount of a chemical substance that appears at an electrode depends on the amount of electricity that passes through the electrode. The second law concerned the relative amounts of different substances released by the passage of a fixed amount of electricity. Faraday observed that the amount depends on the combining power of the substance. This "chemical equivalent" is defined as the amount of the substance that combines with 8g of oxygen. For example, when water is broken up by electrolysis, 1g of hydrogen is produced for every 8g of oxygen: the chemical equivalent of hydrogen is 1:8.

Faraday's laws were of immediate value to chemists in examining the properties of chemical compounds, but it took some time for their full implication to register. It was spelled out in 1881 by the great German scientist, Hermann von Helmholtz (1821-1894) (➧ page 148): "The most startling result of Faraday's law is perhaps this. If we accept the hypothesis that the elementary substances are composed of atoms, we cannot avoid concluding that electricity also is divided into definite elementary portions, which behave like atoms of electricity." In other words, Faraday's study of electrolysis provided the first clue to the existence of a unit of electricity. The existence of this, the electron, was finally established by J.J. Thomson (1856-1940) at Cambridge in 1897, as part of his study of electrical discharges through gas at low pressure. The earliest studies in Britain of such discharges had been carried out by Faraday. He saw this study as a logical parallel to that of electrolysis, the passage of electricity through liquids.

Two early consultancies

Faraday had learnt a great deal on his 1813-1814 tour of Europe, but he did not really enjoy his travels. The main problem was Davy's wife, who insisted on treating him as a servant. However, his salary at the Royal Institution was increased on his return, and he was soon being consulted on a variety of scientific questions. For example, he was called in when the new gas industry encountered problems in bottling gas at high pressure. The process was being affected by the appearance of an oily liquid, which Faraday managed to identify as a hitherto unknown substance – now called benzene. Only later was it realized that this liquid had as much commercial potential as the gas itself. The most intensive of Faraday's consultancy projects came at the end of the 1820s, when he was commissioned by the Royal Society to examine ways of producing better quality glass. His selection for this work indicates the high regard in which he was held as a chemist.

Nonconformist Religion and Science

◄ *In his later life, the Nonconformist scientist Joseph Priestley wrote increasingly against the established political and religious system in Britain. This cartoon from 1792 shows him and Tom Paine, the radical writer, consorting with the Devil. Below it was printed a parody of the National Anthem, "God Save the King", which asserts that "Tom Paine and Priestley wou'd / Deluge the throne with blood". The pictures on the wall include one showing the guillotine, recently introduced by the revolutionaries in France. Both men are surrounded by books containing their writings. Among Priestley's, some refer to electricity. Priestley's house in Birmingham was burnt by a mob in 1791, after which he moved to London. In 1794, feeling against him became so strong that he emigrated, like many other Nonconformists, including Robert Sandeman, to the United States.*

▲ *In the 18th century, John Glas led a breakaway Presbyterian sect in Dundee. After a subsequent move to Perth, he was joined by Robert Sandeman, who later married his daughter. The sect was therefore often called the Sandemanians. Faraday was its most eminent member. It is not surprising that a person of Faraday's Nonconformist background should develop an interest in science. What was unusual was his low level of formal education. The scientists John Dalton and Joseph Henry were also Nonconformists.*

Origins of Nonconformism

The growth of modern science overlapped the dramatic religious changes of the Reformation. Discussion of the question whether there was any connection between these two developments has often concentrated on the English scene, because of the range of scientific and religious attitudes found there (➧ page 62). On the restoration of Charles II in 1660, the Puritans, now relabeled "Nonconformists", were, hardly surprisingly, subjected to a number of legal restrictions. To the extent that interest in science was linked to Puritanism, these restrictions (notably those on education) might be expected to have affected its development. The distinction between Puritans and others was often blurred. Charles II supported the founding of the Royal Society (➧ page 74), which contained many Puritan sympathizers.

The Nonconformist sects were always in a state of flux: it was not uncommon for existing sects to split, or for new ones to appear. The Methodists, for example, arose as an offshoot of the Church of England in the latter part of the 18th century. The Sandemanian sect to which Faraday belonged appeared at much the same time as the Methodists. It brought together one group split away from the Presbyterian church in Scotland, and another from the Church of England in Yorkshire.

Many of the Nonconformist sects continued to hold a favorable view of science and technology, and the industrial revolution in England in the 18th century owed a great deal to them. For example, the rise of the iron industry was dominated by the Quakers, who were responsible for some of the basic advances in the manufacture and use of iron.

Nonconformist education

Nonconformists, teachers and pupils, were often prevented from attending the traditionally church-related schools and universities of England. As a result, Nonconformists themselves established a number of what were called "dissenting academies". The academies trained Nonconformist ministers, but they also accepted laymen. Their curriculum was much wider than in traditional schools and universities – in particular, it contained a significant science component. In this, they reflected contemporary practice in the United States, Scotland and the Netherlands, all of which had been influenced by 17th-century Puritanism. The dissenting academies became an important seedbed of science.

Joseph Priestley (1733-1804) trained in one dissenting academy and taught in another leading academy at Warrington in northwest England. While at Warrington he produced his "History of Electricity", for which he not only reviewed previous work, but repeated and extended the experiments, as Faraday was to do later. Priestley's main fame came from his chemical experiments, and his work on gases was of basic importance for the new chemistry of Antoine Lavoisier (1743-1794) (➧ page 106).

During the 19th century, most of the restrictions on Nonconformists were lifted, and their distinctive contribution to science and technology merged into the mainstream of teaching and research. In some ways, their role was taken over in the 19th and 20th centuries by the Jewish scientists. The Jews, too, were subject to restrictions and saw education as a way of countering them.

The Popularization of Science

The demand for popular science writing
In the years of Galileo and Harvey, all people with a reasonable education could expect to understand contemporary intellectual debates, including those that dealt with scientific topics: there was little need to "popularize" science. However, the expectation of ready comprehension had become questionable by the end of the 17th century. Newton's "Principia" was a notoriously difficult book to understand: attempts to simplify his ideas for the general reader started during his own lifetime. The development of new mathematical techniques during the 18th century, especially in France, led to a continuing need to popularize research into such subjects as astronomy. However, most of the scientific topics of interest to the general public lay in such areas as natural history and geology where mathematics figured hardly at all. Consequently, most science was thought to require little interpretation as late as the 19th century. Darwin, for example, believed that all his books could be understood by nonscientists.

Yet 19th-century science was clearly becoming more difficult for the general reader to follow. The scientific jargon was increasingly impenetrable and the amount of prior knowledge required was growing. At the same time, the reading public was expanding. On the one hand, large-scale production of books and periodicals became much easier, and their national and international distribution had greatly improved. On the other, literacy was on the increase in many countries, and more people were prepared to buy or borrow reading matter. So there was a ready audience for simplified accounts of scientific researches. Books on science, often written as part of a series, proliferated during the 19th century. Most were produced by practicing scientists. This was also true of science articles in the upmarket magazines that became common during the century.

T.H. Huxley – a popular science writer
Some scientists derived considerable income from their writing, and editors competed for their work. For example, Darwin's champion Thomas Huxley (1825-1895) was a highly regarded contributor. A letter he wrote in 1874 to the editor of one magazine, the "Fortnightly Review", gives some idea of the demand for his popular science articles: "Many thanks for your abundantly sufficient cheque – rather too much, I think, for an article which has been gutted by the newspapers. I am always glad to have anything of mine in the 'Fortnightly', as it is sure to be in good company; but I am becoming as spoiled as a maiden with many wooers. However, as far as the 'Fortnightly' which is my old love, and the 'Contemporary [Review]' which is my new, are concerned, I hope to remain as constant as a persistent bigamist can be said to be."

Jules Verne and science fiction
The popular 19th-century interest in science is also reflected in its appearance in fictional form. The science-fiction novel essentially dates from the latter part of the 19th century and, more specifically, from the immensely successful stories

of the Frenchman Jules Verne (1828-1905). These were deliberately planned to present science to the general public. According to Verne's publisher: "The novels of M. Jules Verne have come just at the right time. When an eager public can be seen flocking to attend lectures given at a thousand different places in France, and when our newspapers carry reports of the proceedings of the Academy of Sciences alongside articles dealing with the arts and the theater, it is surely time for us to realize that the idea of art for art's sake no longer meets the needs of the time we live in, and that the day has come when science must take its rightful place in literature."

Popular lectures

Public science lectures were popular, but they necessarily reached a smaller audience. In the early 19th century, itinerant lecturers, including members of the scientific community, often had regular circuits which they traveled each year. Their science lectures typically covered the basics of the subject, as well as referring to more recent developments. As the century progressed, lectures became more often associated with institutions (such as the Royal Institution in London) and they were frequently written up as magazine articles. Faraday's lecture "The Chemical History of a Candle" and Huxley's "On a Piece of Chalk" are famous scientific lectures which were subsequently published and republished many times.

Specialist science correspondents

When massmarket newspapers were created at the end of the last century, scientists began to contribute to these, too. However, the pressures on time and space available meant that most newspaper work had to be done by professional journalists. In the 1920s and 1930s, the first specialist science correspondents appeared. This trend toward media specialists in science was enhanced by the growth of radio and television (which provide the modern equivalent of the 19th-century public lecture and demonstration). This separation of the science reporter from the scientist has led to some disagreement between them over the presentation of science by the media. Scientists' assessment of their work may differ appreciably from that of media specialists. Equally, scientists have always been ambivalent about their own colleagues who become involved in popularization. Unless done in strict moderation, such activities are seen as distracting from the true path of research. Thus Faraday's younger contemporary and his successor as superintendant at the Royal Institution, the physicist John Tyndall (1820-1893), came in for criticism because of his extensive popular science writing and lecturing in Britain and the United States.

▲ The 19th century saw a great growth in popular science journals, such as this German magazine "Der Stein der Weisen" (The Philosopher's Stone).

◄ A French vision of the 20th century which was published in 1883. Note the young lady paying a call in her air car, and the number of electrical cables.

► Humphry Davy lectured to enthusiastic audiences of as many as 1,000 people. Here he is at the Surrey Institution, a lesser-known rival of the Royal Institution of London.

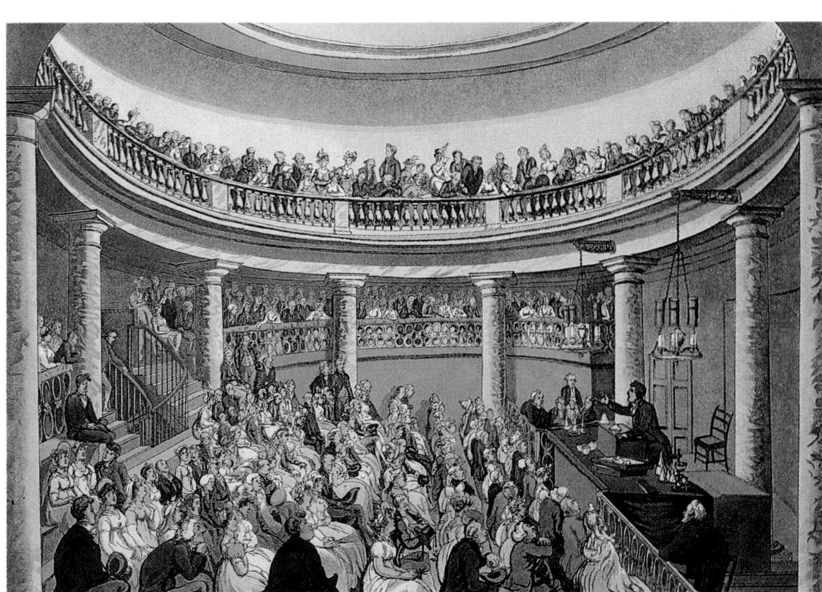

Public and private life

One consultancy helped drive a wedge between Faraday and his erstwhile hero, Davy. The two were retained by the opposing sides of a lawsuit concerned with sugar refining. This exacerbated the problems that had been created by their European tour. Their clash of personalities came to a head in the early 1820s, after Davy had been elected President of the Royal Society in succession to Sir Joseph Banks (1743-1820). When Faraday was proposed for election to the Royal Society, Davy opposed it and, though Faraday was in 1824 eventually elected, relations between the two men were permanently soured.

In later life, Faraday was called in, often by the government of the day, to advise on a wide range of problems. In 1844, for example, he traveled to Durham with the geologist Charles Lyell (1797-1875) (➧ page 152) to investigate a catastrophic explosion in a coal mine. The miner's safety lamp that Davy had invented was supposed to provide light while preventing accidental ignition of the gases that accumulate in mines. As Faraday discovered, however, the miners were frequently ignoring the safety regulations and even used the lamps to light their pipes when they were down the mine. At one stage during the enquiry, Faraday found that he had been given a seat on a sack of gunpowder, as it was the most comfortable place available.

Although he worked well on committees, Faraday was essentially a lone researcher. The Royal Society's funding for his investigation into the properties of glass allowed him to hire a laboratory assistant of his own. The man chosen, Sergeant Anderson, was a soldier, and he remained with Faraday for the rest of his working life. He was the only assistant Faraday had, and had the particular qualification, from Faraday's viewpoint, that he rarely spoke.

Faraday's social life revolved mainly round the religious community to which he belonged – a small Protestant sect called the Sandemanians. Sarah Barnard, whom he married in 1821, was a fellow member of the sect, and Faraday was elected an elder of the church in 1840 (there were no clergy). He was briefly excluded for failing to attend worship one Sunday without adequate reason (he was dining with Queen Victoria), but was subsequently reinvested and remained an elder till ill-health forced him to resign in 1864. Faraday and his wife were childless, but he was particularly interested in the education of children. In 1827 he inaugurated a series of lectures for children to be held at the Royal Institution every Christmas. These famous lectures are still given, and now appear on television. Faraday's most widely read popular writing, *The Chemical History of a Candle*, started life as one of these lecture series.

In 1833 a wealthy member provided the Royal Institution with funds for a new professorship of chemistry, on condition that Faraday was appointed to the post (he was). However, the Sandemanian religion required that money should not be saved, but distributed among the needy, especially if they were members of the sect. Faraday suffered a major illness in 1841; his memory increasingly faded afterwards with a consequent effect on his work at the Royal Institution. His lack of savings then became a source of concern, though the basic problem of accommodation was solved in 1858 when Queen Victoria provided him with a house at Hampton Court to the west of London. By the time of his death, on 25 August 1867, he could hardly speak or move. Typically, though he could have been buried in Westminster Abbey, he preferred a simple burial in Highgate cemetery, where his companions now range from Karl Marx to George Eliot.

▲ *Michael Faraday and his wife, Sarah, recorded by the daguerrotype process. Faraday was no great socializer, happily spending much of his life working in the laboratory or at home with his wife.*

Electricity and magnetism

Faraday's chemical researches would have earned him fame by themselves, but he is actually remembered more for his work on electricity and magnetism. His first important discovery in this field was stimulated by an unexpected result obtained by the Danish scientist Hans Oersted (1777-1851) in 1820. Oersted noted that, when electricity flowed along a wire, it deflected the needle of a magnetic compass held nearby. This was not entirely surprising: a number of scientists, Faraday included, had speculated that there was some connection between electricity and magnetism. What was unexpected was the way the effect depended on the position of the compass relative to the wire. The forces then known – gravitation, as well as electricity and magnetism – all acted along a line joining the two objects together. For example, a stone falls straight toward the Earth's surface under the influence of the Earth's gravitation. However, in Oersted's experiment, the compass needle was not deflected toward, or away from the wire, but at right angles to the compass–wire direction. The implication was that the magnetic force produced by an electric current acted not in a straight line, but in circles round the wire.

▲ *Oersted shows how electricity affects a compass needle.*

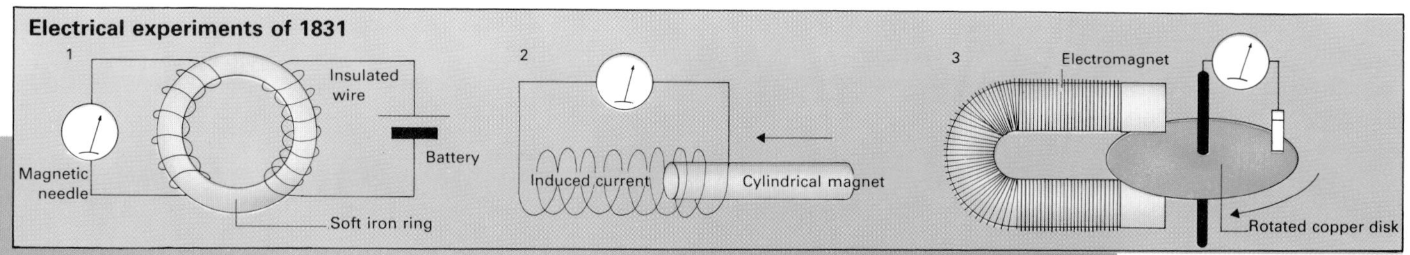

▲◄ *In 1831, experiments carried out by Faraday led to the discovery of electromagnetic induction – a fundamental step forward in understanding the connection between electricity and magnetism. His first piece of apparatus (1) consisted of a metal ring wound with two lengths of insulated wire round opposite sides of the ring. This is one of the surviving items of Faraday's apparatus (left). Faraday found that, when current was started or stopped in one wire coil, a corresponding current was induced in the other coil. In his second experiment (2), a cylindrical magnet was pushed into a wire coil, so producing an electrical current in the wire. He then devised the more complex dynamo (3) in which continuous electric current was produced between the center and edge of a copper disk spun between the poles of a horseshoe magnet.*

Joseph Henry worked on electromagnetism independently of Faraday and at the same time, but Faraday published his results first

It occurred to Faraday that the surprising result of Oersted's experiment might be used to obtain continuous motion. In 1821 he built two pieces of apparatus to demonstrate such motion: in one, a wire carrying a current moved round a magnet, while in the other, a magnet moved round the wire. The importance of these experiments was quickly recognized by the scientific community; Faraday had invented the electric motor converting electricity into mechanical motion.

Since it now appeared that electricity and magnetism were interdependent, the term "electromagnetism" was coined to cover both of them. Various attempts were made to explain this interdependence, the most detailed being in Paris by André Marie Ampère (1775-1836) who tried to interpret all the observed magnetic effects in terms of electric currents flowing in small loops. Faraday thought Ampère's idea was probably right. He deduced that, if a flow of electricity produced magnetism, then magnetism should be capable of producing an electric current. It was not until the 1830s, however, that he had the occasion to look properly for this inverse effect. Meanwhile, at Albany Academy in New York state, the professor of mathematics, Joseph Henry (1797-1878), had in 1830 discovered the phenomenon of self-inductance. Because the American failed to publish his findings, credit for making the discovery went to Michael Faraday.

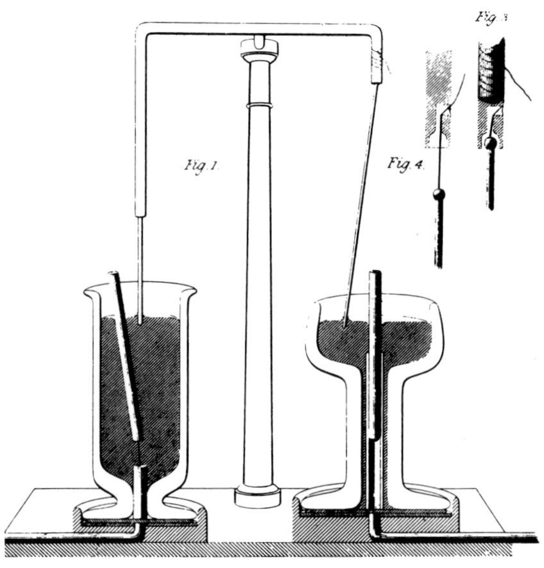

▲ Faraday's apparatus showing that electromagnetic effects can produce motion. On the left, a bar magnet rotates in mercury round a wire carrying an electric current. On the right, the suspended wire moves round the magnet.

◄ In the 1830s, Joseph Henry carried out several experiments on electromagnetism similar to Faraday's. Because Faraday published his results first, Henry is less well remembered today. But he was a major figure in 19th-century physics and played an important role in developing American science.

▲ Faraday's first dynamo comprised a copper disk spun between the poles of an electromagnet.

► By the mid-19th century the interrelationship of magnetism and electricity had been used to build powerful electromagnets for lifting, and the telegraph – both of which had been developed by Henry.

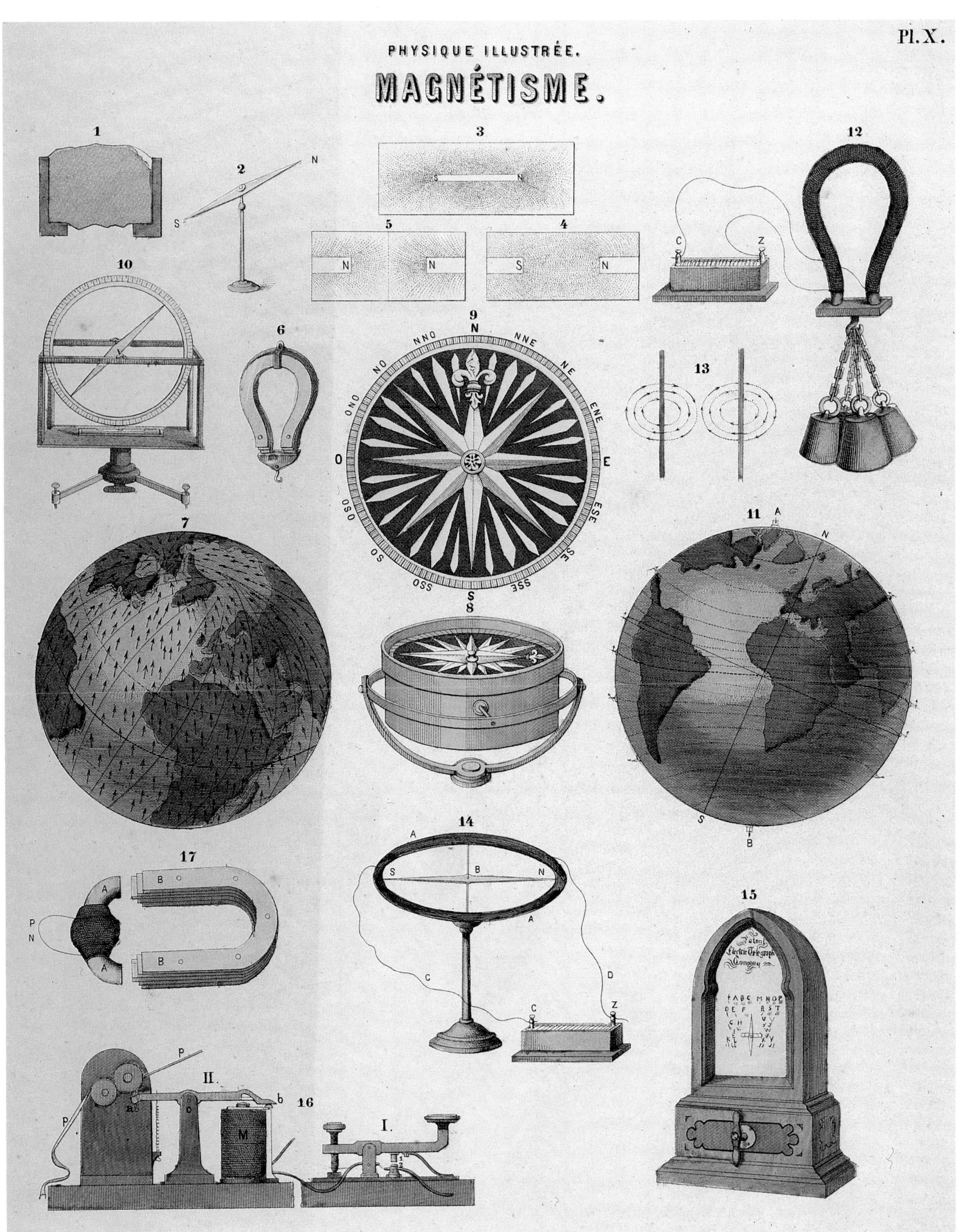
Pl. X.
PHYSIQUE ILLUSTRÉE.
MAGNÉTISME.
1
2
3
4
5
6
7
8
9
10
11
12
13
14
15
16
17
N
S
NNO
NNE
NO
NE
ONO
ENE
O
E
OSO
ESE
SO
SE
SSO
SSE
A
B
C
D
Z
P
M
I.
II.

The Rise of the Professional Scientist

Professionals and amateurs

The word "profession" achieved its present wide application in the 19th century: prior to that it was restricted to those in the law, medicine or the church. When Faraday appeared as a witness in a court case in 1821, it was ruled that scientists did not have professional status (and, therefore, could not claim expenses). By the time he died, scientists and engineers were coming to be seen as having the status of professionals in both competence and qualifications. However, the term "professional" was also used to describe someone who was paid to do his job – in his case, science. By this second definition, Faraday was a professional scientist throughout most of his life, even though he possessed no formal qualifications.

Faraday was unusual in his day in being paid to carry out scientific work. Most of his scientific contemporaries obtained their money from other activities. Such men must be called amateurs, in terms of pay, but were obviously professionals in terms of expertise. The word "amateur" had none of the connotations of lower-quality work then that it may have now.

Payment and specialization

Today virtually all contemporary scientists of importance are paid to practice science. This development got under way in Faraday's lifetime, and was partly related to the growth of specialization in scientific research. Before the last century, scientists frequently covered a range of the sciences. Faraday himself was equally at home in physics and chemistry. As the 19th century progressed, such wide-ranging interests became increasingly rare. Thus, Faraday's Cambridge correspondent, William Whewell, suggested that the word "physicist" should be introduced to distinguish scientists who specialized in physics research. Faraday would have none of it: he preferred the older, broader term "natural philosopher". Specialization was a consequence of an increase in both the amount and the complexity of scientific research. The amount of research findings that had accumulated by the 19th century meant that newcomers had a great deal of background to absorb. At the same time, the sciences were gradually acquiring a theoretical base, in terms of which new results could be interpreted. This was true, for example, of chemistry at the end of the 18th century. All this meant not only that intending scientists required much more training and time for study than before, but also that they had to restrict the range of topics in which they could be interested. To make progress in scientific research, it became increasingly important to work at it fulltime, and to be paid for doing so. This point is reflected in the changing composition of the Royal Society in Britain. By 1900, almost half of the members were paid scientists. (Today, nearly all of them are.)

Support from state and industry

Certain types of scientific activity had traditionally been given state support. For example, most countries supported some astronomy and geology. The reasons for this support were practical.

▼ The "amateur" British geologist Roderick Murchison (1792-1871) was a man of wealth who, after serving in the Peninsular War in Spain against Napoleon, devoted his life to geology. He invented the Silurian and (with Adam Sedgwick) Devonian stratigraphic systems.

▼ The great American inventor Thomas Edison (1847-1931) exemplified the scientifically-minded entrepreneur in industry. His "invention factory" employed 50 researchers, and at his death he held over 1,000 patents. Most new scientific posts were, however, in education.

▲ Henri Sainte-Claire Deville (1818-1881), who combined the roles of research chemist and university teacher. Best known for his work on bulk production of aluminum, Deville was professor in turn at Besançon, and at the École Normale and University in Paris.

Astronomy provided time and position measurements, and geology assisted mining. During the 19th century, the growth of new industry created further scientific posts, most obviously in chemistry. The rapidly growing chemical industry supported, directly and indirectly, an increasing number of jobs for fulltime researchers. However, the most important new source of jobs for scientists was education. Many countries saw a major growth in education at all levels, and a considerable emphasis on science in the curriculum.

The appearance of paid posts in two sectors – industrial and academic – led to some problems for the development of the profession of scientist. Specialist scientific societies became common in the 19th century; but academic and industrial scientists found that what they wanted from such societies increasingly diverged. Academic members saw the societies as places for the discussion and publication of research. Industrial scientists saw them as professional bodies protecting standards of competence, salary scales and related matters. In Britain, for example, a national Chemical Society was set up in 1841. By the latter part of the century, the industrial chemists who were members found the academic orientation of the society unacceptable, and split off to form a separate Institute of Chemistry. In the 20th century academic and industrial research have grown back together, so that the two organizations recombined in 1980 to form the present united Royal Society of Chemistry.

The basic components for an electrical industry

In 1831, Faraday constructed a simple apparatus consisting basically of an iron ring about 15cm across. Around one side of the ring he wound a number of turns of wire insulated from the ring and connected to a battery. Round the opposite side of the ring, he wound wire in a similar way and extended it some distance to run near a compass needle. His idea was that current in the first coil of wire would produce magnetic effects, as Oersted had found. Iron was a magnetic material, so the magnetism produced would be guided round the ring and through the second coil of wire. If this magnetism was converted to electricity in the second coil, it would be observed via a deflection of the compass needle. Faraday found that a steady current from the battery did not affect the compass, but every time he connected, or disconnected the battery, the compass needle jumped. This explained why the conversion of magnetism to electricity had not been observed before: it only occurred when the amount of magnetism was changing, not when it was steady.

Faraday's ring was actually the first electrical transformer. The electrical pressure – that is, the voltage – in the second coil of wire depended not only on the power of the battery, but also on the relative number of turns of wire in the first and second coils. So high-voltage electricity could be converted into low-voltage, or vice versa, simply by changing the number of turns in each coil.

The discovery of electromagnetic induction (the current in the second coil was said to have been "induced" by the magnetism) established the possibility of going from electricity to magnetism and back again. It did not demonstrate the production of electricity from a permanent source of magnetism, such as a bar magnet. Faraday soon remedied this now it was clear that the essential requirement was change. His simplest apparatus consisted of a coil of wire into which a bar magnet could be pushed. Every time the magnet was pushed in, or pulled out, a current flowed in the wire. At a rather more complex level, Faraday showed that a metal disk rotated between the ends of a horseshoe magnet produced a continuous electric current between the center and edge of the disk. What he had invented was a dynamo – a method of generating electricity from motion. So his work on electromagnetism led to the electric motor, transformer and dynamo – the basic elements of the modern electrical industry.

Faraday's last important researches on electricity and magnetism were carried out in the 1840s. By then he had become certain that there must be some connection betwen electromagnetism and light. He used polarized light in his experiments (that is, light that vibrates in one direction only, not in all directions like ordinary light). When such light passes through an ordinary piece of glass, the direction in which the light vibrates is unchanged. But Faraday discovered that the direction of vibration can be changed if a powerful magnet is brought close to the glass. Since this "Faraday rotation", as it is now called, only occurred when the polarized light was passing through glass, Faraday wondered whether the glass itself might have hitherto unsuspected magnetic properties. He suspended a piece of glass like a compass needle near a magnet, and noted that it preferentially set itself at right angles to the magnet (whereas ordinary magnetic material, such as iron, would set itself pointing toward the magnet). He found that a number of other materials besides glass possessed this property of "diamagnetism", as he labeled it; magnetism was much more widespread and complicated than had previously been supposed.

Growth of the Electrical Industry

Competition with steam

Although Faraday's researches provided the basic components required to establish an electrical industry, it took some time for this industry to become viable. The reasons were economic, rather than technical. Steam power was already well established, and electricity could not compete for many applications. For example, an electric locomotive was tried out in Scotland in the early 1840s: it reached a maximum speed of 6km/h. So electricity attracted more attention where steam power offered problems, as in the development of a submarine. The first attempt to propel a boat by electricity seems to have been made by Moritz Jacobi (1801-1874) in Russia. He wrote to Faraday in 1838 telling him how a paddle boat with a dozen passengers had been driven in this way. The problem was power to drive the motors had to come from batteries, and these were both heavy and expensive. The general use of electricity for transportation had to await the availability of other sources from which electricity could be picked up while the vehicle was in motion. As these sources appeared during the last 20 years of the 19th century, so streetcars, trains and subways or underground lines began to run on electricity.

Electroplate and lighting

The new sources of electrical power were based on another of Faraday's inventions, the dynamo. Early commercially-produced dynamos were built for fairly small-scale applications, especially in electroplating. In electrolysis, if the cathode consisted of objects made of a cheap metal – brass, say – use of a solution containing a precious metal, such as silver, would produce a coating of this metal that made the objects look much more expensive. On an industrial scale, this process takes more electricity than batteries can readily provide. The first large dynamo for electroplating was installed in Birmingham, England, in the 1840s.

Increasing demand for electricity led to further development of the dynamo, so that, by the 1880s, electricity could be provided to the general public as well as to industry. The great demand was for lighting. The problem was to develop a cheap, durable electric light. In the 1870s both Thomas Edison (1847-1931) in the United States and Joseph Swan (1828-1914) in England produced lamps containing filaments of carbon which satisfied those requirements. By the 1880s, buildings from New York to Moscow were lit by electricity.

The transmission of electric power

Initially, dynamos were installed close to the activities they served, but the need to transmit electricity over quite long distances soon became evident. For example, it was decided toward the end of the 19th century that the power of the Niagara Falls should be used to drive dynamos. The question was how the electricity produced could best be transmitted elsewhere. It soon became clear that the voltage required for transmission over long distances was not the same as that needed for domestic use. So the third of Faraday's inventions – the transformer – came into increasing use.

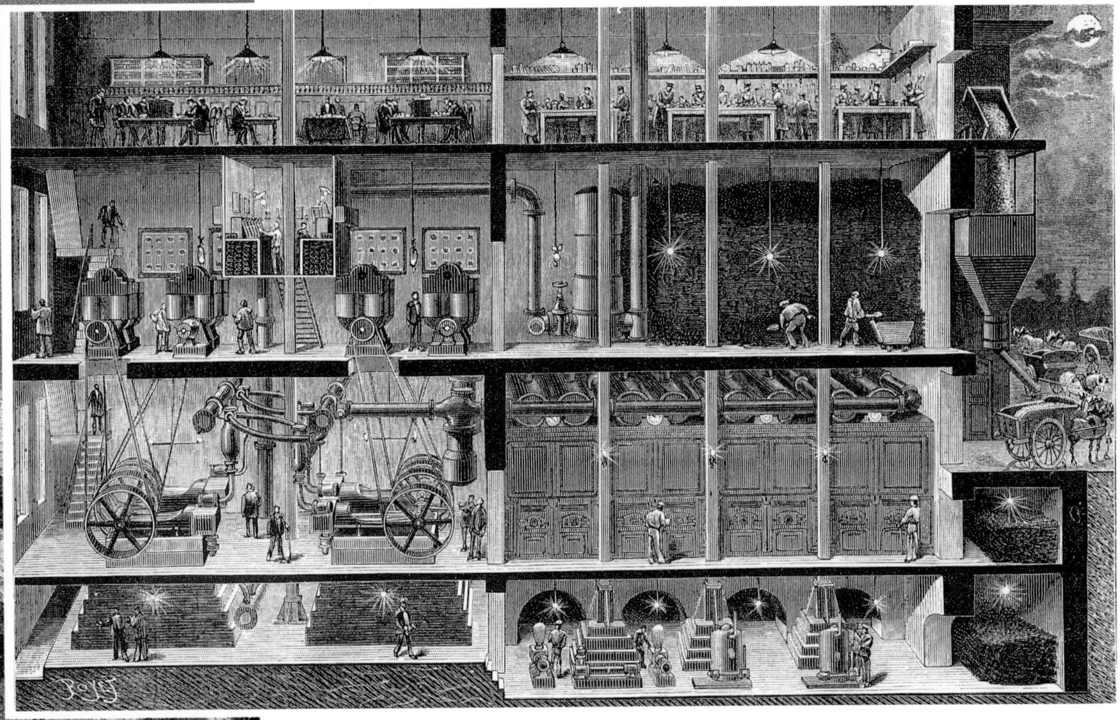

◄▲ By the 1880s, electricity was increasingly used for lighting. Left: the 1885 International Invention Exhibition at Crystal Palace, London. The demand for electricity led to the construction of power stations, such as (above) the Edison Company's one at Brooklyn, New York (1890).

▼► Cheap electrical appliances for the home have had a great impact on our lives in the 20th century. Those shown here include a vacuum cleaner (pre-World War I), a fire (from the 1920s) and a radio (1930s).

◄ Electricity provided useful power for transport by the end of the 19th century, especially within cities. In May 1903, the Prince of Wales (later King George V) inaugurated the first electric tram service in London.

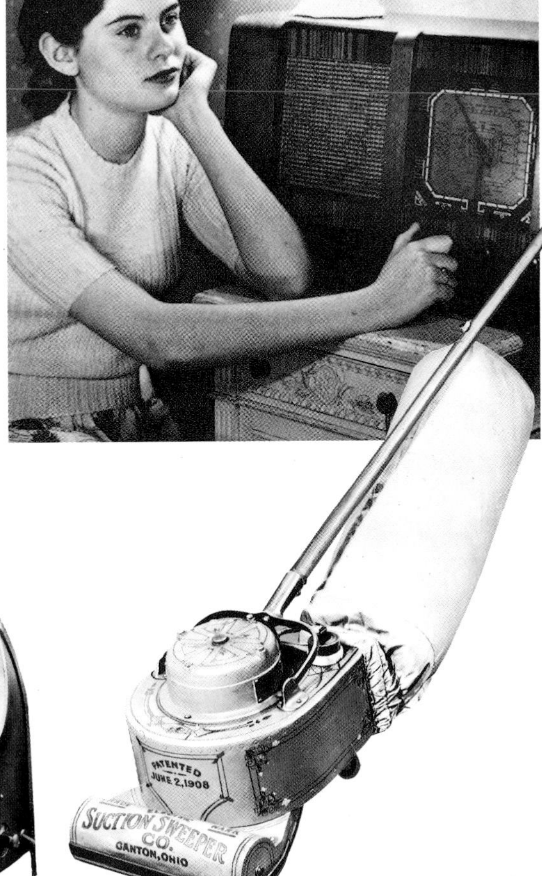

Birth of the radio industry

Not all industrial applications of electricity stemmed from Faraday's work. Electromagnets which could lift heavy loads were built very early on by the American physicist Joseph Henry (◆ page 141). But Faraday was indirectly responsible for a totally different area of commercial activity – the radio. James Clerk Maxwell's prediction that electromagnetic waves existed was followed up in 1887-1888 by the German physicist Heinrich Hertz (1857-1894) who, as a result, discovered radio waves. He showed not only how to produce and detect such waves, but also that they could be reflected by appropriate "mirrors" in just the same way as light waves. His experiments were taken up throughout Europe and North America, and the possibility of using radio waves for signaling was soon being investigated. The most persistent experimenter was the Italian Guglielmo Marconi (1874-1937), who in 1901 managed to send radio signals from England to North America. The resulting enormous publicity ushered in the birth of a new radio industry.

Faraday initiated the "Christmas Courses of Lectures Adapted to a Juvenile Auditory" at the Royal Institution in 1827 – they still run

▲ The work of the Englishman William Gilbert (1544-1603) established many of the basic ideas of magnetism. He understood that the Earth itself is a giant magnet (♦ page 127), and investigated its field in terms of dip and variation.

Faraday's "field theory"

From the 17th century onward, scientists had believed that a theoretical description of the world must ultimately involve mathematics. Faraday knew hardly any mathematics, but this did not prevent him from constructing a theoretical framework which he used to guide his experimental work. He believed from early on in his career that there was an underlying unity between the different types of activity he investigated. Magnetism could be converted into electricity which could be used to produce chemical reactions via electrolysis, and so on. By the 1840s, this viewpoint was becoming widely accepted. It led to one of the fundamental laws of modern science – the conservation of energy. Scientists, such as the physicist James Joule (1818-1889) in Manchester, showed that not only are different types of activity interconvertible, but the amount of energy involved at each conversion remains constant (once allowance has been made for heat losses, which are, themselves, a form of energy). Thus, by the end of Faraday's life, his belief in the unity of all sorts of physical and chemical activity had been largely vindicated.

Acceptance of Faraday's ideas on electricity and magnetism came more slowly. Faraday's contemporaries saw these as being similar to gravitation: each body had its own share, and forces between bodies simply depended on their distance apart. Faraday, on the contrary, came to believe that what mattered was the properties of the space between the bodies. He developed a picture in which the space between electric or magnetic bodies is filled with "lines of force" – observable, for example, when small pieces of iron are scattered

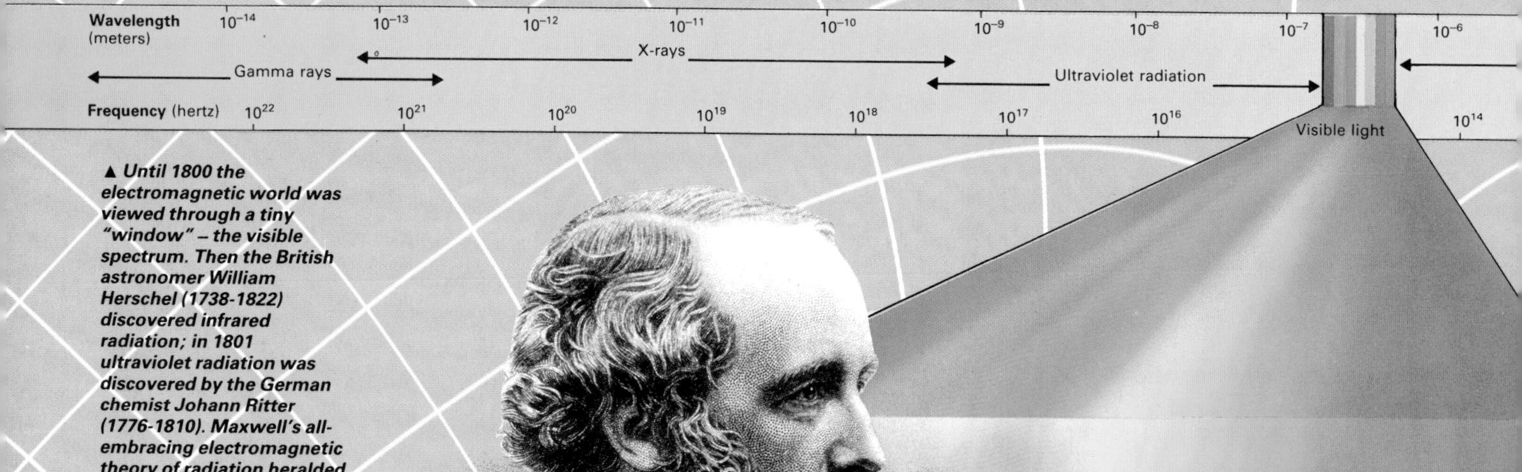

Wavelength (meters) 10^{-14} 10^{-13} 10^{-12} 10^{-11} 10^{-10} 10^{-9} 10^{-8} 10^{-7} 10^{-6}

Gamma rays

X-rays

Ultraviolet radiation

Visible light

Frequency (hertz) 10^{22} 10^{21} 10^{20} 10^{19} 10^{18} 10^{17} 10^{16} 10^{14}

▲ Until 1800 the electromagnetic world was viewed through a tiny "window" – the visible spectrum. Then the British astronomer William Herschel (1738-1822) discovered infrared radiation; in 1801 ultraviolet radiation was discovered by the German chemist Johann Ritter (1776-1810). Maxwell's all-embracing electromagnetic theory of radiation heralded discovery of other parts of a single spectrum: long-wavelength, low-energy radio waves (1887), then short-wavelength, high-energy X-rays (1895) and gamma-rays from natural radioactivity (1896) – usually described in terms of particles rather than waves. The hertz unit (1 cycle per second) is named for the discoverer of radio waves.

► Faraday's ideas on electromagnetism were taken up by James Clerk Maxwell, who represented the electromagnetic field mathematically.

round a magnet – which determine what interactions take place.

Here Faraday introduced a fundamental shift in the way of looking at the world. It is similar to thinking of a football game in terms of the limitations imposed by the pitch, rather than those imposed by the players. By analogy, Faraday's sort of picture is called a "field theory". What Faraday did for electricity and magnetism, Einstein was subsequently to do for gravitation (➧ page 238).

At the time, Faraday's ideas were of little use to his contemporaries, who wanted a quantitative (that is, mathematical) picture of the world. The first step toward this was taken by William Thomson (1824-1907), subsequently to become better known as Lord Kelvin. Thomson showed that Faraday's picture could be expressed in mathematical terms, but took the matter no further. It was left to James Clerk Maxwell (1831-1879) to demonstrate that Faraday's approach, when expressed mathematically, led to important new predictions. Maxwell, like Thomson a brilliant mathematician from Scotland, published his first work on Faraday's lines of force in 1855, but his great *Treatise on Electricity and Magnetism* only appeared in 1873, after Faraday's death. In this he publicized his conclusion that light is a form of electromagnetic vibration.

Unknown to his contemporaries (or to anyone, for a century), Faraday had reached the same conclusion long before. In 1832 he deposited a sealed envelope at the Royal Society which was not opened until 1937. It was then found to contain a proposal that electromagnetic waves exist which are analogous to waves on the surface of water. This described qualitatively what Maxwell predicted quantitatively.

▲ Michael Faraday in 1863, toward the end of his life, when his mental powers were already in decline. He is shown with some of his electromagnetic apparatus, though it was some years since he had last used it.

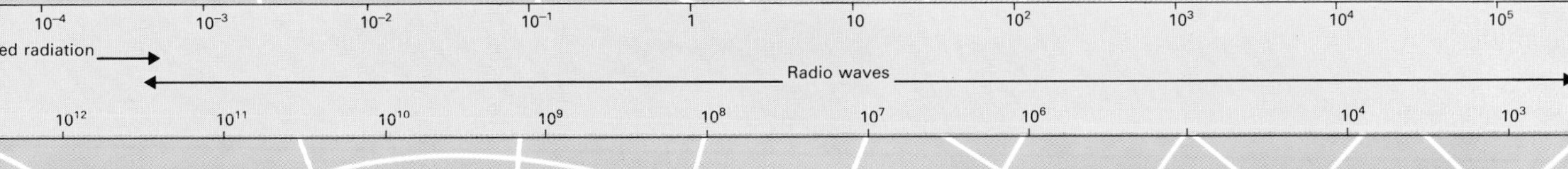

▲ Faraday was the driving force behind the children's lectures given at the Royal Institution each Christmas. He believed strongly that children should be taught science and here (above) at this 1855 lecture the Prince Consort and his two sons are listening on the left. Prince Albert, like many Germans, attached a high importance to science and worked hard to interest the British public in it.

► James Joule determined the mechanical equivalent of heat in 1843. Faraday's belief in the unity of nature was vindicated in work by Joule, Thomson, Helmholtz, Clausius and Maxwell.

"Decompose an element and tell us what it is made of – that would be a discovery indeed worth making" – Faraday

Considering the ultimate nature of matter

Faraday's activities as a chemist meant that he was also forced to consider from early on the ultimate nature of matter. Here he was influenced by Davy. Most of their contemporaries believed that matter consisted of atoms, which could be thought of as very tiny billiard balls. Neither Davy, nor Faraday, thought this picture could explain the complex interactions that occurred beween atoms. They therefore adopted the view that atoms were not material in the ordinary sense at all, but should be regarded as points from which various kinds of force emanated. This idea of "immaterial" atoms fitted in with his belief that a scientist's task is to examine the forces which fill space throughout the Universe. It also led him to realize that atoms were not necessarily the ultimate form of matter. A few years before his death, he wrote, "To discover a new element is a very fine thing, but if you could decompose an element and tell us what it is made of – that would be a discovery indeed worth making." The identification of radioactive elements some 30 years later (➧ page 193) showed that such decomposition occurred in nature, and led on to our present understanding of what atoms are made of.

► The great German scientist Hermann von Helmholtz drew important parallels between fluid flow and the electromagnetic field. He demonstrated how Faraday's work pointed to the existence of "atoms of electricity". (The electron was discovered in 1897.)

A SCIENTIFIC CENTENARY.

Faraday (returned). "WELL, MISS SCIENCE, I HEARTILY CONGRATULATE YOU; YOU HAVE MADE MARVELLOUS PROGRESS SINCE MY TIME!"

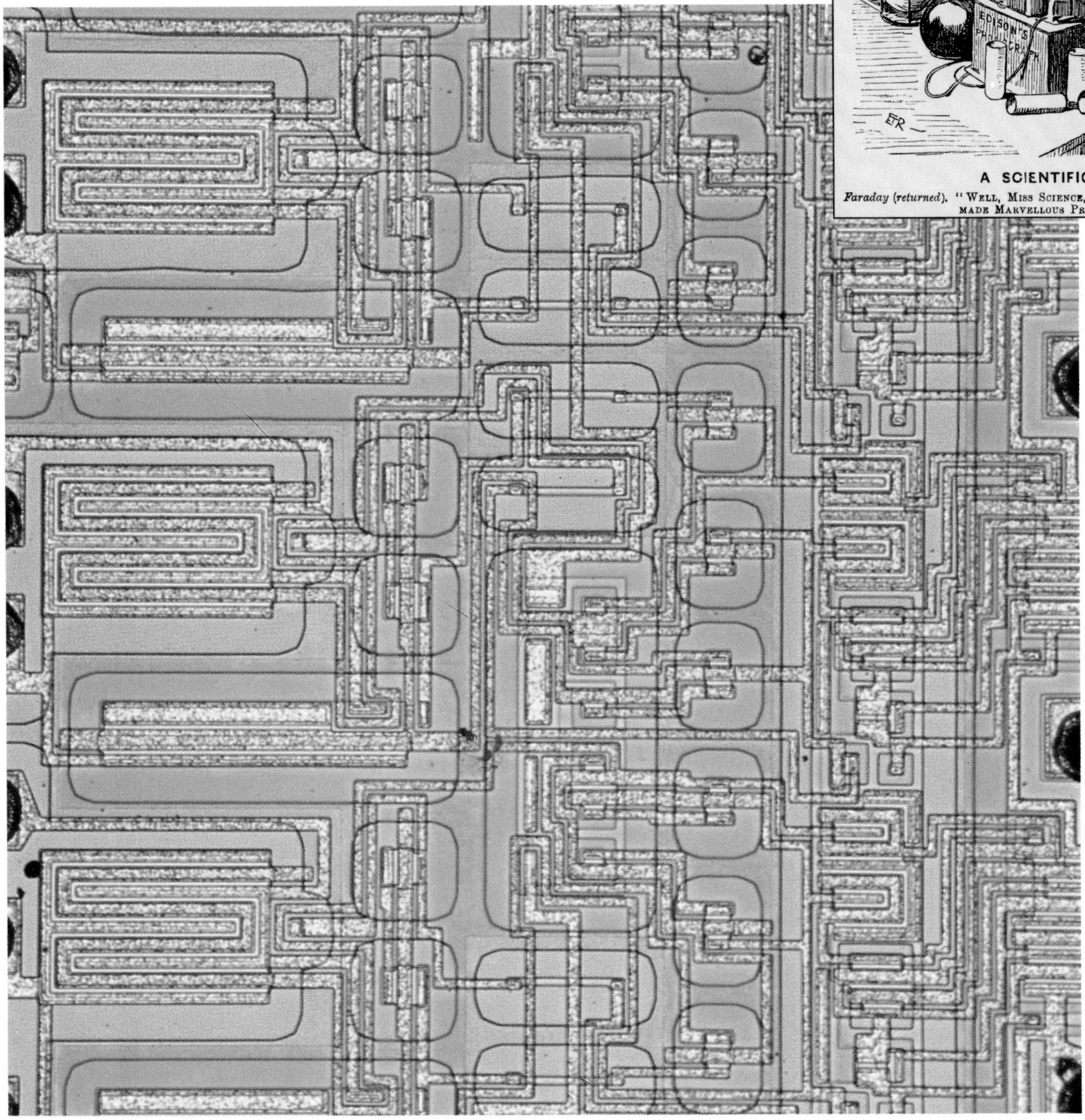

▲◄ This cartoon of Faraday appeared in 1891 in the British weekly "Punch" to celebrate the centenary of his birth. Four applications of electromagnetism subsequent to his death are shown. He is listening to Edison's phonograph – a forerunner of the gramophone. Behind him, on the left, is a box labeled "microphone", while, on the right, is an early telephone. The chart at the back shows telegraph cables criss-crossing the world. Today's integrated circuit (left) has affected all these devices. By Faraday's bicentenary in 1991, every type of communication will use such circuits.

Charles Darwin 1809-1882

8

The Darwin pedigree...The voyage of the "Beagle"... Uniformitarian geology...Recording evidence of transmutation...Variation and selection...The "Origin of Species"...Wallace's letter...Other writings...Objections to evolution...Altruism and sexual selection... PERSPECTIVE...A captain of the old school...Scientific societies...Religion and science at war?...Literary aspects of evolution

At the end of the *Origin of Species*, Darwin deliberately echoes Shakespeare's *Midsummer Night's Dream* and Oberon's

"...bank where the wild thyme blows,
Where oxlips and the nodding violet grows,
Quite over canopied with luscious woodbine,
With sweet musk roses and with eglantine."

Darwin invites us to "contemplate an entangled bank clothed with many plants of many kinds, with birds singing on the bushes, with various insects flitting about, and with worms crawling about through the damp earth." But, instead of leading us toward a sweet-sleeping Titania, he asks the reader "to reflect that these elaborately constructed forms, so different from each other, and dependent on each other in so complex a manner, have all been produced by laws acting around us." It was Charles Darwin's life work to understand, explain and elaborate these natural laws and the mechanism behind them.

The evolution of a theory

Until the end of the 18th century, educated Christians believed that the Earth and all its creatures had been created by God in about 4,000 BC. Between 1794 and 1809 Erasmus Darwin and Jean Baptiste Lamarck proposed that simple forms of life were spontaneously generated, then made increasingly complex over time by a vital force. Acquired characteristics were inherited.

By the middle of the 19th century a number of philosophers and naturalists found the idea of evolution a simpler way to explain geographical, geological and biological facts. It was Charles Darwin's triumph to invent a plausible mechanism for evolution, natural selection. The discovery, made independently also by Wallace, was publicly announced in 1858. By the time Darwin died, few naturalists denied that evolution had occurred.

By 1930 population geneticists, such as R.A. Fisher (1890-1962), had shown with mathematical models that natural selection could work on minute variations generated by mutations to produce evolutionary change. Whether evolution is gradual or not is still debatable, but Darwin's theory remains the cornerstone of biology.

▼ In ancient Greece followers of Thales believed that all living creatures had their origins in water. Aristotle argued that all nature reflects the purposes of a final cause: spiny lobsters have tails because they swim about, so a tail is useful. In Darwin's day such "teleology" was used to argue for the existence of God, the grand designer whose "will" was the final cause, against ideas of evolution.

Charles Darwin – His Life, Work and Times

For Charles Darwin, the flora and fauna of the "entangled bank" of nature had not, as many educated Christians believed, been specially created in their present forms by God some 6,000 years ago, but had slowly developed, or evolved, from different forms over a period of hundreds of millions of years by a mechanism of natural selection. Whenever a particular kind of variation in a plant or animal enabled it to survive longer and more easily in a particular environment, the variation could be perpetuated in order to produce divergence from the original form – a "transmutation" of one species into another. In publishing this viewpoint in 1859 in his book *On the Origin of Species by Means of Natural Selection* and again in 1871 in *The Descent of Man*, Darwin was to alter mankind's self-perception and to bring about a new view of man's place in nature.

Darwin's origins

Charles' pedigree can be traced back to a family of 16th-century Darwins who lived in Lincolnshire. His grandfather Erasmus Darwin (1731-1802), a Derby doctor, believed that plants and animals had developed from a few primitive species by an innate vital force. Beginning with Erasmus, there has been a Darwinian fellow of the Royal Society continuously until the present day.

Charles Robert Darwin was born at Shrewsbury on 9 February 1809. His father was a distinguished local physician; his mother was a daughter of the famous potter, Josiah Wedgwood (1730-1795), who had been a close scientific friend of Erasmus Darwin's. In 1825, it being by then a well-established family tradition, Darwin began medical studies at the University of Edinburgh; however, like his older brother, Erasmus, he soon abandoned the course when he found that he had no stomach for human suffering. Two years later he entered Christ's College, Cambridge, with the intention of taking an ordinary degree prior to becoming an Anglican clergyman.

Although there was then no science degree at the University of Cambridge, there were professors of both geology – the Rev. Adam Sedgwick – and botany – the Rev. John Henslow. They held regular lectures and field excursions which undergraduates might attend on a voluntary basis. Despite Darwin's dissipated behavior as a student (he belonged to a hunting and drinking set), both Sedgwick and Henslow took to him and he to them. The result was that the Cambridge experience focused Darwin's long-standing countryman's interest in the world of nature and the science of botany, geology and natural history.

The minimal examination requirements for the pass degree (which he took successfully in 1831) also forced Darwin to study two books by Archdeacon William Paley (1743-1805), *View of the Evidences of Christianity* (1794) and *Natural Theology* (1802). The *Evidences* showed him how a thesis for which the evidence was indirect (e.g. the resurrection of Christ) could be shown to be probable by the cumulative effect of many different lines of argument. Paley's *Natural Theology* was the argument for the existence of God from design; everything in nature seemed to be contrived and adapted to its immediate use and surroundings, just as a watch bears all the hallmarks of having had a maker. Although Darwin was eventually to reject this argument, Paley's text did provide him with an enormous range of examples and awkward puzzles. How could an eye originate by chance? Would it not be simpler to suppose that a benevolent creator had designed it specifically for seeing?

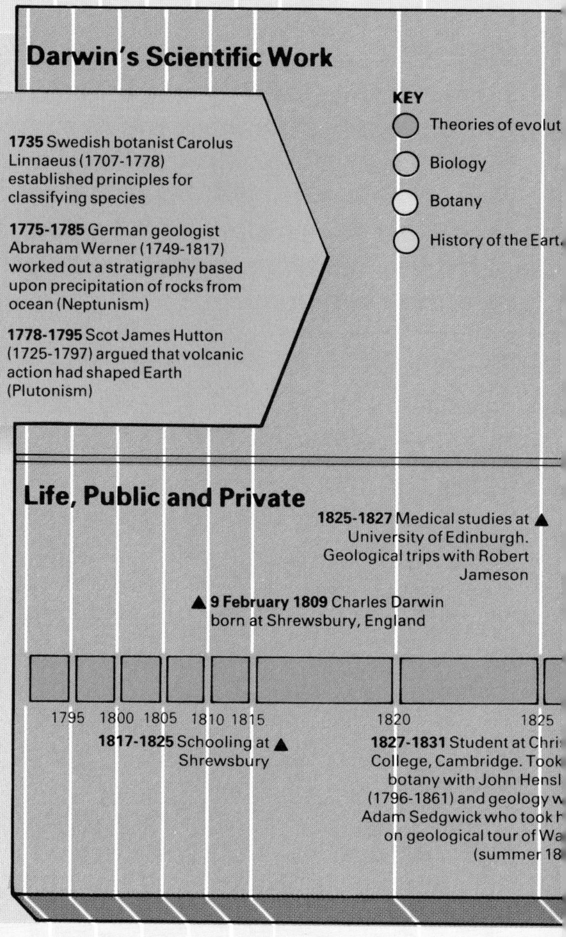

Historical Background
1815 Final defeat of Napoleon, Waterloo
Darwin's Scientific Work
KEY
Theories of evolut
Biology
Botany
History of the Eart
1735 Swedish botanist Carolus Linnaeus (1707-1778) established principles for classifying species
1775-1785 German geologist Abraham Werner (1749-1817) worked out a stratigraphy based upon precipitation of rocks from ocean (Neptunism)
1778-1795 Scot James Hutton (1725-1797) argued that volcanic action had shaped Earth (Plutonism)
Life, Public and Private
1825-1827 Medical studies at University of Edinburgh. Geological trips with Robert Jameson
9 February 1809 Charles Darwin born at Shrewsbury, England
1795 1800 1805 1810 1815 1820 1825
1817-1825 Schooling at Shrewsbury
1827-1831 Student at Chri College, Cambridge. Took botany with John Hensl (1796-1861) and geology w Adam Sedgwick who took h on geological tour of Wa (summer 18

Scientific Background

1800-1812 French naturalist Georges Cuvier (1769-1832) reconstructed extinct animals from fossil bones and suggested "revolutions or catastrophes" in Earth's prehistory

1798 Briton Thomas Malthus (1768-1834) argued that since food supplies cannot keep pace with rates of reproduction, a struggle for existence is inevitable

1794 Darwin's grandfather, Erasmus Darwin (1731-1802), used embryological evidence to argue that species are transformable

1802 William Paley (1743-1805) proved existence of God from evidence of adaptation in nature in *Natural Theology*

1809 French biologist Jean Baptiste de Lamarck (1744-1829) developed vitalist theory of transformation of species and inheritance of acquired characteristics

1807 Geological Society, London, founded

1815-1822 Lamarck published *Histoire naturelle des animaux*

1814-1829 German Alexander von Humboldt (1769-1859) described his voyages to South America in his *Personal Narrative*

1820-1840 Emerger "catastrophist" viev geological history b geologists William (1784-1836) and Ad Sedgwick (1785-187

1790 1795 1800 1805 1810 1815 1820 182

1837 Accession of Queen Victoria (reigned until 1901)

1839 First British Opium War with China

1845-1849 Potato famine in Ireland

1848 Revolutions in France, Germany, Italy, Hungary

1853-1855 Crimean War

1855 D. Livingstone discovered Victoria Falls, heralding European expansion in Africa

1857-1858 British suppression of Indian Mutiny

1861-1865 Civil War in United States

1861 Unification of Italy

1870-1871 Franco-Prussian War

1837 Opened notebook on "transmutation hypothesis" and read Thomas Malthus' *Essay on Population*

1839 Published *Journal of Researches during Voyage of the Beagle*

1842 *Structure and Distribution of Coral Reefs*

1842 First draft of transmutation theory, enlarged (1844) into 230pp essay

1846-1854 Studies of barnacles

1856 Darwin advised by Lyell to publish his views; he began his "big book"

1860-1862 Work on plant fertilization by insects

18 June 1858 Letter from Wallace announcing a theory of natural selection

1 July 1858 Theory of evolution announced at Linnean Society

November 1859 *Origin of Species*

1868 *Variations of Animals and Plants*

1871 Argued for human evolution by natural and sexual selection in *Descent of Man* and in *Expression of Emotions* (1872)

1874-1880 Several books on botanical subjects

1881 *Formation of Vegetable Mould through Action of Worms*

September 1835 Visited Galapagos Islands

1836 Returned to London

1842 Moved to Down House, Kent, living life of semi-invalid

Birth of son George Darwin (1845-1913), physicist

Birth of son Leonard Darwin (1850-1943), engineer and scientist

Birth of son, Horace Darwin (1851-1928), scientific instrument maker

1881 Obtained civil-list pension for Wallace

19 April 1882 Died at Down House

1835 1840 1845 1850 1855 1860 1865 1870 1875 1880

December 1831–2 October 36 On *HMS Beagle* under mmand of Robert FitzRoy as paid naturalist

1838-1841 Secretary of Geological Society

1839 Elected Fellow of Royal Society. Married cousin, Emma Wedgwood

Birth of son Francis Darwin (1848-1925), botanist

1849 Took first of several water cures at Malvern to improve health

1854-1856 On Council of Royal Society

1864 Awarded Copley Medal by Royal Society

1876 Composed an autobiography for his children

1831-1833 British geologist Charles Lyell (1797-1875) published "uniformitarian" view of Earth's history in *Principles of Geology*. Attacked Lamarckianism in vol. 2 (1832)

1830 First survey of South ca by *HMS Beagle*

1831 British Association for Advancement of Science founded

1836 William Buckland, *Geology and Mineralogy considered with reference to natural theology*

1839 Animal and plant cells described by Theodor Schwann (1810-1882) and Matthias Schleiden (1804-1881)

1840 Penny Post in Britain made scientific communication easier

1844 Robert Chambers (1802-1871) published theory of development of species anonymously in *Vestiges of Creation*, which was bitterly attacked by Sedgwick

1845-1862 Humboldt published *Cosmos*

1846 Smithsonian Institution founded, Washington DC

1848 American Association for Advancement of Science founded

1848 British paleontologist Richard Owen (1804-1892) developed notion of ideal types

1848-1852 British naturalist A. R. Wallace (1823-1913) colleted specimens in Amazon jungle

1852 British social philosopher Herbert Spencer (1820-1903) published a general theory of evolution

1855 Wallace in Malaysia published essay in English journal supporting the development of species

1858 British zoologist T. H. Huxley (1825-1895) published theory of vertebrate skull, challenging views of Owen

1860 Oxford debate on evolution between Bishop Wilberforce, T. H. Huxley and botanist J. D. Hooker (1817-1911)

1863 Lyell and Huxley examined evidence for antiquity of man and affinity with apes in *Antiquity of Man* and *Man's Place in Nature*

1864-1868 Wallace rejected possibility that man's higher faculties have arisen by chance

1865 Austrian botanist Gregor Mendel (1822-1884) published law of inheritance, which was not influential until 1900

1868 British physicist William Thomson (Lord Kelvin) (1824-1907) reduced the age of the Earth by thermodynamic arguments

1869-1910 Darwin's cousin, Francis Galton (1822-1911), developed eugenics, the breeding of human beings for evolutionary improvement

1872-1876 Oceanographic survey voyage by *Challenger*

1889 Wallace published *Darwinism* and received first Darwin Medal (1890)

1890-1896 Rejection of Lamarckianism by German biologist August Weismann (1834-1919) who developed idea of the germ plasm

1900 Revival of Mendel's theory of inheritance

1835 1840 1845 1850

Only after intervention from his uncle – father of his future wife Emma – was Darwin allowed by his father to join the "Beagle"

▲ Darwin at 31, 4 years after the voyage, and 19 before publication of "The Origin of Species".

▼ Georges Cuvier, founder of paleontology and opponent of evolutionary ideas before Darwin.

The lessons of the *Beagle*

It was through John Henslow, professor of botany at Cambridge, that Darwin, despite initial opposition from his father, was invited to sail, both as a naturalist and as personal companion to Robert FitzRoy. A strange, brooding, but highly efficient naval officer, FitzRoy had been given command of HMS *Beagle* for a long voyage to complete a survey of the coasts of South America. And so it came about that from December 1831 until October 1836, Darwin was given a unique opportunity to study first-hand a wide range of geological, biological and anthropological phenomena – an opportunity denied to all the well-known previous advocates of the transmutation of species, like his grandfather Erasmus Darwin and Jean Baptiste Lamarck (1744-1829), whose speculative notions had been based on the study of books and museum specimens.

On his voyage of discovery, Darwin took with him *Principles of Geology*, the influential book by the British geologist Charles Lyell (1797-1875), who had published the first of three volumes in 1830. In this Lyell argued for a "steady state" view of geological history. The shape and disposition of the present-day landscape – its mountains and valleys, volcanoes, coastlines, rivers, lakes and cliffs – could all be explained as having developed through the long-term uniform action of the forces of erosion and change which were seen in action today: heat, ice, wind, rain and sea. Lyell contrasted this (often referred to as the "uniformitarian" theory) to the more "catastrophist" explanations of most other British geologists. They preferred to interpret fossils as signs of the violent extinction of life during several past epochs by floods, earthquakes and other violent cataclysms.

According to Lyell's opponents, such as the French naturalist Georges Cuvier (1769-1832), the Earth's history suggested "direction", rather than "steady-state", for, following each catastrophe, it seemed that the Creator must have restocked the Earth with plants and animals of an "improved" or more advanced pattern. An Age of Invertebrates had been followed by ones of fishes, reptiles, birds and mammals. It seemed, then, that God's most recent and "highest" creation, man, was the intended result of past violent, but progressive, creativity. While Lyell agreed that the world had been designed for man, he disagreed that it showed any signs of progress. He may well have feared that if "progress" was admitted it would be only a small step to argue that animals had "evolved" from one another; and Lyell's notebooks clearly show that the idea that man was linked with the animals was abhorrent to him.

Darwin's problem was to reconcile Lyell's uniformitarian geology, which he soon found made good sense when he studied the South American landscape, with the equally clear progressive nature of paleontological evidence. As he noted later in his *Autobiography*:

"During the voyage of the Beagle I had been deeply impressed by discovering in the Pampean formation great fossil animals covered with armour like that on the existing armadillos; secondly by the manner in which closely allied animals replace one another in proceeding southward over the Continent; and thirdly by the South American character of most productions of the Galapagos archipelago, and more especially by the manner in which they differ slightly on each island of the group, none of these islands appearing to be very ancient in a geological sense. It was evident that such factors as these, as well as many others, could be explained on the supposition that species gradually became modified; and the subject haunted me."

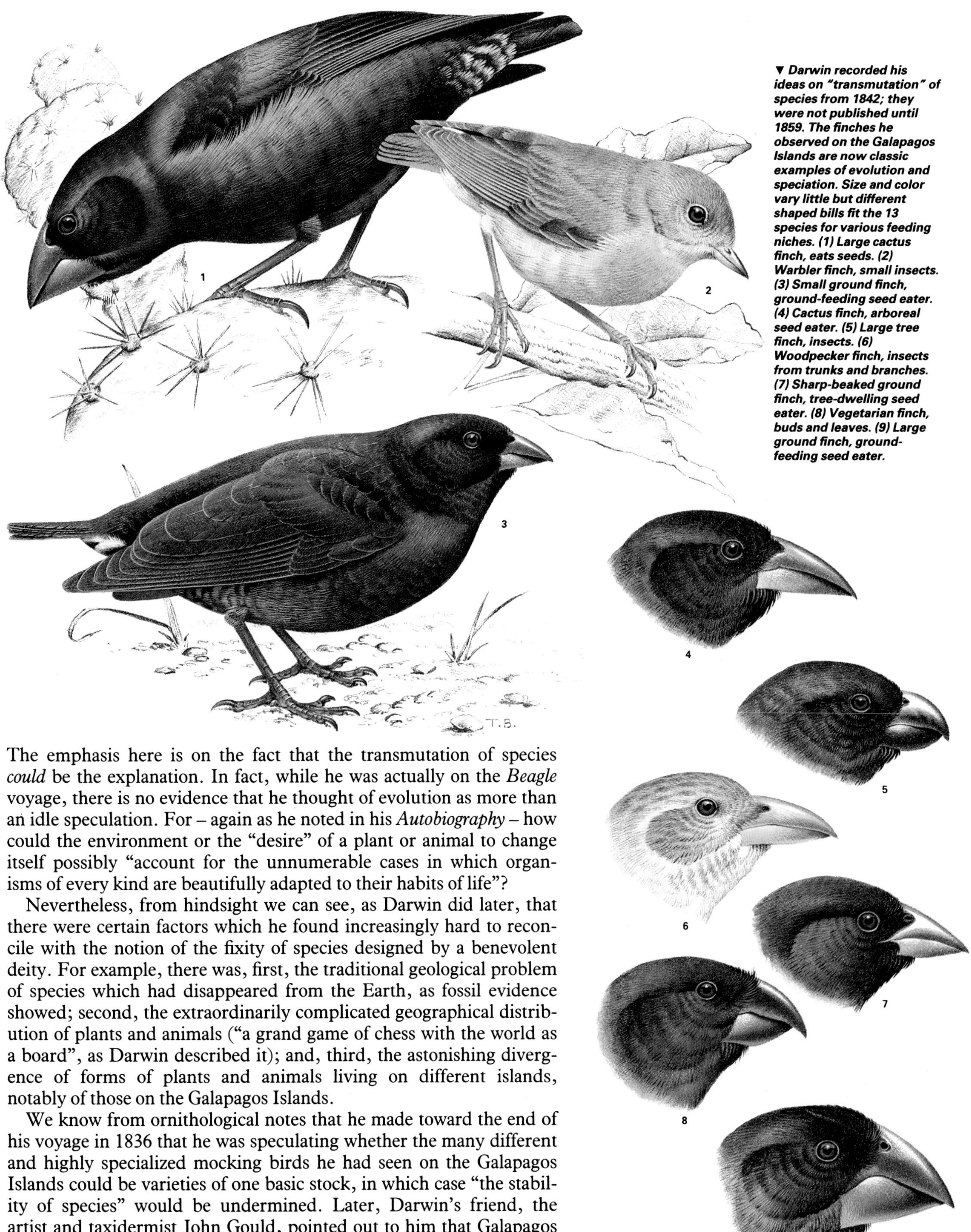

▼ *Darwin recorded his ideas on "transmutation" of species from 1842; they were not published until 1859. The finches he observed on the Galapagos Islands are now classic examples of evolution and speciation. Size and color vary little but different shaped bills fit the 13 species for various feeding niches. (1) Large cactus finch, eats seeds. (2) Warbler finch, small insects. (3) Small ground finch, ground-feeding seed eater. (4) Cactus finch, arboreal seed eater. (5) Large tree finch, insects. (6) Woodpecker finch, insects from trunks and branches. (7) Sharp-beaked ground finch, tree-dwelling seed eater. (8) Vegetarian finch, buds and leaves. (9) Large ground finch, ground-feeding seed eater.*

The emphasis here is on the fact that the transmutation of species *could* be the explanation. In fact, while he was actually on the *Beagle* voyage, there is no evidence that he thought of evolution as more than an idle speculation. For – again as he noted in his *Autobiography* – how could the environment or the "desire" of a plant or animal to change itself possibly "account for the unnumerable cases in which organisms of every kind are beautifully adapted to their habits of life"?

Nevertheless, from hindsight we can see, as Darwin did later, that there were certain factors which he found increasingly hard to reconcile with the notion of the fixity of species designed by a benevolent deity. For example, there was, first, the traditional geological problem of species which had disappeared from the Earth, as fossil evidence showed; second, the extraordinarily complicated geographical distribution of plants and animals ("a grand game of chess with the world as a board", as Darwin described it); and, third, the astonishing divergence of forms of plants and animals living on different islands, notably of those on the Galapagos Islands.

We know from ornithological notes that he made toward the end of his voyage in 1836 that he was speculating whether the many different and highly specialized mocking birds he had seen on the Galapagos Islands could be varieties of one basic stock, in which case "the stability of species" would be undermined. Later, Darwin's friend, the artist and taxidermist John Gould, pointed out to him that Galapagos finches were individually adapted to their different island habitats.

FitzRoy and the *Beagle*

Advances in hydrography
Hydrography, the science which deals with the study and mapping of seas, coasts, ocean depths, tides and currents, has always been important for maritime nations. Although the British Admiralty's Department of Hydrography was only founded in 1795, the first accurate maps of coasts and inshore waters date from the reign of Henry VIII. In the 17th century several Fellows of the Royal Society, including the astronomer Edmund Halley, helped to perfect the trigonometrical methods and instruments of marine surveying which are still in use today. The ability to determine longitude using chronometers standardized to London time, was an 18th-century development and allowed long voyages round the world to be undertaken with confidence. Between 1829 and 1855, under the aegis of the government hydrographer, Sir Francis Beaufort, not only was a "grand survey" of British coastal waters completed, but survey ships began detailed mapping of all the world's shipping lanes.

The captain of the "Beagle"
Robert FitzRoy, the captain of HMS "Beagle" during Darwin's circumnavigation of the globe between 1831 and 1836, was a Tory aristocrat, being descended from both the Duke of Grafton and the Viscount Castleraugh. He entered the Royal Navy in 1818, receiving what was in effect a first-class scientific education at the Royal Naval College in Portsmouth, before ascending rapidly through the ranks until he took command of the "Beagle" following her captain's suicide in the Magellan Straits in 1828.

The "Beagle", a 10-gun brig of 242 tons launched in 1820, together with its mother ship the "Adventure" had already spent two years surveying the southern coasts of South America for the Admiralty under the overall command of the hydrographer Captain Philip King.

◄ Robert FitzRoy, captain of the "Beagle", pioneer hydrographer and meteorologist. Later, as head of the country's meteorological services, he invented a barometer, introduced a system of storm warning signals and compiled the first weather charts. FitzRoy's resistance, during discussions on the voyage, to the new "uniformitarianism" helped shape Darwin's views on evolution.

FitzRoy and the "scientific person"
In June 1831 the Admiralty appointed FitzRoy to complete King's South American survey in the "Beagle". "Anxious that no opportunity of collecting useful information during the voyage should be lost", FitzRoy recommended to the Admiralty that "some well-educated and scientific person should be sought for who would willingly share such accommodation as I have to offer, in order to profit by the opportunity of visiting distant countries yet little known." And so it came about that the landlubber, Charles Darwin, shared FitzRoy's tiny cabin from 27 December 1831, until 2 October 1836. Because he traveled as a private gentleman, not as a regular crew, Darwin had to pay his own way at a cost of some £500 per annum.

Darwin's conversations and observations obviously disturbed FitzRoy's sincerely-held belief that the world was designed. In 1860, at the famous Oxford meeting of the British Association (▶page 163) he publicly declared that "he regretted the publication of Mr Darwin's book 'Origin of Species' and denied Professor Huxley's statement that it was a logical arrangement of facts."

A failed conversion

It was the survey of the Magellan Straits by the "Beagle" which awakened FitzRoy's interest in the primitive peoples of Tierra del Fuego. When the first South American survey ended in 1830, FitzRoy returned to England with four Fuegians – a girl aged nine he named Fuegia Basket, together with Jemmy Button (14), York Minister (26) and Boat Memory (20). The latter soon died of smallpox, but the other three were educated at FitzRoy's expense, his idea being to return them to their cold and uninviting country as Christian missionaries. When sent back on the "Beagle", however, they soon reverted to their former life-style, as FitzRoy found on a return visit to the settlement.

◄▲ Jemmy Button, "civilized" and "savage".

▲ The "Beagle" and observing Fuegians, south Tierra del Fuego: the painting was made by the expedition's artist Conrad Martens.

After the voyage

Although the voyage was a brilliant personal success for FitzRoy, it seems to have brought out his hereditary tendency to depression.

After briefly serving as a member of parliament for Durham, FitzRoy was governor of New Zealand from 1843 to 1845. His governorship ended in failure when settlers disagreed with his liberal policies toward the Maoris. In retirement in London during the 1850s, and awarded a fellowship of the Royal Society for his services to the science of navigation, FitzRoy began to develop a keen interest in weather forecasting for ships in British coastal waters. By issuing simple barometers (the FitzRoy barometer) to fishing ports, composing a "Barometer Manual", and by exploiting the electric telegraph, he collected centrally in London meteorological information from 24 coastal weather stations and issued synoptic charts which began to be published in "The Times" in 1861. Then as now, weather forecasting was unreliable and FitzRoy soon became censured for making mistakes. Despite support from insurers and people who lived on stormy coasts, official forecasting was abandoned, not to be revived until the 1880s. FitzRoy was never one to take criticism lightly. This public censure, together with his inability to accept Darwin's theory or the growing archaeological evidence for the antiquity of man, unhinged his mind. He cut his throat on 30 April 1865.

After 1836 the "Beagle" was used for hydrographical surveys of Australia but in 1851 it was turned into a customs' watch vessel for British coastal waters. The famous ship was sold to the Japanese as a training vessel before eventually being broken up in 1888.

It was not until May 1837, after his return to England, by which time he had been lionized in London as a brilliant geologist, that Darwin opened a notebook to record evidence for the transmutation of species. Into this and other notebooks he recorded queries and miscellaneous facts and information from his extensive reading and from the many zoological specimens he had brought back with him from South America. He kept up this systematic recording for some 15 months until, in September 1838, a reference in a review prompted him to reread *An Essay on the Principle of Population* by the Englishman Thomas Malthus (1766-1834). Malthus had explained away a crucial problem for theologians – the presence of hunger, vice and misery in a divinely-constructed world – by suggesting that it resulted from an inevitable law of nature. This was that food supplied could never match the rate of reproduction of animal populations and particularly of human beings. Although food supplies could be made to increase, they did so far more slowly than the population they were supposed to

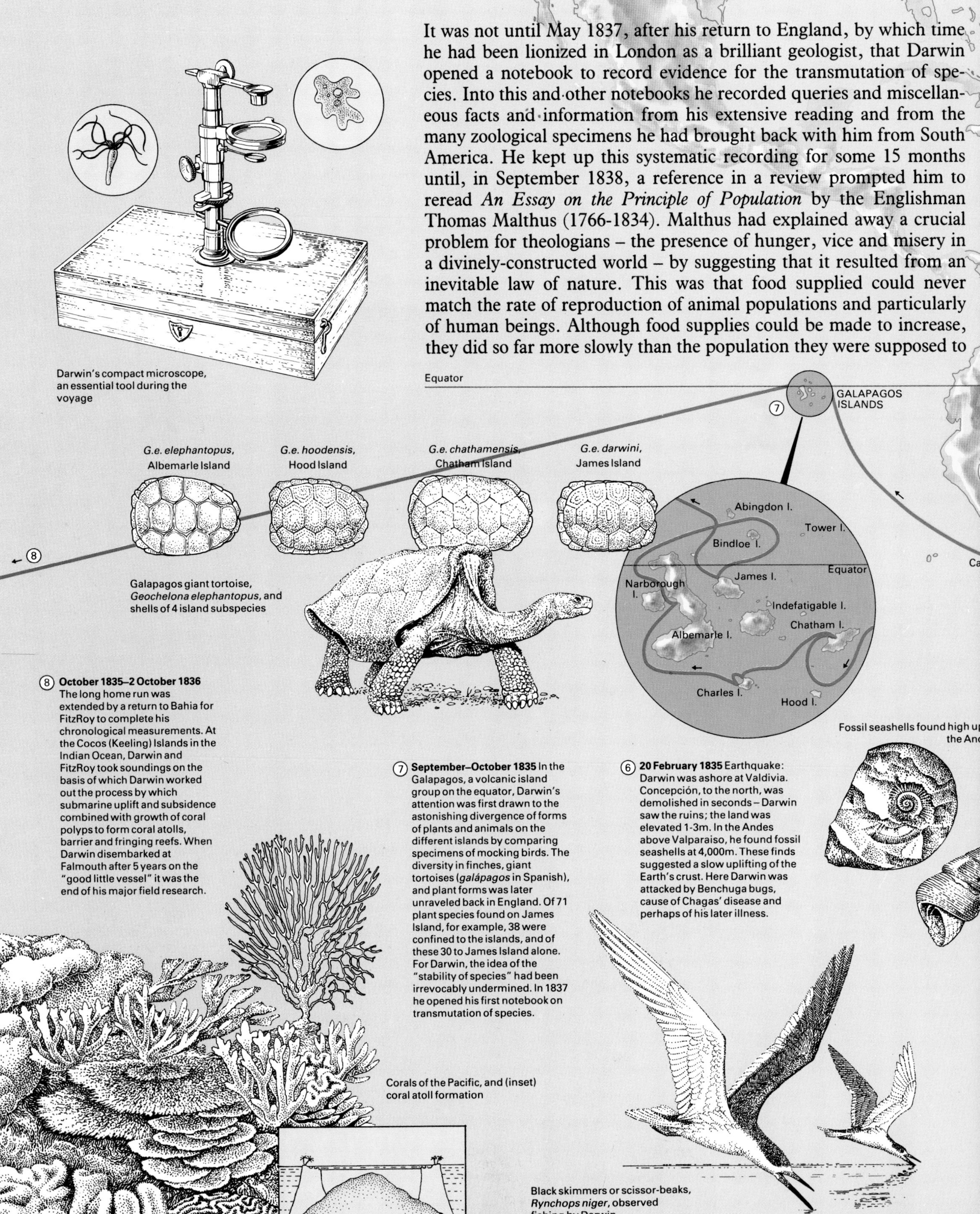

Darwin's compact microscope, an essential tool during the voyage

Galapagos giant tortoise, *Geochelona elephantopus*, and shells of 4 island subspecies

Fossil seashells found high up the And

(8) **October 1835–2 October 1836** The long home run was extended by a return to Bahia for FitzRoy to complete his chronological measurements. At the Cocos (Keeling) Islands in the Indian Ocean, Darwin and FitzRoy took soundings on the basis of which Darwin worked out the process by which submarine uplift and subsidence combined with growth of coral polyps to form coral atolls, barrier and fringing reefs. When Darwin disembarked at Falmouth after 5 years on the "good little vessel" it was the end of his major field research.

(7) **September–October 1835** In the Galapagos, a volcanic island group on the equator, Darwin's attention was first drawn to the astonishing divergence of forms of plants and animals on the different islands by comparing specimens of mocking birds. The diversity in finches, giant tortoises (*galápagos* in Spanish), and plant forms was later unraveled back in England. Of 71 plant species found on James Island, for example, 38 were confined to the islands, and of these 30 to James Island alone. For Darwin, the idea of the "stability of species" had been irrevocably undermined. In 1837 he opened his first notebook on transmutation of species.

(6) **20 February 1835** Earthquake: Darwin was ashore at Valdivia. Concepción, to the north, was demolished in seconds – Darwin saw the ruins; the land was elevated 1-3m. In the Andes above Valparaiso, he found fossil seashells at 4,000m. These finds suggested a slow uplifting of the Earth's crust. Here Darwin was attacked by Benchuga bugs, cause of Chagas' disease and perhaps of his later illness.

Corals of the Pacific, and (inset) coral atoll formation

Black skimmers or scissor-beaks, *Rynchops niger*, observed fishing by Darwin feed. Malthus's somewhat gloomy conclusion was that it was up to individuals to look after themselves ("self help") and that this was God's way of encouraging work, thrift and reverence. If individuals became paupers, it was their fault. But those who "selected" to struggle to work hard would survive and gain their due reward on Earth, as well as in Heaven.

"Selection" was the key to the process Darwin had already referred to in his notebooks as "descent" or "modification". "One may say", he now concluded, "there is a force like a hundred wedges trying [to] force every kind of adapted structure into the gaps in the economy of nature, or rather forming gaps by thrusting out weaker ones." This idea of a "struggle for existence" was not new; it had been present in the writings of Malthus, Paley, Lyell, the Swiss botanist Augustin de Candolle (1778-1841) and others. (The English social philosopher Herbert Spencer (1820-1903) was to refer to "the survival of the fittest" as lying at the heart of evolutionary progress as early as 1852.)

▼ In possibly the most important scientific voyage ever made, HMS "Beagle" undertook a survey of the coastline of South America and set up time stations around the world for precise longitude determinations. Darwin was able to study coral reefs, confirm Lyell's views on geology, and assemble information on the distribution of plants and animals which convinced him that they had evolved by natural selection.

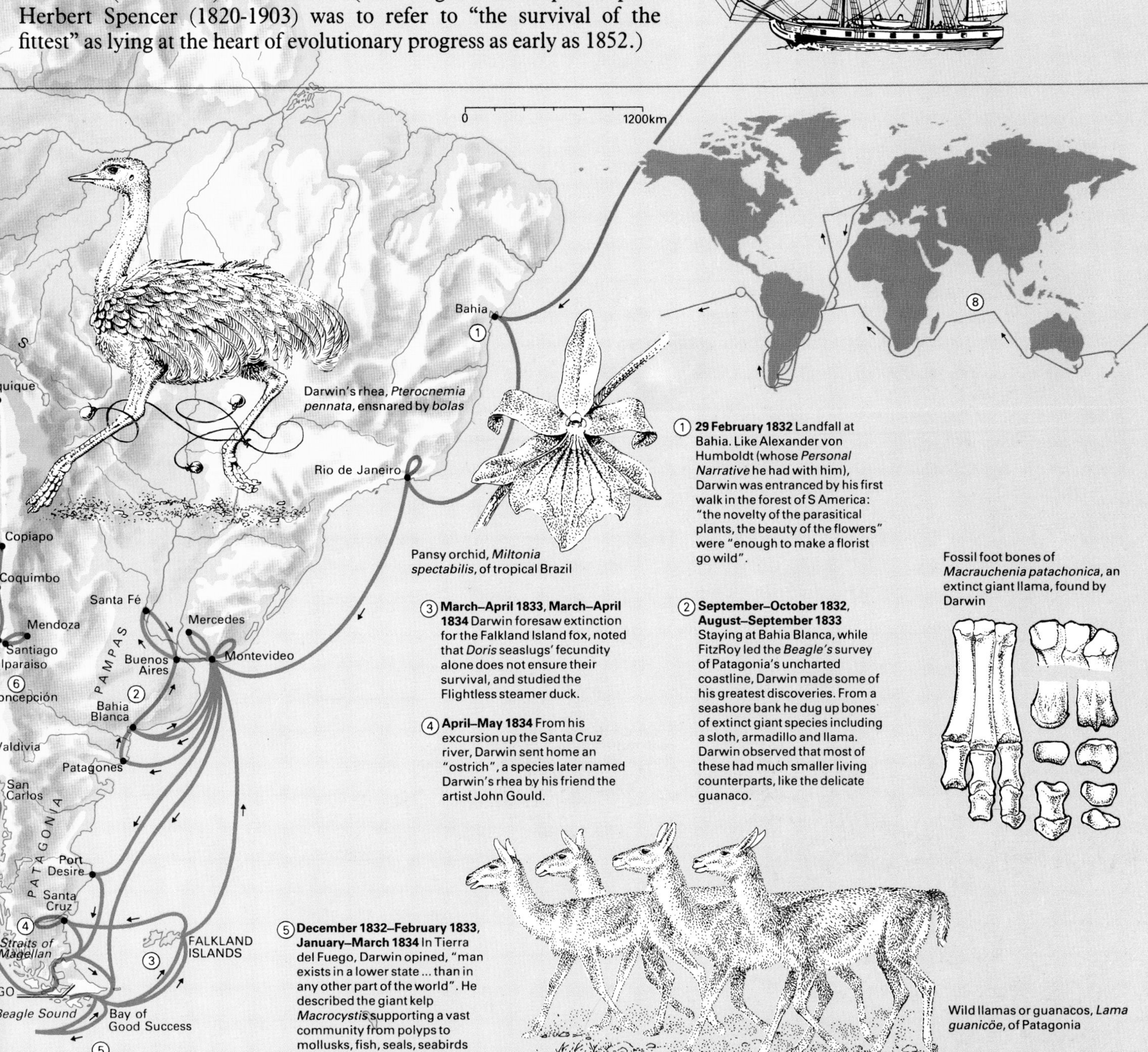

0
1200km
Bahia
1
Rio de Janeiro
Santa Fé
Mercedes
Buenos Aires
Montevideo
PAMPAS
2
Bahia Blanca
Patagones
PATAGONIA
Port Desire
Santa Cruz
4
Straits of Magellan
FALKLAND ISLANDS
3
Beagle Sound
Bay of Good Success
5
Copiapo
Coquimbo
Mendoza
Santiago
Valparaiso
6
Concepción
Valdivia
San Carlos
8
Darwin's rhea, *Pterocnemia pennata*, ensnared by *bolas*
Pansy orchid, *Miltonia spectabilis*, of tropical Brazil
Fossil foot bones of *Macrauchenia patachonica*, an extinct giant llama, found by Darwin
Wild llamas or guanacos, *Lama guanicöe*, of Patagonia
(1) **29 February 1832** Landfall at Bahia. Like Alexander von Humboldt (whose *Personal Narrative* he had with him), Darwin was entranced by his first walk in the forest of S America: "the novelty of the parasitical plants, the beauty of the flowers" were "enough to make a florist go wild".
(2) **September–October 1832, August–September 1833** Staying at Bahia Blanca, while FitzRoy led the *Beagle's* survey of Patagonia's uncharted coastline, Darwin made some of his greatest discoveries. From a seashore bank he dug up bones of extinct giant species including a sloth, armadillo and llama. Darwin observed that most of these had much smaller living counterparts, like the delicate guanaco.
(3) **March–April 1833, March–April 1834** Darwin foresaw extinction for the Falkland Island fox, noted that *Doris* seaslugs' fecundity alone does not ensure their survival, and studied the Flightless steamer duck.
(4) **April–May 1834** From his excursion up the Santa Cruz river, Darwin sent home an "ostrich", a species later named Darwin's rhea by his friend the artist John Gould.
(5) **December 1832–February 1833, January–March 1834** In Tierra del Fuego, Darwin opined, "man exists in a lower state ... than in any other part of the world". He described the giant kelp *Macrocystis* supporting a vast community from polyps to mollusks, fish, seals, seabirds and the Fuegian people.

Darwin was familiar with the ways in which plant and animal breeders artificially "selected" traits which they wished to cultivate, and when he read Malthus's essay in 1838 it reminded him how prolific nature is, and how rapidly population can grow. This greatly increased the chances of *variation* occurring within populations on a wide scale. The fossil record, geographical distribution and adaptation of species might then be explained in terms of the environment, which changed over millions of years by the sort of mechanisms Lyell had suggested, so acting as an agent which selected varieties for survival. The original ancestor of the giraffe, which strained its neck to feed off the succulent upper branches of trees did not in consequence develop a longer neck, as suggested by Lamarck. Instead, those which had longer necks would survive, reproduce and perpetuate the trait of long-neckedness when ordinary food resources were scarce. The giraffe had developed by a mechanism of "natural selection". Thus, although in 1838 he still lacked an explanation of why interbreeding *varieties* of plants and animals became separate nonbreeding *species*, Darwin had become a convinced believer in transmutation.

This did not mean that he abandoned the view that the world was harmonious and designed. Although he was to lose his belief in God – mainly because he thought the God of the Old Testament a tyrant and the doctrine of everlasting punishment of unbelievers a "damnable doctrine" – Darwin's essentially religious approach to evolution was to see it as producing a grander view of nature: a world in which God, or Purpose writ large, governed through natural law. As the sublime ending of Darwin's *Origin of Species*, already quoted, suggests, Darwin never abandoned a belief in the ultimate perfection and harmony of the Universe. Just as Paley had before, Darwin consoled his readers "that the war of nature is not incessant, that no fear is felt, that death is generally prompt, and that the vigorous, the healthy, and the happy survive and multiply". But would his religious readers be so consoled and would his scientific readers be satisfied with the slow and random mechanism of natural selection? Would he be laughed at, derided and attacked? For this had been the fate of the anonymous author (Robert Chambers, the publisher, 1802-1871) of *The Vestiges of Creation* when, in 1844, he had argued that everything in the universe, including man, had "developed" from a nebulous gas because the Creator had endowed matter with such potentialities.

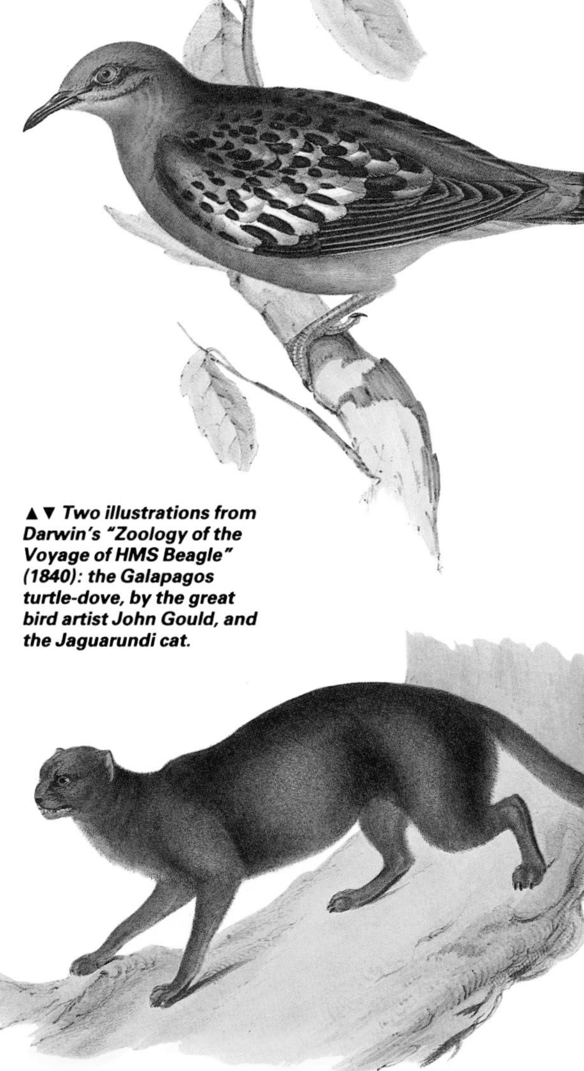

▲▼ Two illustrations from Darwin's "Zoology of the Voyage of HMS Beagle" (1840): the Galapagos turtle-dove, by the great bird artist John Gould, and the Jaguarundi cat.

▼ The ruins of Concepción, as they were found by Darwin after Chile's most violent earthquake for decades. "The world, the very emblem of all that is solid, had moved beneath our feet like a crust over a fluid," wrote Darwin.

Specialized Societies

Specialized societies
Until science found a firm place in universities in the second half of the 19th century, scientists tended to find their homes in learned societies rather than in academic institutions. This was particularly true of Great Britain, the clubbable character of whose nation was brilliantly captured by Charles Dickens in the Pickwick Club, to which Samuel Pickwick Esq. delivered his memorable paper on the theory of tittlebats. By the time "Pickwick Papers" appeared in 1837, the inn dining room and the coffeehouse had helped produce, in the preceding decades, literary and philosophical societies and many ephemeral associations for scientific conversation in London and provincial towns. Until its reform in 1847, even the Royal Society of London to which Darwin was elected in 1839 as a geologist and much-traveled expert on natural history, continued to bear the hallmarks of an elite gentleman's club. However, the growing need to collect, identify and classify the natural objects of the Universe – its stars, minerals, chemicals, plants, animals, languages and races of men – soon proved too great a task for individuals yet one too highly detailed and specialized to interest all the fellows of the Royal Society. Hence, despite the unease of Sir Joseph Banks, president of the Royal Society from 1778 until his death in 1820 (♦page 100), specialized scientific societies began to be founded from the 1780s onward. The Linnean (1788), Geological (1807), Astronomical (1831) and Chemical (1841) societies soon became forums for the reporting and support of individual specialized research, but initially they were created so that members could collaborate in handling large quantities of data. For example, the Linnean Society was founded in order to encourage the classification of the living world. It took its name from the Swedish taxonomist, Carolus Linnaeus (♦page 112), whose collections of plants and other materials had been purchased by the Society's founder and first president, Sir James Edweard Smith. Oddly, Darwin did not choose to become a fellow until 1854, and he appears to have joined mainly so that he could borrow books from its excellent library by post! He was too unwell to attend the Society's meeting in 1858 at which Hooker and Lyell read his and Wallace's joint announcement of the theory of evolution by natural selection.

Developments in Europe
The Royal Society and the French Academy of Sciences (♦pages 75, 94) had quickly become models for national academies in Russia (1725), Sweden (1739), America (1743) and Switzerland (1815). Each produced its own journal. By 1800 specialist periodicals, such as Crell's "Chemisches Journal" and Lavoisier's "Annales de chimie", had begun publication. However, in continental Europe specialist learned societies were slower to develop. Although separated into many separate states before 1870, Germany possessed a quasi-national scientific society in the Berlin Academy, which the philosopher Gottfried Leibniz (1646-1716) founded in 1700. However, the fact that the rival governments of her separate states came increasingly to support scientific teaching and research within Germany's many universities (♦page 120), made specialized scientific societies rather less important than they were in Britain. For example, the German Chemical Society was only founded in 1868. On the other hand, the annual meetings of doctors, scientists and laymen in different German towns which began in 1822 (♦page 124) were to be widely copied.

The Geological Society of London
The Geological Society, founded in 1807 to map Britain's mineral resources, was the liveliest and most prestigious of the London societies. Its meeting room was arranged in parliamentary fashion with seats facing each other, so that debate was encouraged. Other learned societies sat facing the speaker in theater, or church-congregation, style. By the 1870s, comment and discussion had come to be seen as an essential ingredient in the presentation and promotion of scientific advance – a ritual which both the Geological Society and the controversial nature of Darwin's ideas did much to promote. By then, too, specialized scientific societies had become essential features of the structure and communication of modern science.

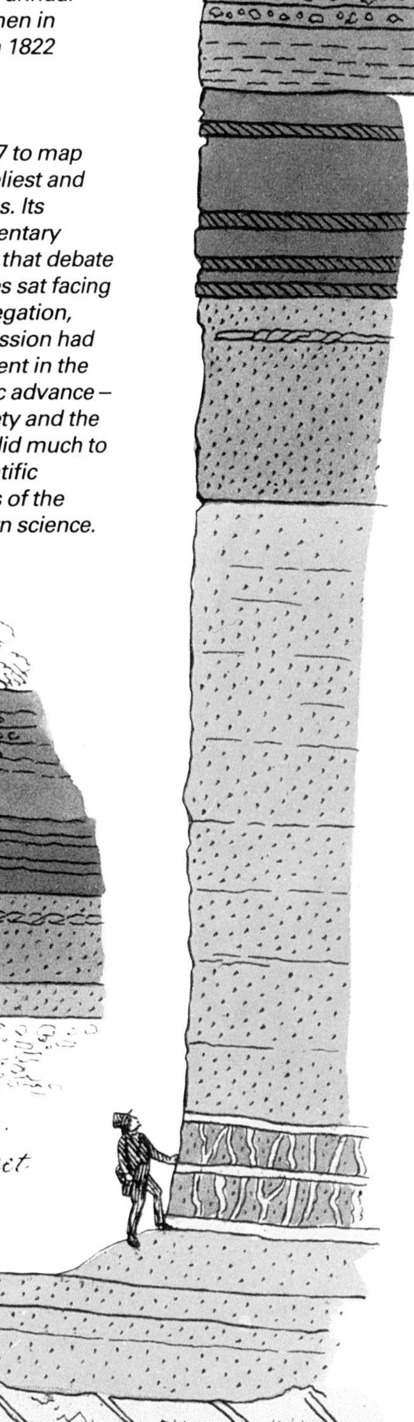

▼ **Many societies ran research projects. The Geological Society (where Darwin was secretary, 1838-1841) mapped Britain's rock strata. These elegant sections of Aust Cliff, River Severn, were drawn in 1824 by Henry De la Beche (1798-1855), first director of the Geological Survey.**

► *Giant tortoises on Isabela (Albemarle), one of the Galapagos islands. The fact that species differed from one island to another contributed to Darwin's ideas on evolution.*

The Origin of Species

By June 1842 Darwin had penned a 1,500-word sketch of his views, which he expanded into a 52,000-word essay in July 1844. This he showed to the botanist Joseph Hooker (1817-1911) but to no one else. (Contrary to popular belief, Darwin was much closer to Hooker than to the zoologist T.H. Huxley (1825-1895), who was 16 years his junior and whom he befriended later.) The last of the *Beagle* reports was completed and published by 1846: these, incidentally, had revealed Darwin to be a talented travel writer.

Darwin then devoted the next eight years of his life to testing the strengths and weaknesses of the evolutionary hypothesis by classifying fossil and living barnacles. His study embraced 10,000 specimens and culminated in four large monographs, published between 1852 and 1854. From the experience he gained in, for example, the problem of taxonomy, Darwin found this laborious task immensely valuable.

Looking back on his life, he was to recognize three stages in his professional development: "the mere collector at Cambridge; the collector and observer on the *Beagle* and for some years afterwards; and the trained naturalist after, and only after, the Cirripede [barnacle] work." Whatever his scientific peers made of his theory, they would have to recognize him as a professional biologist and could not attack his ideas as coming from a geologist straying outside his area of professional competence.

It was during this period that Darwin hit upon the final key to his theory: the principle of divergence. To the question "how do the lesser differences between varieties become augmented into the greater differences between species?" he was able to answer that "the more diversified the descendants from any one species became in structure, constitution and habits, by so much will they be better enabled to seize on many and widely diversified places in the polity [organization] of nature." By branching away into their own minute adaptations and ecological niches, varieties slowly diverged into nonbreeding species.

Encouraged by Hooker, Lyell and his elder brother, Erasmus, Darwin now began to draft a "big book" which would document in great detail how species evolved principally by the mechanism of natural selection.

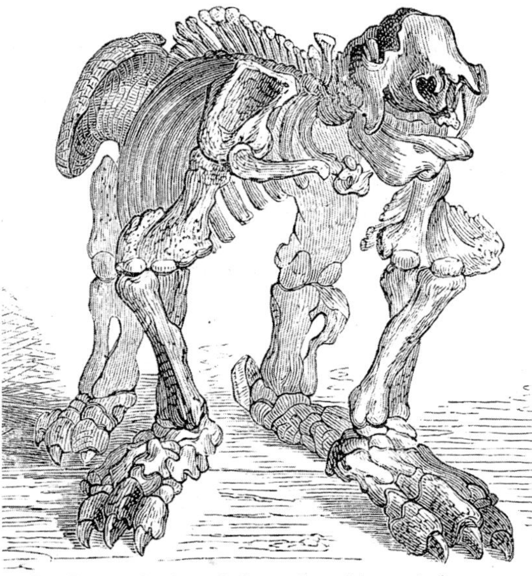

▲ *Fossil megatherium skeleton, from "Journal of Researches". Darwin sent home quantities of fossils.*

◄ Down House, where for 40 years Darwin followed a routine that combined prodigious work – he wrote 16 books and 152 other publications – a happy family life and visits from friends including Hooker and Huxley.

► The co-discoverer of evolution A.R. Wallace. He later called Darwin "the Newton of Natural History".

▼ Galapagos land iguana. Darwin discovered over 200 new species in just one month in the Galapagos.

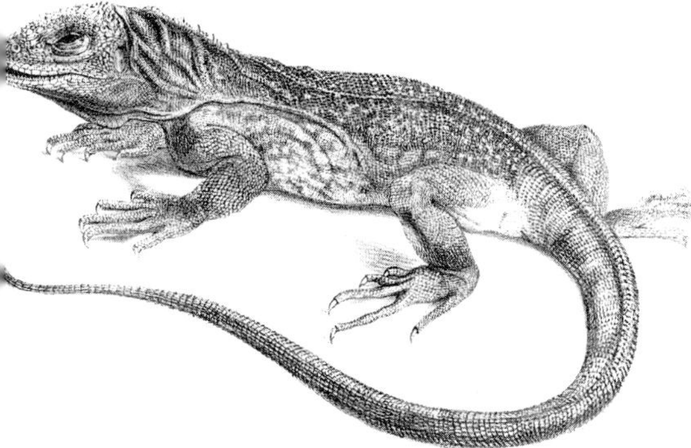

► Sir Joseph Hooker (seated, left) seen here on the Kew expedition he led to the Rockies in 1877. Britain's leading botanist of the time, he organized the herbarium at Kew Botanical Gardens, and was keenly involved in economic botany.

Science *versus* Religion

An image of warfare

The imagery of a militant, campaigning Christian church took its inspiration from Hebrew psalms, the Roman military metaphors of St. Paul's Epistles, the evangelical tradition of seeing life as a struggle between good and evil, and the revivalist hymnody produced in the aftermath of the American Civil War (1861-1865). From the 1860s onward, Christian churches, sects and denominations became increasingly engaged in "missions" and "campaigns". Such imagery, of course, implied a foe, and a casual reading of history has led to the incorrect assumption that the enemy was science, and Darwin in particular.

In 1874, in response to Pope Pius IX's announcement of the doctrine of papal infallibility, and following the Roman Catholic Church's publication of a "syllabus" of modern errors in 1864, the American chemist John William Draper (1811-1882) published a polemical "History of the Conflict between Religion and Science". (The "Syllabus" had condemned the doctrine that a Pope "can and ought to reconcile himself with progress, with liberalism, and with modern civilization".) Draper claimed that conflict between science and religion (meaning in practice Roman Catholicism) had been a continuous theme in the history of science. This claim was echoed by another American historian, A.D. White (1832-1918), in "A History of the Warfare of Science with Theology in Christendom" (1896). Although neither Draper nor White suggested that Darwin had been the special target of religious opposition (they scarcely mentioned him), their works put the science-versus-religion warfare metaphor into circulation. This metaphor was considerably reinforced by the development in the United States after World War I of the religious movement known as Fundamentalism.

In February 1858 Darwin wrote to a cousin:

"I am working very hard at my book, perhaps too hard. It will be very big; and I have become most deeply interested in the way facts fall into groups. I am like Croesus overwhelmed with my riches in facts, and I mean to make my book as perfect as I can. I shall not go to press at soonest for a couple of years."

Clearly Darwin was in no hurry. By 18 June 1858 he had written a quarter of a million words of the book; but on that day work was arrested when he received a letter from Malaysia by the English naturalist, Alfred Russel Wallace (1823-1913). Wallace asked Darwin to read an enclosed essay and to arrange for its publication in a biological journal if the argument seemed sound. The essay explained the origin of species by natural selection! Darwin, already depressed by ill-health and the imminent death of a sick child, was appalled. As he wrote immediately to Lyell, "I never saw a more striking coincidence; if Wallace had my Ms sketch written out in 1842, he could not have made a better abstract!"

Although some historians have seen the advice which Lyell and Hooker gave to Darwin as somewhat unethical and unfair to Wallace – who was scarcely in a position to protest – the decision to read extracts from the writings of both men at a meeting of the Linnean Society on 1 July 1858 attracted little attention. Meanwhile, Lyell approached the publisher John Murray, who had profited from the excellent sales of Darwin's *Journal of Researches* (1839) which described the *Beagle* voyage, and asked him to publish an abstract of Darwin's "big book". Murray agreed, generously offering Darwin two-thirds of the profits without first seeing the manuscript. The "abstract" of 155,000 words, with the more eye-catching title of *On the Origin of Species by Means of*

▲ *Judge and jury at the 1925 Dayton trial of John Scopes for teaching evolution theory, against Tennessee state law.*

◄ *Protagonists in the legendary Oxford debate on evolution in 1860: Bishop Samuel Wilberforce and biologist T.H. Huxley.*

Fundamentalism in the United States

The Fundamentalists insisted, among other things, on the absolute truth and inspiration of the Bible. American historians who grew up during the period of maximum publicity for Fundamentalism – including the Scopes trial – tended to assume erroneously that the "religion-versus-evolution" skirmishing of the 1920s and 1930s had been the situation in the 1860s. In reality, throughout Christian history, men and women have found innumerable ways of combining a wide variety of ideas and beliefs about nature (the subject of science), God, Christ and the Bible. Indeed, late 19th-century religious doubts were probably engendered more by the historians' and theologians' decision to treat the Bible as an historical text (the "higher criticism") than by Darwinism. In any case, most Protestant theologians, scientists and laymen had experienced little difficulty in interpreting evolution as a natural mechanism through which God had chosen to design the world.

The "Oxford debate" on evolution

What then of the notorious debate over evolution and religion held between Bishop Samuel Wilberforce (1805-1873) and the biologist T.H. Huxley (1825-1895) at a meeting of the British Association held at Oxford in 1860? The first versions of the story only began to appear after Draper's book had been published. Historians have therefore tended to see the Oxford debate as possessing legendary and symbolic features. Contemporary evidence reveals that Wilberforce did not attack evolution on religious grounds at all; instead (primed by the paleontologist Richard Owen, 1804-1892) he pointed to scientific difficulties that Huxley simply could not answer. It was Joseph Hooker who rescued Huxley from intellectual embarrassment. Huxley's purported reply to the Bishop "that he would rather have a miserable ape for a grandfather" than a bishop who used his intelligence to ridicule a scientific matter was perhaps rhetoric and bluster to cover his own nakedness! Only in the 1900s, when science and technology had come to preeminence, were science and Huxley seen to have been on the winning side.

Creationism today

Fundamentalism has more recently reemerged in the United States as "creationism" or "creationist science", which claims the right of equal time with evolutionary biology in the school curriculum. Creationism derives from a literal interpretation of the book of Genesis, and rejects evolution as atheistic speculation. It seems to be based not only upon the misconception that science is necessarily against religion (the Draper-White thesis again), but also on ignorance of history.

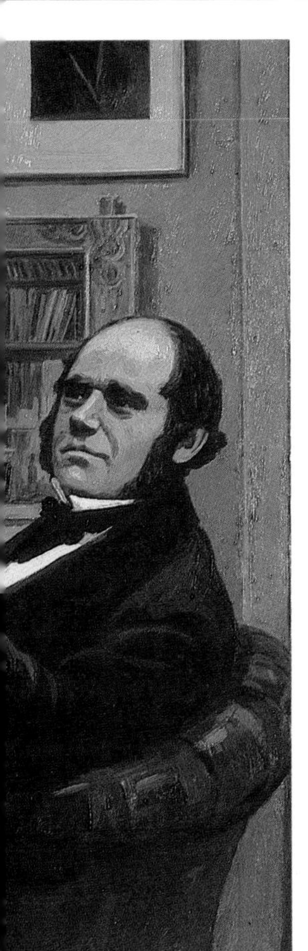

▲ *American plant taxonomist and pioneer plant geographer Asa Gray (1810-1888). Darwin outlined his theory of evolution in a letter to Gray in 1857, and Gray became Darwin's chief advocate in America.*

◄ *Hooker (left) and Lyell advise Darwin what to do about Wallace's letter, received in June 1858.*

► *In his book "Insectivorous Plants" (1875) Darwin showed the astounding adaptations that enable the sundew to capture, digest and absorb its prey.*

Natural Selection, or the Preservation of Favoured Races in the Struggle for Life, appeared on 24 November 1859.

Darwin's "big book" was never published (though the portion which he had written was published as an aid to Darwin scholarship in 1975). Revised six times, its abstract, *On the Origin of Species*, is best viewed as the forerunner to a dazzling sequence of books which both provided further evidence for evolution and drew out its consequences. These included the botanical masterpieces *On the Various Contrivances by Which British and Foreign Orchids are Fertilised by Insects* (1862), *On the Movements and Habits of Climbing Plants* (1865) and *The Power of Movement in Plants* (1880); a long treatise on artificial selection, *The Variation of Animals and Plants under Domestication* (1868); and two controversial books which examined the evidence for man's animal nature and descent, *The Descent of Man* (1871) and *The Expression of the Emotions in Man and Animals* (1872). Darwin was an extraordinarily successful science writer. In the last year of his life he published the factual, but poignant *Formation of Vegetable Mould through the Action of Worms* (1881). He was convinced that no one would read it; yet it went through six printings in less than a year and had sold 11,000 copies by 1888.

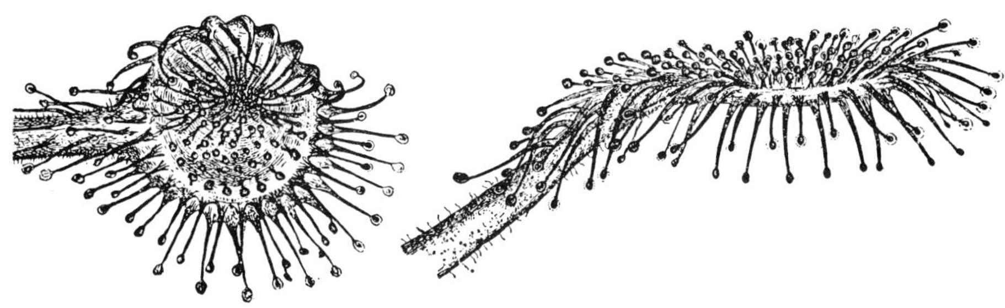

Darwin and Literature

▲► A water-baby is examined by professors Owen and Huxley; and (right) Tom converses with a lobster – from Kingsley's "The Water-Babies".

"The Origin" as literature

In her book "Darwin's Plots" (1983), the Cambridge literary historian Gillian Beer has argued persuasively that not only was Darwin's "Origin of Species" filled with literary allusions (e.g. to Milton's "Paradise Lost") and the kind of metaphorical language used by poets such as Wordsworth whom he had read greedily as a young man, but that the book actually resembled a Victorian novel in its construction. "Darwin was seeking to create a story of the world – a fiction – which would not entirely rely upon the scope of man's reason nor upon the infinitesimally small powers of observation he possesses." Darwin's sensational plot was natural selection, which he wove into a web of narrative and arguments supported by illustrations from nature. Just as authors such as Dickens, Thackeray and Eliot, personally intervene in their novels, so Darwin (the narrator) occasionally addresses the reader directly to make a moral or philosophical point. By viewing Darwin's book as a literary work, historians can gain a better insight into how the Victorians read him and why his work had such a cultural impact.

Evolutionary fantasy – "The Water-Babies"

Evolution and its implications also had direct cash value for conventional novelists. Charles Kingsley (1819-1875), who was one of Darwin's earliest admirers and himself a keen naturalist, saw evolution as leading to a more dignified view of God as "a living, immanent, ever-working" creator who improved creatures through a natural law of development. In his children's evolutionary fantasy, "The Water-Babies" (1863), Tom (the former child chimney sweep) asks Mother Carey directly whether she makes new beasts out of old, and is

▼ In "Tess of the D'Urbervilles", Thomas Hardy explored the idea of sexual selection: a still from Polanski's 1980 film starring Nastassia Kinsky.

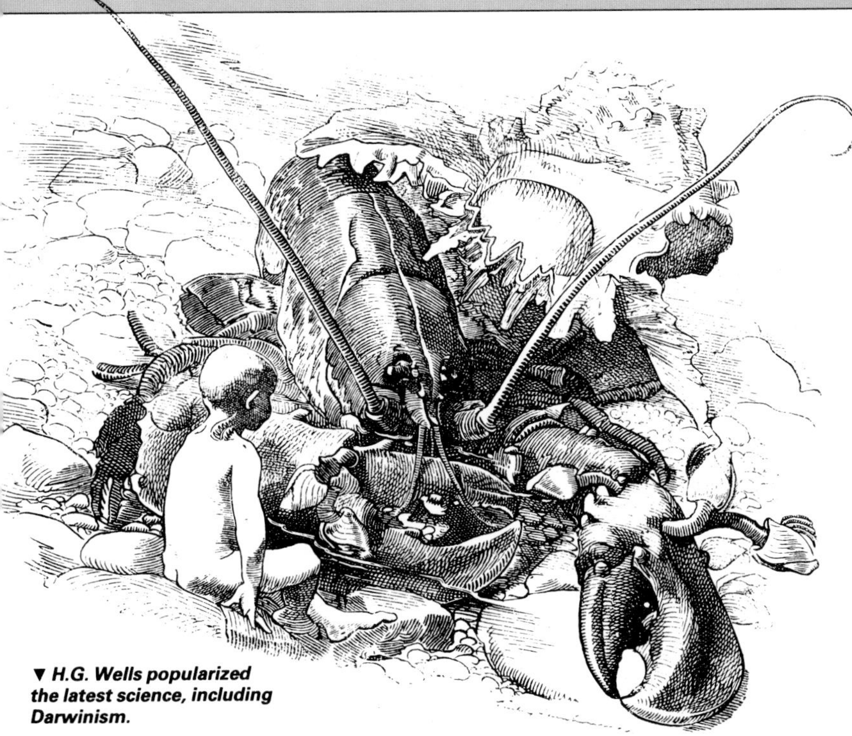

▼ **H.G. Wells popularized the latest science, including Darwinism.**

told "So people fancy. But I am not going to trouble myself to make things, my little dear. I sit here and make them make themselves."

In the story, Tom, one of the abandoned Malthusian masses, drowns and begins life again in an evolutionary river as a water baby. At the end, after witnessing many transformations, and having performed a morally edifying task of doing what he did not like doing, Tom grows into "a great man of science, and can plan railroads, and steam-engines and electric telegraphs, and rifled guns...and all this from what he learnt when he was a water-baby, under the sea."

The implications of evolution

If Kingsley, more than other writers, perceived the allegorical or fairy-story possibilities of evolution, others like Samuel Butler (1835-1902), George Eliot (1819-1880), Thomas Hardy (1840-1928) and H.G. Wells (1866-1946) explored its moral and philosophical implications.

In the satire on Victorian pretentiousness, stuffiness and religiosity, "Erewhon" (i.e. "nowhere", written backwards) which Butler published in 1872, the hero stumbles upon a utopian community. He is arrested for possessing a watch, for here is a society which had abandoned industrialization some 400 years previously when a "Professor of Unreason" had proved "that machines were ultimately destined to supplant the race of men, and to become instinct with a vitality as different from, and superior to, that of animals, as animals to vegetable life." Although meant as a parody of analogical reasoning, Butler later took seriously the idea that machines were only modified human limbs. If humans had evolved machines as extra limbs to conquer their environment, might they not have earlier developed their real limbs through conscious effort also? Butler worked out this idea in a series of books, making in the process an enemy of Darwin.

More conventionally, in "Middlemarch" (1872), George Eliot brilliantly explored the idea of variation of human character in outwardly similar people, and, in "Daniel Deronda" (1876), those basics of Victorian fiction – courtship, matchmaking, the innocence and purity of women, male aggression and dominance, and "inheritance in all its forms" – took on new meaning because Eliot had read the "Descent of Man". Thomas Hardy who acknowledged Darwin as an influence on his pessimism, also explored the idea of sexual selection on "Tess of the D'Urbervilles" (1891). In this, the natural sexual law which dictates that Tess and Angel Clare should select one another as mates, is undermined by the unnatural social law which dictates that brides had to be virginal. The result is extinction. Ultimate extinction of the human race was the bleak forecast of H.G. Wells' early science fiction, "The Time Machine" (1895).

However, not all literary assimilations of Darwin were pessimistic. In "A Modern Utopia" (1905) Wells himself described a perfect future world – a theme which was adopted by many science fiction writers like Olaf Stapleton, whose "Last and First Man" (1930) examined evolution optimistically in a cosmic setting.

▲ *Sharing their later years: Emma Darwin plays the piano while the invalid Charles is resting. Behind them are trees in the garden of their family home Down House.*

◄ *"Man is but a worm", a fanciful interpretation of Darwin's theory of evolution from a "Punch" cartoon of 1881.*

► *In 1881 Darwin published yet another pioneering work, the "Formation of Vegetable Mould through the Action of Worms". The engraving (right) from a cast photographed in Calcutta illustrates the range of material Darwin received from collaborators abroad.*

► *Darwin could not reconcile the existence of social insects with his theories. Modern kin selection theory explains "altruism" in insects like bees and ants.*

The reception of Darwin's theory

Darwin had foreseen most of the scientific objections to his theory. These he discussed frankly in the *Origin*. This did not, of course, prevent critics from raising them publicly. Why, if evolution had occurred by insensible gradations, was the fossil record incomplete? Where were the intermediate fossilized forms? How could such perfect adaptations as the eye have ever developed? Darwin freely admitted that "to suppose that the eye, with all its inimitable contrivances for adjusting the focus to different distances, for admitting different amounts of light, and for the correction of spherical and chromatic aberration, could have been formed by natural selection [seemed] absurd in the highest possible degree." Again, how could sterile worker ants and bees, whose social role and behavior were quite different from those of male and female ants, have arisen by natural selection?

Darwin never pretended that he had answers to these and other particular problems; but given that the theory of evolution was compatible with the facts of the geological record, geographical distribution, comparative anatomy and embryology, Darwin found the overall case, like Paley arguing for the feasibility of Christ's miracles, highly probable.

One objection that Darwin could not have foreseen came from the leading British physicist William Thomson (1824-1907) (Lord Kelvin as he became in 1892, ◆ page 146). Thomson made calculations of the Sun's heat in the 1860s, on the assumption that this came from the contraction of the Sun, and concluded that the Earth could not possibly be as old as Lyell's uniformitarian geology demanded. The Earth, Thomson asserted authoritatively, could not be older than 20 million years; Darwin had confidently estimated the Earth's age at 300 million years old! Only with the discovery of radioactivity at the end of the 19th century did an unsuspected mechanism for the generation of the Sun's heat become available. Darwin and the geologists were then vindicated.

Man's place in nature

Although Darwin had deliberately avoided the question of man's evolution in the *Origin*, much of the objection to the book hinged on its implicit rejection of man's miraculous creation and spiritual endowment. Instead, it looked for a natural origin, implying that man's higher powers had developed from those already present in a less developed form in animals. In *The Descent of Man* Darwin argued:

"The great difference in mind between men and the higher animals, great as it is, is certainly one of degree and not of kind. The senses and institutions, the various emotions and faculties, such as love, memory, attention, curiosity, imitation, reason, etc, of which man boasts, may be found in an incipient, or even sometimes in a well-developed condition, in the lower animals."

This indomitable naturalism had its own intrinsic difficulties, quite apart from the unease it undoubtedly generated in the minds of religious people. If natural selection were the mechanism whereby qualities of human character such as altruism and skill in art, music and mathematics had developed, of what evolutionary advantage could they have been initially? Altruistic behavior, such as laying down one's life for someone else, would surely have been a character trait opposing survival. Darwin's answer, which his colleague Wallace refused to adopt, was to suggest mechanisms other than natural selection whereby such changes might occur.

Darwin suggested that sexual selection explained display and color in animals and the emergence of man's aesthetic senses, and correlations of growth. Since organisms were integrated systems, the alteration of one part necessarily led to changes in others. These nonadaptive correlated changes might have latent uses which only became apparent at a later evolutionary stage. Wallace agreed that the human brain contained latent powers, for he had observed directly that "savages" could be taught mathematics or to play western musical instruments – skills which were of no use to them in the jungle! However, for Wallace, a strict natural selectionist, such latent abilities must have been put there by design through some supernatural intervention. Wallace, who had become a spiritualist in the 1860s, agreed with Darwin that man had physically evolved from higher animals; but he preferred a supernatural explanation for the existence of man's higher powers.

Darwin was genuinely appalled by Wallace's backsliding from naturalism. Most of the scientific and religious communities had accepted with surprising ease that Darwin's position was not incompatible with the idea of God working through a natural mechanism rather than a supernatural intervention.

Darwin's century

Darwin died from a heart attack at his home at Down House in the village of Downe in Kent (today a Darwin museum) on 19 April 1882. He had expected to be buried without fuss in the village churchyard. Instead, in response to an appeal from 20 members of parliament, Darwin's body was taken to Westminster Abbey in London for a public funeral – borne through the nave by his sons, his scientific friends Hooker, Huxley, Wallace and Sir John Lubbock, and watched by representatives of every scientific society in Great Britain, and by ambassadors from all over the world. In this honorable Christian way, accompanied by Beethoven's funeral march, a specially-composed anthem by the abbey's organist, and two memorial sermons by a canon and a bishop, did this erstwhile "scourge of orthodoxy" and supposed catalyst of 19th-century secularism come to rest beside the graves of Sir Isaac Newton and Sir Charles Lyell in the northeastern corner of the abbey's nave.

Today, despite modifications to Darwin's theory from the development of genetics, it is still an essential ingredient of biology. To the historian of the 19th century, Darwin's name symbolizes that century's intellectual doubts. Although Darwin did not, and could not, have caused such a major change single-handedly, he will always stand for the transformation and enrichment of man's understanding of his place in nature.

▲▼ *In establishing the basis for modern genetics, the Austrian botanist Friar Gregor Mendel (1822-1884) found a mechanism of inheritance and causes of variation within a species which vindicated Darwin's theory of evolution and solved difficulties recognized by Darwin himself. In his own work of about the same time, Darwin concluded that "man and all other vertebrate animals have been constructed on the same general model..." The drawings (below) of affinities between human and dog embryos are from Darwin's "The Descent of Man" (1871).*

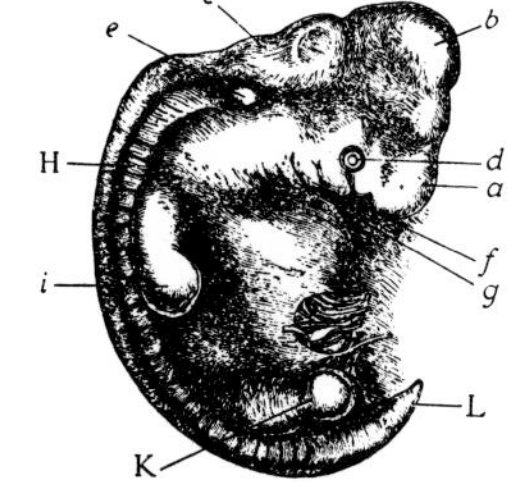

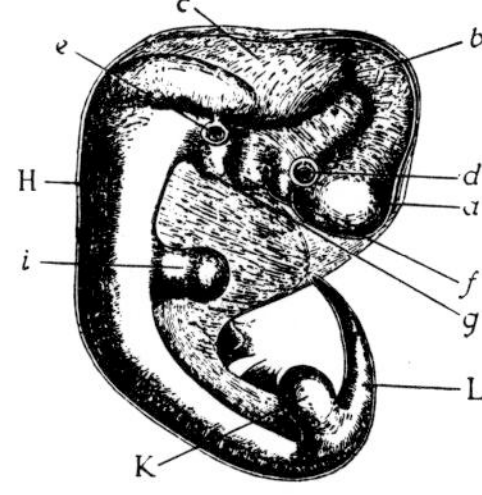

◄ *Denied state honors in his lifetime, Darwin was buried in London's Westminster Abbey, close to Isaac Newton.*

► *Mementos of the "Beagle's" voyage: marine and Galapagos land iguanas, and the Falkland Island fox (warrah) described by Darwin which became extinct within 50 years.*

Louis Pasteur 1822-1895

At the École Normale, Paris...A tradition of crystallographers...The "handedness" of nature... Two kinds of bacteria...The spontaneous generation controversy...Diseases of wine, beer...and silkworms ...Fowl cholera, anthrax and rabies vaccines... PERSPECTIVE...Roman Catholicism and science... The technology of bacteriology...Medicine orthodox and unorthodox...Medical treatments

With the growth of bottle-feeding of babies on cow's milk at the end of the 19th century there was a marked rise in infant mortality rates and in infant diarrhea. Pasteurization and bottling of milk after 1900, and the elimination of tuberculosis-ridden cattle from dairy herds, led to a dramatic decline in infant mortality and in some forms of human tuberculosis. These advances were all consequences of the emergence of a new science of microorganisms – "bacteriology".

The technique of milk pasteurization, whereby milk is heated to a temperature of 60-65°C in order to destroy any bacteria which it may contain, is named after Louis Pasteur, who introduced the method in 1865 to prevent the souring of wine. The fact that a chemist, as Pasteur was, helped establish the germ theory of disease and began the conquest of certain diseases by vaccine therapy, is revealing about the process of scientific discovery and the pressures which can mold and shape a scientist's career.

A practical and theoretical scientist

From the time of Leeuwenhoek's discovery of bacteria in 1677, microbiology was concerned with practical problems arising from the study of alcoholic fermentation, medicine and sanitation, and with the theoretical question whether organized beings could be spontaneously generated. The French scientist Louis Pasteur, who first gained his reputation as a chemist and crystallographer, made fundamental contributions to all of these concerns.

In the late 1830s, Theodor Schwann argued that fermentation was due to living yeast cells, but this was denied by chemists such as Justus von Liebig. Pasteur's demonstration that the process was firmly linked to life helped revive an ancient idea that diseases were caused by living "contagia". In the 1860s and 1870s both Pasteur and his German rival Robert Koch laid the foundations for this germ theory of disease and for the development of preventive vaccination techniques which led in the 20th century to the development of chemotherapy. Pasteur's elegant experiments also helped to disprove the idea that present-day organisms can arise spontaneously. In the 1880s, the discovery of heat-resistant germs by Koch and J. Tyndall, together with the development of genetics by A. Weismann (1834-1914) confirmed Pasteur's conclusions. This has not prevented scientists from speculating that "life" must have evolved from inorganic materials.

► Pasteur made fundamental contributions to chemistry and biology. Work on stereochemistry led him to the study of fermentation and the germ theory of disease. In his École Normale laboratory (right) he developed the anti-rabies vaccine in 1885.

▼ Pasteur is much revered in France. The stamp shows the swan-necked flask with which he conducted a famous experiment, crystals, a microscope and bacteria.

Louis Pasteur – His Life, Work and Times

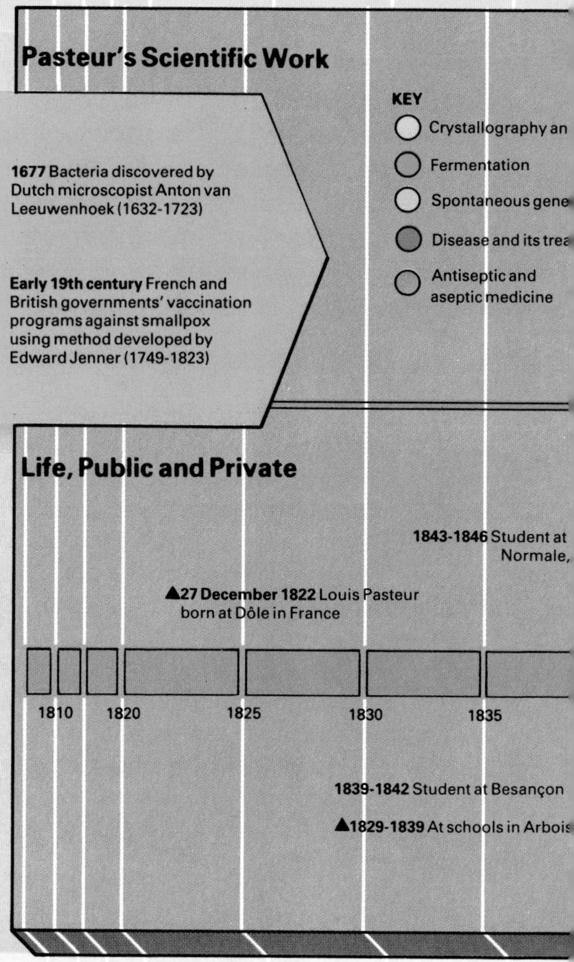
Historical Background
1830 France began
Algeria
1830 July Revolution
autocratic monarchy
France
Pasteur's Scientific Work
KEY
Crystallography
Fermentation
Spontaneous
Disease and its
Antiseptic and
aseptic medicine
1677 Bacteria discovered by Dutch microscopist Anton van Leeuwenhoek (1632-1723)
Early 19th century French and British governments' vaccination programs against smallpox using method developed by Edward Jenner (1749-1823)
Life, Public and Private
1843-1846 Student at
Normale,
27 December 1822 Louis Pasteur born at Dôle in France
1810
1820
1825
1830
1835
1839-1842 Student at Besançon
1829-1839 At schools in Arbois

Pasteur the chemist

Pasteur was born on 27 December 1822, at Dôle, near Dijon in the east of France. Most of his childhood was spent 25km to the southeast at Arbois, amid the vineyard-covered slopes of the Jura mountains, where his father kept a tannery. In 1843 Pasteur began studies at the École Normale in Paris. He had passed the entrance examination the year before, but he was so disgusted at coming 14th that he reentered the examination a second time, and took fourth place. Pasteur's biographers have seen this as an example of his extraordinary perseverance; but it also illustrates his pride and egotism.

The École Normale was France's leading institution for preparing students for the state examination of *agrégation* – a teaching qualification which guaranteed its graduates the highest rate of salary in French secondary schools. Here, under the teaching of France's leading organic chemist, J. B. Dumas (1800-1884), and of A.-J. Balard (1802-1876), the discoverer of the element bromine, Pasteur became fired by an interest in chemistry, particularly in crystallography.

French science had a long tradition of research into the nature and properties of crystals. In 1815, the French physicist, J. B. Biot (1774-1862), found that many organic liquids were, like some crystals, capable of rotating the direction of polarization of a beam of polarized light that was passed through them. That is, the beam emerged at an angle to the plane of vibration of the original polarized beam. Since these organic molecules were not crystalline, Biot concluded that their power to alter the light must be inherent in the structure of their molecules. By 1844 it was known that there were two different kinds (isomers) of the organic compound tartaric acid. One turned the plane of transmitted polarized light to the right, the other produced no effect at all. Since their crystalline shapes and therefore, by implication, their molecular forms, seemed to be identical, it was a puzzle as to how their solutions could have such different effects on polarized light. This was the problem that Pasteur tackled successfully for his *agrégation* doctorate in 1847.

Pasteur noticed that the polarizing, or optically-active, tartaric acid crystals possessed some distinctive facets which seemed to be absent in the non-polarizing crystals. Detailed comparisons made with a magnifying glass showed that the inactive crystals were symmetrical, whereas the active crystals were asymmetrical. The polarization effect seemed, therefore, to depend upon the existence of asymmetric crystals. Upon preparing larger optically-inactive crystals, Pasteur noticed by eye that there were, in roughly equal proportions, two different kinds of crystals in the sample; both were asymmetric, but one lot were mirror images of the other. Pasteur then boldly concluded that since the former rotated the plane of polarization to the right (the "dextro" form), the latter ought to turn it to the left (the "laevo" form). Because each group of crystals was present in equal proportions in the inactive sample, polarization effects normally cancelled out. Pasteur had prepared a new form of tartaric acid which did not occur naturally. His discovery was to form the basis of an important branch of chemistry, called stereochemistry.

The work on tartaric acid, which was highly praised by Biot, assured Pasteur of an academic career. In 1848 he was appointed physics teacher at the *lycée* (secondary school) in Dijon, but within three months he had moved to the faculty of science at what was later the University of Strasbourg, in industrialized Alsace, as professor of chemistry. Here he married the daughter of a local headmaster.

Scientific Background

1808 Étienne Malus (1775-1812) discovered reflection of light causes polarization

1815 French physicist J. B. Biot (1774-1862) found plane of polarization is rotated by many organic solutions

1830 Achromatic micro… developed by J. J. Lister (… of the surgeon)

1837 Charles Cagniard de Latour in France and Theodor Schwann (1810-1882) in Germany argued that fermentation is a living process

1839 Schwann developed …

1839 J. von Liebig (1803-…) German chemist, argued that fermentation and putrefaction are purely chemical processes

1840 A "germ theory" of … suggested by … pathologist Jakob Henle (…)

1805 1815 1825 1830 1835

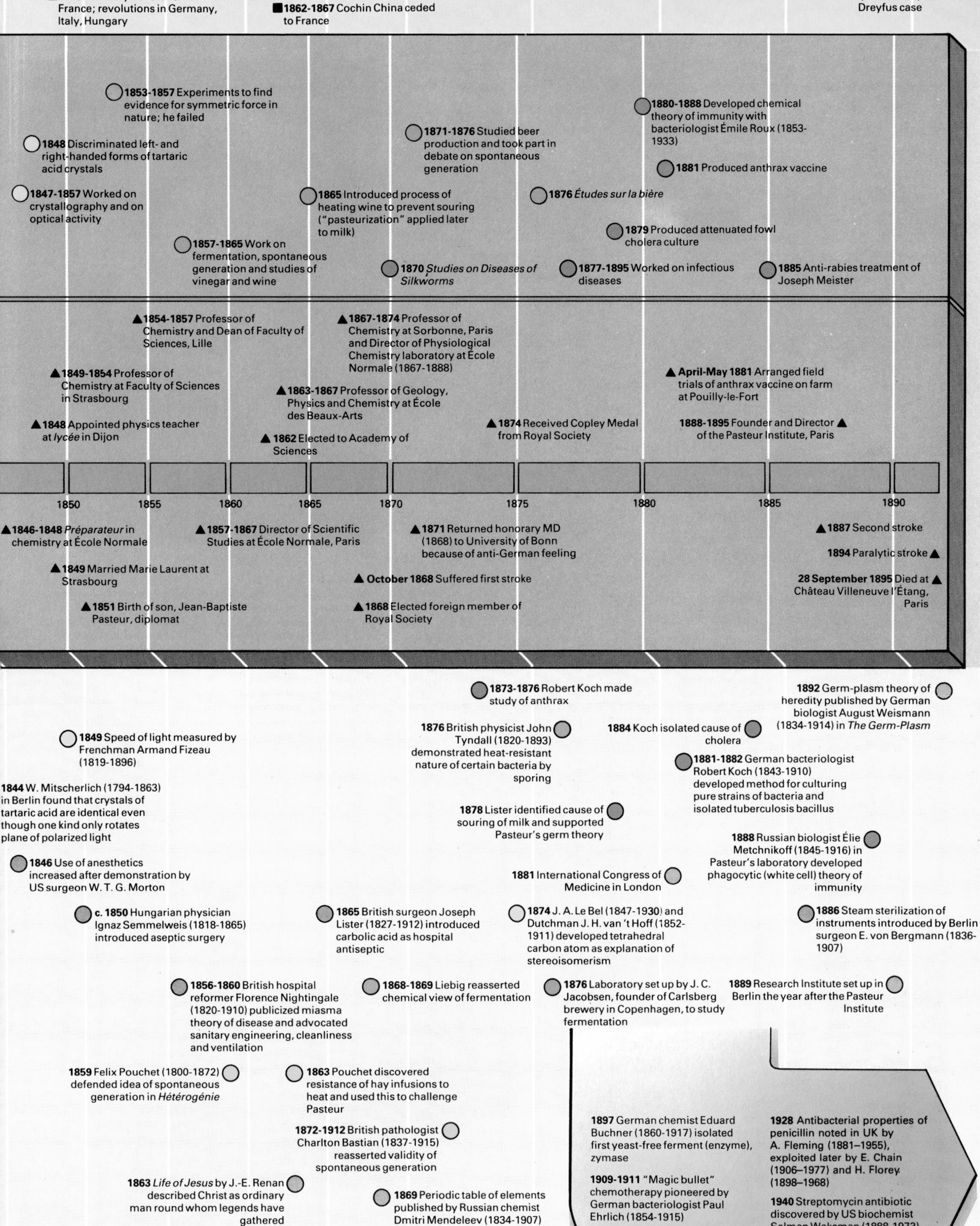

1852 Louis Napoleon became Napoleon III of a Second Empire
1870-1871 France defeated in Franco-Prussian War; Third French Republic founded; Paris Commune suppressed
1848 February Revolution in France; revolutions in Germany, Italy, Hungary
1862-1867 Cochin China ceded to France
1894-1905 France polarized by Dreyfus case
1853-1857 Experiments to find evidence for symmetric force in nature; he failed
1880-1888 Developed chemical theory of immunity with bacteriologist Émile Roux (1853-1933)
1871-1876 Studied beer production and took part in debate on spontaneous generation
1848 Discriminated left- and right-handed forms of tartaric acid crystals
1881 Produced anthrax vaccine
1847-1857 Worked on crystallography and on optical activity
1865 Introduced process of heating wine to prevent souring ("pasteurization" applied later to milk)
1876 Études sur la bière
1879 Produced attenuated fowl cholera culture
1857-1865 Work on fermentation, spontaneous generation and studies of vinegar and wine
1870 Studies on Diseases of Silkworms
1877-1895 Worked on infectious diseases
1885 Anti-rabies treatment of Joseph Meister
1854-1857 Professor of Chemistry and Dean of Faculty of Sciences, Lille
1867-1874 Professor of Chemistry at Sorbonne, Paris and Director of Physiological Chemistry laboratory at École Normale (1867-1888)
1849-1854 Professor of Chemistry at Faculty of Sciences in Strasbourg
April-May 1881 Arranged field trials of anthrax vaccine on farm at Pouilly-le-Fort
1863-1867 Professor of Geology, Physics and Chemistry at École des Beaux-Arts
1848 Appointed physics teacher at lycée in Dijon
1874 Received Copley Medal from Royal Society
1888-1895 Founder and Director of the Pasteur Institute, Paris
1862 Elected to Academy of Sciences
1850
1855
1860
1865
1870
1875
1880
1885
1890
1846-1848 Préparateur in chemistry at École Normale
1857-1867 Director of Scientific Studies at École Normale, Paris
1871 Returned honorary MD (1868) to University of Bonn because of anti-German feeling
1887 Second stroke
1894 Paralytic stroke
1849 Married Marie Laurent at Strasbourg
October 1868 Suffered first stroke
28 September 1895 Died at Château Villeneuve l'Étang, Paris
1851 Birth of son, Jean-Baptiste Pasteur, diplomat
1868 Elected foreign member of Royal Society
1873-1876 Robert Koch made study of anthrax
1892 Germ-plasm theory of heredity published by German biologist August Weismann (1834-1914) in The Germ-Plasm
1849 Speed of light measured by Frenchman Armand Fizeau (1819-1896)
1876 British physicist John Tyndall (1820-1893) demonstrated heat-resistant nature of certain bacteria by sporing
1884 Koch isolated cause of cholera
1881-1882 German bacteriologist Robert Koch (1843-1910) developed method for culturing pure strains of bacteria and isolated tuberculosis bacillus
1844 W. Mitscherlich (1794-1863) in Berlin found that crystals of tartaric acid are identical even though one kind only rotates plane of polarized light
1878 Lister identified cause of souring of milk and supported Pasteur's germ theory
1888 Russian biologist Élie Metchnikoff (1845-1916) in Pasteur's laboratory developed phagocytic (white cell) theory of immunity
1846 Use of anesthetics increased after demonstration by US surgeon W. T. G. Morton
1881 International Congress of Medicine in London
c. 1850 Hungarian physician Ignaz Semmelweis (1818-1865) introduced aseptic surgery
1865 British surgeon Joseph Lister (1827-1912) introduced carbolic acid as hospital antiseptic
1874 J. A. Le Bel (1847-1930) and Dutchman J. H. van 't Hoff (1852-1911) developed tetrahedral carbon atom as explanation of stereoisomerism
1886 Steam sterilization of instruments introduced by Berlin surgeon E. von Bergmann (1836-1907)
1856-1860 British hospital reformer Florence Nightingale (1820-1910) publicized miasma theory of disease and advocated sanitary engineering, cleanliness and ventilation
1868-1869 Liebig reasserted chemical view of fermentation
1876 Laboratory set up by J. C. Jacobsen, founder of Carlsberg brewery in Copenhagen, to study fermentation
1889 Research Institute set up in Berlin the year after the Pasteur Institute
1859 Felix Pouchet (1800-1872) defended idea of spontaneous generation in Hétérogénie
1863 Pouchet discovered resistance of hay infusions to heat and used this to challenge Pasteur
1872-1912 British pathologist Charlton Bastian (1837-1915) reasserted validity of spontaneous generation
1897 German chemist Eduard Buchner (1860-1917) isolated first yeast-free ferment (enzyme), zymase
1928 Antibacterial properties of penicillin noted in UK by A. Fleming (1881–1955), exploited later by E. Chain (1906–1977) and H. Florey (1898–1968)
1909-1911 "Magic bullet" chemotherapy pioneered by German bacteriologist Paul Ehrlich (1854-1915)
1863 Life of Jesus by J.-E. Renan described Christ as ordinary man round whom legends have gathered
1869 Periodic table of elements published by Russian chemist Dmitri Mendeleev (1834-1907)
1940 Streptomycin antibiotic discovered by US biochemist Selman Waksman (1888-1973)
1845
1850
1855
1860
1865
1870
1875

While at Strasbourg Pasteur asked himself why nature appears to be "handed". Why do naturally occurring crystals always uniquely possess either the dextro- or laevo- form? Pasteur speculated that there was an asymmetric force present in nature – perhaps light, heat, or electricity. He experimented with the effects of giant magnets on chemical reactions to see if optically-active rather than optically-neutral crystals were produced. They were not. In 1854 he arranged for plants to be grown in sunlight whose rays were reversed by means of an optical device, to see whether they produced dextro rather than laevo compounds in their leaves. The results were inconclusive. In any case, Pasteur saw that there was no experimental way he could remove the influence of any asymmetric materials that were already present in the plant seeds themselves. (DNA, deoxyribonucleic acid, which we know now to be the key molecule in the genetic structure of living systems, is, in fact dextro-rotatory.) It seems that chemical syntheses affected by the "handedness" of the compound both in nature and in the laboratory can only be achieved through the intervention of an optically-active intermediate in the synthesis. It is interesting to recall that in the detective novel, *The Documents in the Case* (1930), Dorothy L. Sayers has a victim die apparently from mushroom poisoning. Murder is proved by the fact that the poison turns out to be racemic, optically-inactive and therefore synthetic muscarine stolen from a laboratory, while the muscarine that occurs in the fly agaric mushrooms only exists in dextro-rotatory form!

For Pasteur, nature's asymmetry was the barrier between inorganic and organic life. Dextro-tartaric acid is a normal constituent of grapes and had long been known to form the principal ingredient of the sludge which separates from wine during fermentation. In 1852 a German firm of manufacturing chemists informed Pasteur (who was in the habit of visiting such establishments when he traveled) that if solutions of impure tartrates were left out in warm weather in contact with organic matter, they underwent fermentation, producing various products in the process. Pasteur tried it out using both the dextro- and racemic forms of tartaric acid, and found that fermentation brought about laevo-rotatory activity in the latter. In other words, during

▲ As a youth, Pasteur's chief interest and talent lay in painting. This charming pastel portrait of his mother shows her dressed in a pensioner's costume.

► Light rays undergo double refraction and polarization if passed through asymmetric crystals such as Iceland spar.

▼ The University of Strasbourg where Pasteur worked on crystallography, 1849-1854. Following Strasbourg's occupation by Germany (1871) Pasteur argued for improvements in French science, industry and education.

▲ *Building upon Pasteur's work on optical effects, the Dutch chemist J.H. van't Hoff (above) (1852-1911) and Frenchman J.A. Le Bel (1847-1930) independently advanced the theory of the asymmetric carbon atom in 1874.*

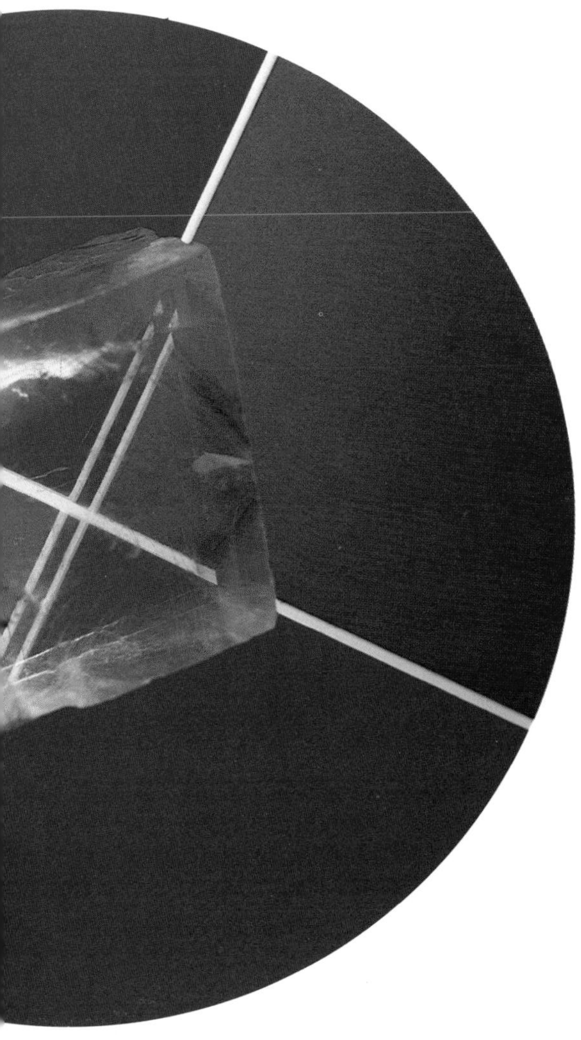

fermentation, the dextro-tartaric acid in the neutral racemic acid was consumed by some living organism, leaving behind only the laevo-tartaric acid. Not only did this – as we have seen – give Pasteur a new method of resolving mirror-image isomers, but it revealed that they were different physiologically. Bacteria would feed upon the dextro, but not on the laevo- acid. It was this discovery which led Pasteur into the study of fermentation as a chemical and biological process, and slowly shifted his career from chemistry to biology.

The study of fermentation

In 1854 Pasteur moved to Lille, another French industrial center, as dean of the faculty of science. Faculties were special schools for higher education which Napoleon had established in the provinces in 1802 as part of his educational reforms (◆ page 103). Since one of Lille's leading industries was the fermentation of beetroot and grain to make alcohol, Pasteur decided to offer courses on fermentation.

The predominant view of alcohol fermentation in the mid-1850s was that it was a purely chemical process in which sugar is transformed into alchohol. To the German chemist Justus von Liebig (1803-1873), who had put his great authority behind this view, the yeast cells which brewers used promoted the fermentation through their death and decomposition; living, biologically-active yeast had nothing to do with the process. Liebig (◆ page 174) held this opinion despite the fact that Theodor Schwann (1810-1882), the German microscopist who discovered animal cells in 1839, had already shown that yeast cells reproduce and grow during fermentation.

Although as a chemist Pasteur might have been expected to have shared Liebig's "pro-chemical" view of fermentation, his experience of the asymmetric nature of fermentation products inclined him to the biological interpretation. He was also much struck by the continuous nature of the yeast's growth, and he did not know of any analogy with other chemical reactions.

By 1857 he had succeeded in showing that the souring of milk – lactic acid fermentation – was caused by a microscopic rod-like plant which, when introduced into a solution of sugar containing a nitrogen source, such as ammonia, caused the formation of lactic acid. Pasteur had demonstrated, as had Schwann before him, that fermentation was dependent upon the presence of microscopic forms of life, with each fermenting medium serving as a unique food for a specific microorganism. Although Liebig was worsted in this exchange, it was later realized that yeast and other microorganisms promote fermentation by the secretion of cell chemicals called ferments or enzymes. The compromise between these two extreme views was one of the factors in the development of a new science, biochemistry, in the 1880s.

Meanwhile Pasteur's work on fermentation also helped open up another new science, bacteriology. (This was the German word for what the French for nationalistic reasons, following Pasteur, called microbiology.) In 1857, after only three years in Lille, Pasteur was called to Paris to become director of scientific studies at his *alma mater*, the École Normale. He was sacked from the post ten years later because students and staff found him authoritarian, inflexible and overpowering as an administrator; but he stayed on at the college as its director of physiological chemistry until his retirement in 1888. Prestigious though his 1857 appointment was, it carried the disadvantage that the director had no laboratory. Pasteur therefore had to pay for one to be erected in the institution's attics at his own expense.

Pasteur identified "anaerobic" bacteria inactive in air and "aerobic" microbes active only in the presence of oxygen

▼ When Pasteur cultured microbes he discovered that some were inactive in air, or more specifically oxygen, while others were only active in the presence of oxygen. Identification of anaerobic bacteria (which obtain their oxygen indirectly from other compounds) suggested to Pasteur that decay or putrefaction, like fermentation, was a biological process. Anaerobic bacteria (above, Clostridium perfringens) begin the processes of fermentation and putrefaction, and the products they produce are then acted on by aerobic bacteria (below, Lactobacillus bulgaricus from live yoghurt).

Aerobic and anaerobic microbes, or bacteria

It was in Paris that Pasteur developed techniques for culturing microbes in liquid broths (➧ page 181) and he discovered that some bacteria were inactive in air, or more specifically, oxygen, while others were only active in the presence of oxygen. He called the former anaerobic, the latter aerobic.

The identification of anaerobic bacteria (which obtain their oxygen indirectly from other compounds) suggested to Pasteur that decay or putrefaction, like fermentation, was another biological process. Anaerobic bacteria began the processes of fermentation and putrefaction, and the products they produced were then acted on by aerobic bacteria. One consequence of this was that microorganisms were of great significance, for without them the world would become saturated with undegraded organic molecules. There would be no recycling of materials.

It was to demonstrate this point that Pasteur designed a classical experiment. In the first of two flasks he placed a solution of yeast with some pre-heated sugar or milk and pure, dust-free air; the second flask contained the yeast and unheated sugar or milk and unpurified air. Both flasks were kept for several days at a temperature of about 30°C, following which the air in both flasks was analyzed. In the first flask, the air still contained a large quantity of oxygen; but in the second, where the access of microscopic organisms had been unimpeded, the oxygen was exhausted and replaced by carbon dioxide. Clearly fermentation, whether in the anaerobic or aerobic mode, only proceeded in the presence of microorganisms.

The experiment initially had nothing to do with the question whether life could originate by spontaneous generation. Yet, in 1861, when Pasteur first made his demonstration, this question was a particularly live issue in France, both in academic circles and among the general public. When, therefore, Pasteur realized that his work on fermentation was relevant to the debate, he entered into it with enthusiasm. By 1864 Pasteur would be able, on the basis of the results of series of further experiments, to declare with certainty that the minute organisms causing fermentation came from similar organisms to be found in ordinary air.

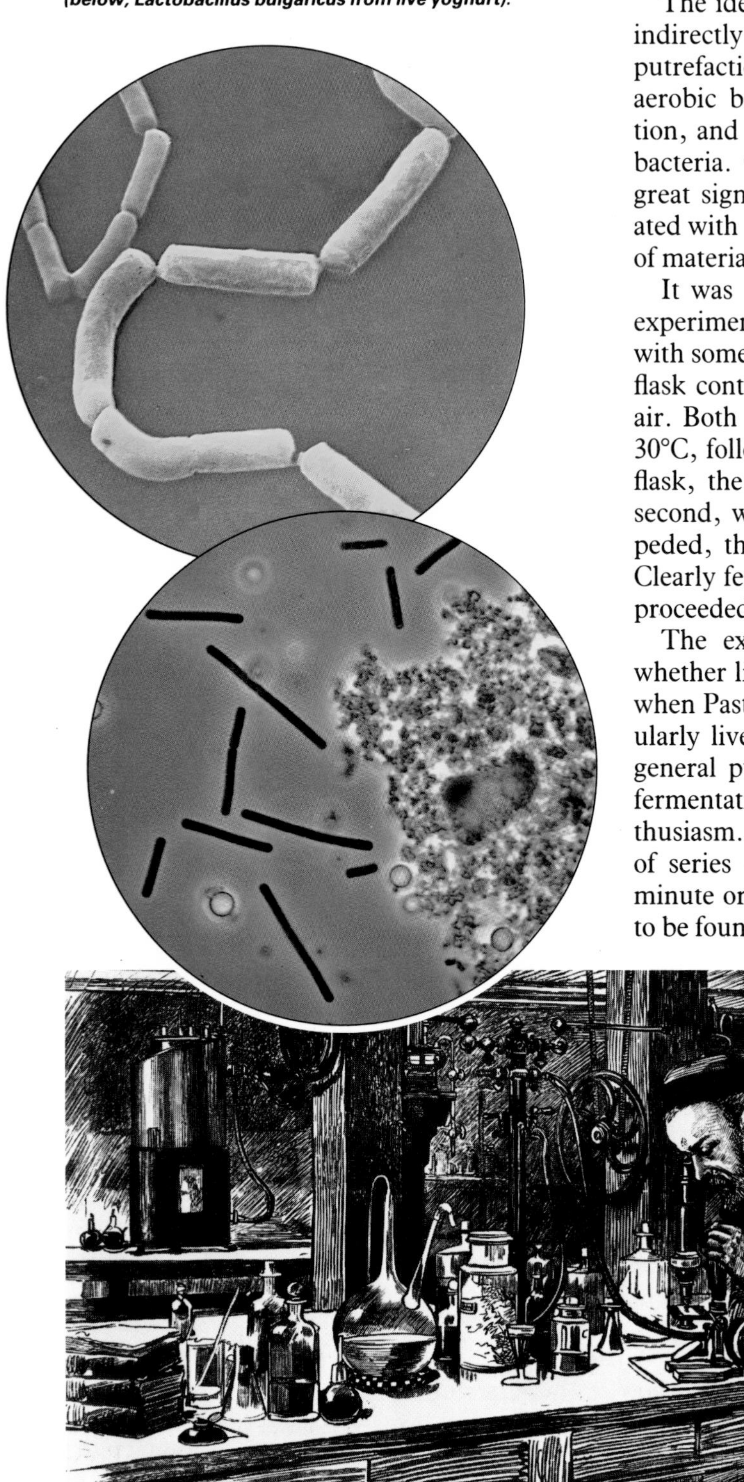

◄ Pasteur using the compound microscope to study bacteria in his laboratory at the Ecole Normale. His "germ theory" proposed that diseases are spread by germs (bacteria).

▲ The German chemist Justus von Liebig believed that fermentation and putrefaction were chemical processes, whereas Pasteur argued that living organisms were implicated. The discovery of enzymes in the 1890s showed that both men were right.

The "Bankruptcy" of Science

▲ **The anti-vivisection movement began in the 1880s in response to the development of experimental physiology, pharmacology and vaccine therapy. Typical of experiments which led to protests from laymen was the one above, in which a chloroformed dog was resuscitated by electric shocks. While Pasteur had no moral objections to vivisection – the preparation of the rabies vaccine involved live rabbits and that for smallpox, young calves – he usually made his assistants carry out such operations. The anti-vivisection controversy led to legislation which permitted experimentation, but only under license.**

The secularization of knowledge

Catholic scientists and intellectuals, among them Pasteur, generally found it especially difficult to come to terms with the increasing secularization of knowledge in the 19th century. These difficulties were particularly acute in France, where Catholicism had ceased to be the official state religion. In order to restore its authority many French Catholics sought political and religious support from the Pope and the freedom to run their own church schools in opposition to the state schools established at the time of Napoleon. Although church schools were allowed from 1850, they were inspected by state officials who laid down the curriculum. Thus many Catholics became antagonistic toward tendencies of modern thought which, correspondingly, caused anticlerical sentiments among more liberal Frenchmen.

Renan, Darwin and Catholic idealists

In 1863, Joseph-Ernest Renan (1823-1892), a French philosopher who had abandoned clerical training because of his growing scepticism, published a "Life of Jesus", describing Christ as an ordinary man about whom legends had accumulated. There followed a heated debate in France and Britain between those who followed Renan and Darwin in upholding that human reason could, through the application of scientific method, understand everything ("scientific naturalism") and those who felt that such a creed was intellectually bankrupt. To Catholic idealists science was "bankrupt" because it was powerless to save souls and ultimately harmful to man because it threatened to destroy the ethical foundations of European civilization. In his novel, "Le disciple" (1889), the French writer Paul Bourget told the story of a man obsessed with the "scientific method" who seduced a woman as a psychological experiment. When the victim learned this she committed suicide. The novel was seized upon as showing the moral irresponsibility of scientific naturalism. Other critics pointed to the growing amount of vivisection which lay at the heart of the new physiology (and which the devout Pasteur was also criticized for supporting) and at the new bacteriological medicine which treated the body as a mere mechanism.

The claim that science was bankrupt was denied in France by the physiologist Charles Richet (1850-1935), who was to win a Nobel prize in 1913 for work on serum therapy. Richet, who was a keen critical investigator of psychic phenomena, was prepared to admit that science did not have an answer to everything. He admitted, too, that the scientist did not always care about morality; but looking for the truth without regard for the consequences was good in itself since it dispelled human ignorance! The French chemist Pierre Berthelot (1827-1907) also placed his faith firmly in a future in which science would continue to transform the material and moral conditions of human existence. Catholic scientists including Pasteur and the physicist Pierre Duhem (1861-1916) responded by founding the Catholic Scientific Society of Brussels and an international journal dedicated to the premise that knowledge of God was as certain as that of the physical world. However, the Franco-Prussian War of 1870-1871 led to the foundation of a French Third Republic dedicated to the cult of science. Catholic scientists were frequently snubbed. In 1892 the republican government created the world's first chair of history of science but appointed an elderly rationalist rather than the better-qualified Catholic historian of mathematics, Paul Tannery (1843-1904). The latter was again deliberately over-looked in 1903 when the chair became vacant.

A degree of reconciliation

Intellectual reconciliation was made easier by Pope Leo XIII's recommendation (1879) of the writings of Thomas Aquinas (♦page 26) who, in the 13th century, had been faced with the problem of reconciling Aristotle's science and philosophy with Christianity. Nevertheless, succeeding popes again turned their face against science. Only in the 1950s, during the pontificate of Pius XII, was an "historical" reading of the Bible officially permitted. At the same time Darwin's theory of evolution was accepted for man's physical condition, God having endowed man with his spiritual "soul" at a specific moment in evolutionary history.

► **Darwin's "Origin of Species" and Renan's "Life of Jesus" were among the many works which Pio Nono, Pope Pius IX, had in mind when he condemned rationalism, the liberty of the press and liberty of religious thought in an encyclical of 1864 to which he attached a "Syllabus" listing 80 "errors" of the age.**

A hundred years ago the great controversy over spontaneous generation of living things was drawing to a close

Spontaneous generation theory discarded

Spontaneous generation refers to the belief that living entities can arise either from inorganic elements (abiogenesis) or from a reconstitution of dead organic matter (heterogenesis) without any intervention by a living "parent". The idea dated back to Aristotle – the Flemish physician and early chemist J.B. van Helmont, for example, had given directions for the growth of mice from old rags – and it had still not been disproved.

One of the strongest advocates of spontaneous generation in France was Felix-Archimède Pouchet (1800-1872), director of the natural history museum at Rouen. In *Hétérogenie* (1859), Pouchet argued that there were three factors necessary for the generation of lower forms of life – decaying organic matter, air and water; light and electricity were also helpful. Realizing that contamination might be used to explain his results (as, in effect, was done by Pasteur), Pouchet took great pains to purify the air used in his tests. He pointed out that microscopic examination of his air filters showed mainly mineral contamination and occasionally fungal spores (a reminder that bacteria were extremely difficult to "see" before the advent of staining techniques in the 1870s). Even if solutions were heat-sterilized, Pouchet claimed, germination occurred nevertheless. By 1860, the French Academy of Sciences was sufficiently interested in Pouchet's claims to make spontaneous generation the subject of a prize paper competition. Always on the lookout for finance to fund his research, Pasteur entered with an essay which impressively undermined Pouchet's position. Although Pasteur's argument was experimental, as a devout Roman Catholic and opponent of evolution he also had religious and philosophical reasons for disliking a doctrine which reduced life to chemical reactions.

Pasteur first showed that, by using guncotton (cellulose nitrate) instead of raw cotton wool as an airfilter, an enormous number of microorganisms could be trapped which bypassed Pouchet's filter. Secondly, he pointed out that Pouchet opened his flasks to the "purified" air under mercury, which was another vehicle for introducing life into the sterile flask. Pouchet accepted this criticism, but argued that the real point was that oxygen, not Pasteur's supposed life-bearing contaminants, was essential for spontaneous generation. This argument led Pasteur to an elaborate and elegant refutation. He put broth into a large number of flasks with elongated necks. The contents were boiled (i.e. sterilized) and the air driven out. The necks were then heat-sealed. Pasteur took the samples into the countryside of Arbois, where he opened 20 flasks; more were opened on a mountainside some 2,000m above sea level. All the flasks were resealed, carried back to Paris, and deposited with the Academy of Sciences. The results were variable: of the 20 flasks opened in the countryside, eight developed life, but only one flask exposed upon the mountain did so. Hence, despite the access of oxygen into the flasks in all cases, life only appeared if germs were present in the air.

But supposing, Pouchet objected, that in boiling the infusions Pasteur had decomposed the chemicals crucial for life to emerge? Pasteur's answer was to place infusions in swan-necked flasks, the necks of which were left open to the air. The infusions were heat sterilized in the usual way, but no life emerged despite the air's open access. The reason for this, Pasteur argued, was that the germs were trapped in the U-bend of the flask's neck. If the neck was broken, or the flask tilted so that the contents came into contact with the U-bend,

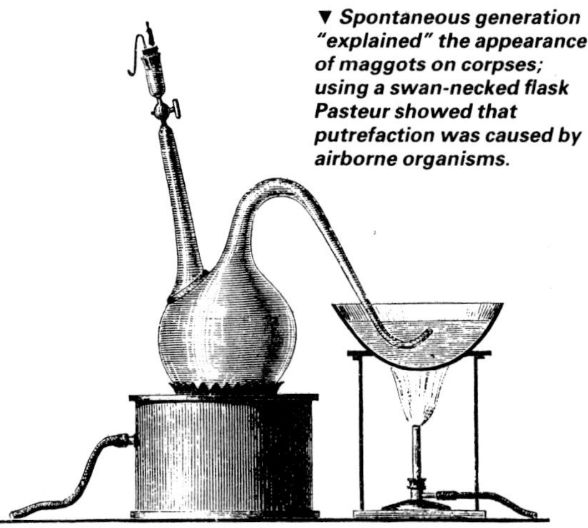

▼ ***Spontaneous generation "explained" the appearance of maggots on corpses; using a swan-necked flask Pasteur showed that putrefaction was caused by airborne organisms.***

JEBELLE (Monastère de la Trappe-Drôme)
LA LOI DES FERMENTS

◄ *In 1863 the Emperor Napoleon III commanded Pasteur to investigate the diseases of French wines. Using an improvised laboratory in the back room of a cafe in the wine region of Arbois, he showed that the souring of wine was caused by the development of the organism Mycoderma aceti at the expense of Mycoderma vini, and could be cured by neutralizing the acetic acid (vinegar) formed with potassium hydroxide. Other wine diseases were cured by heat treatment to 55°C ("pasteurization") and by excluding excess air from the bottles.*

▲ *Pasteur's apparatus for cooling and fermenting wort (the liquid from mashing malt in water) during his work on beer in 1871. The contents are seen through a spyhole.*

▼ *A modern plant for the pasteurization of milk using a heat-exchange system. Originally Pasteur invented the process to improve storage qualities of wine and beer.*

organisms rapidly developed. Clearly, suggested Pasteur, the infusion's chemical power had not been altered by the pre-heating.

This brilliantly-conceived experiment did not close the argument. Later, in the mid-1870s, it was realized that many germs, including those in the hay infusions which Pouchet used, are heat-resistant and can withstand very high temperatures because they form inert spores. Under favorable conditions, the spores regenerate the rod or spiral bacterial form. Neither Pasteur nor Pouchet knew this during the 1860s and their dispute was settled on ideological and religious grounds rather than those of experimentation. Pouchet suggested in 1864 that the issue should be settled experimentally by a special Committee of the French Academy of Sciences. The Committee was packed with Roman Catholic scientists who strongly supported Pasteur. Not surprisingly, its proceedings were farcical, since it contrived to view only Pasteur's experiments. This deliberate official rejection of evidence that was embarrassing to Pasteur's case delayed the understanding of bacterial life cycles by a decade.

In 1876 (Pouchet had died in 1872), a new exponent of the cause of heterogenesis came forward, the English pathologist, H. Charlton Bastian (1837-1915). In the process of demolishing Bastian's case, the heat resistance of bacterial spores became understood and accepted as the explanation of the anomalous evidence that both Pouchet and Bastian had produced. An important consequence of the debate was a growing awareness of the need for heat treatment, or "pasteurization", as it came to be called, in the food and dairy industries. Equally important, the skirmishes over spontaneous generation led to the science of bacteriology and the germ theory of disease.

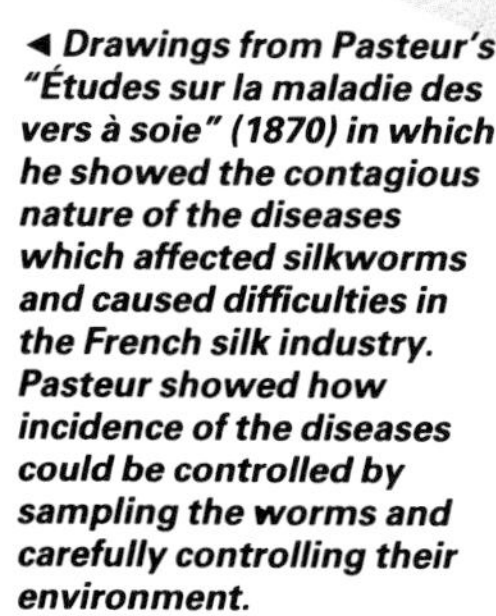

◄ *Drawings from Pasteur's "Études sur la maladie des vers à soie" (1870) in which he showed the contagious nature of the diseases which affected silkworms and caused difficulties in the French silk industry. Pasteur showed how incidence of the diseases could be controlled by sampling the worms and carefully controlling their environment.*

▼ *Already before the establishment of the germ theory, the idea that diseases were spread by a miasma arising from human dung- and waste-heaps encouraged the introduction of the water closet and piped sewerage systems. In London sewers were constructed beneath roads by "cut and cover" and the sewage piped to Barking in Essex.*

The germ theory

Pasteur's work on fermentation and putrefaction demonstrated the role of airborne microorganisms in these chemical processes. The predominant theory of disease in the mid-19th century was the "miasmatic" theory which associated disease, and epidemics like cholera and fever, with poisonous fumes given off from dung heaps and decaying matter. These fumes, or miasma, were blown by prevailing winds from one area to another.

However, once Pasteur had shown that decay was due to microorganisms, and not the often obnoxious effluvia they produced as a byproduct of their actions, it became possible to picture them as organisms which "germinated" in animal and human hosts and to regard diseases as an interference with the normal chemistry of the body by a parasitic invasion. This suggested defensive strategies. For example, in 1865, believing from Pasteur's work that the reason why surgical wounds often turned septic was because of dangerous organisms in the hospital air, the surgeon Joseph Lister (1827-1912) introduced carbolic acid as an antiseptic agent. In 1874 Pasteur declared that if he had been a surgeon he would "never introduce an instrument in the human body without having passed it through boiling water".

Despite Lister's advocacy of antisepsis – that is, the use of agents (antiseptics) which prevent the growth of microorganisms – Pasteur's

◄▲ ***Best known for curing hydrophobia (rabies) using a rabbit culture, Pasteur also investigated anthrax, a disease fatal to both cattle and people, caused by Bacillus anthracis (left).***

► ***Madame Pasteur taking dictation from her husband in the garden of their home at Pont-Gisquet following Pasteur's stroke in 1868. Madame Pasteur died in 1910.***

◄ ***The realization that epidemic diseases such as cholera and typhoid fever were spread by water contaminated with microbes, rather than by noxious fumes, led to detailed microscopic analysis of water supplies and made possible effective preventive action. This 19th-century drawing shows microorganisms living in drinking water.***

idea of "asepsis" (the removal of microorganisms altogether), took a surprisingly long time to introduce into hospital practice (➧ page 186).

One reason why aseptic routines were resisted was that it was extremely difficult to obtain direct evidence that microorganisms were present in large numbers in the atmosphere and, hence, to establish the germ theory over and above sensible sanitary and ventilation practice. The germ theorists overcame this difficulty both by arguing that germs were too small to be seen in contemporary microscopes and (quite correctly) by turning their attention to the presence of microorganisms on solid surfaces, in water and in other potentially dangerous fluids. Only in 1878, when Lister identified and cultured the bacteria which soured milk, was the single-celled nature of microorganisms realized.

Pasteur's contribution to the establishment of the germ theory came about through his practical work on the diseases of wine and beer. In 1862 he investigated the manufacture of vinegar (acetic acid) from wine. He showed that the scum on a vinegar vat consisted of rod-like microorganisms which promoted the oxidation of the alcohol in the wine. The presence of these organisms helped explain the souring of wine and led to Pasteur being consulted about other malfunctions of wine-making and brewing. In all cases, he found that the yeasts employed to cause fermentation had become contaminated with foreign organisms. This was why, in 1865, Pasteur recommended the heat treatment of wines and beers to 55°C before leaving them to stand. (Pasteur later discovered that "pasteurization" had been recommended for wines as early as 1795 by Nicolas Appert, 1749-1841, the inventor in 1804 of the method of preserving food in bottles and tin cans.)

From 1863 to 1867, while still at the École Normale, Pasteur also acted as professor of geology, physics and chemistry at the École des Beaux-Arts in Paris; in 1867 he was appointed professor of chemistry at the Sorbonne. During this period the French silk industry was almost destroyed by a disease of silkworms which made their eggs sterile or the larvae unable to feed. Between 1850 and 1866 French silk producers had lost nearly 20 million francs. In 1865 the chemist J.B. Dumas (also a former minister of agriculture) prevailed on his former student to help. On visiting Alès, the center of the silk industry in the south of France, Pasteur found that the bodies of the silkmoth larvae were infected by bright oval bodies.

The Art of Pure Culture

The importance of the microscope

*Technical difficulties inherent in the lens systems of microscopes and in the preparation of tissues for study meant that for much of the 19th century it was possible to maintain conflicting theories concerning the nature of, and relationships between, cells, bacteria and disease. Until Joseph Jackson Lister (the wine merchant father of the surgeon) developed the achromatic microscope in 1830, microscope images were blurred by multiple-colored images because light was not being brought to a common focus. Once this technical frontier had been crossed, progress was rapid. The German botanist Matthias Schleiden (1804-1881) and Theodor Schwann (**▸ page 173**) quickly developed the idea that the principal organizing unit of vegetables and animals was the cell.*

Koch's postulates

As early as 1840, the German pathologist Jakob Henle (1809-1885) had speculated from the available evidence concerning animal diseases caused by fungal infections (for example, the silkworm disease, muscardine) that different diseases were caused by specific living parasites which "germinated" in their hosts (the "germ theory" of disease). Henle suggested that this hypothesis would be proved if: 1) specific organisms were found constantly (not merely occasionally) in pathological samples from the host exhibiting the disease; 2) these organisms could be physically isolated and shown to live independently of their hosts. Implicit in Henle's argument were two further criteria: 3) the isolated organisms would have to be tested on healthy host tissue; 4) inoculation into healthy tissue would have to reproduce the original disease symptoms exactly.

These four criteria are the fundamental tenets of bacteriological pathology. They are usually known as "Koch's postulates", after Robert Koch, who had attended Henle's lectures as a student, first used them in his research on tuberculosis in 1881. Henle had not been in a position to use them, for they depend upon the bacteriologist's ability to prepare pure cultures – a technique pioneered by Koch and, to a lesser extent, Pasteur.

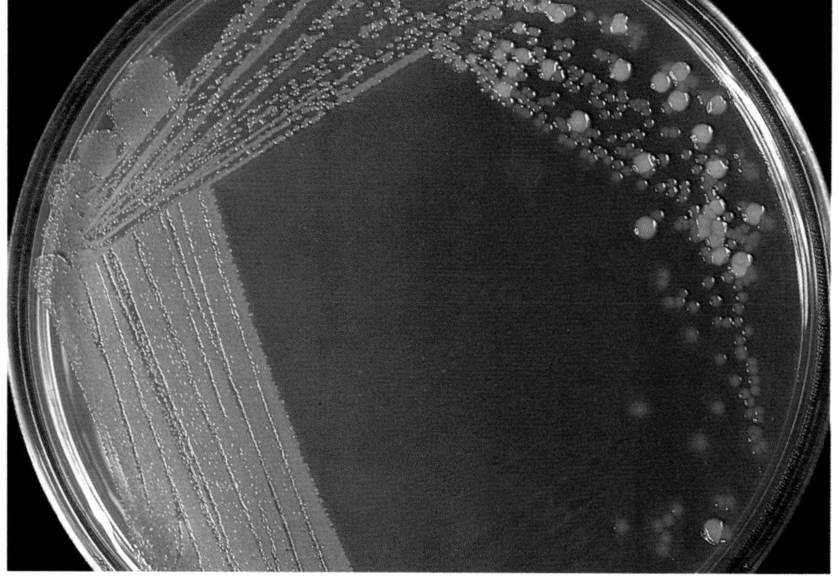

▲ Perfection of culturing techniques in the 1880s led to study of normally harmless bacteria which live in the human gut. Using a solid gelatine medium in a petri dish mixed colonies can be cultivated.

◄ Robert Koch (1843-1910), who was Pasteur's great German rival, established the techniques of bacteriology and its institutionalization as a public health discipline. He discovered the cause of tuberculosis, but a cure which he announced in 1890 proved unsuccessful, though useful in diagnosis.

▼ Growth of the German dyestuffs industry proved invaluable for microscopic detection of bacteria. This early example of staining shows Salmonella.

◄► Microscopy was popular among amateurs and professional scientists in the 19th century. Instruments made by the firm of Powell & Lealand (left) had objectives with a resolving power of 0.5 microns – about half-way to the limit achievable by the optical microscope. In the modern scanning electron microscope (right) data are analyzed by computer.

Growth mediums and culturing techniques
Pasteur's technique for cultivating pure, homogeneous strains of bacteria was to use a liquid growth medium, such as sugar solution laced with yeast ash. Once growth had begun, Pasteur used a minute amount to inoculate a fresh medium, and so on, until he was personally satisfied by microscopic analysis that he had a pure culture. About 1880 the commercial meat extract (a beef broth) developed by Justus von Liebig was found to be an ideal nutrient medium, but by then it had become clear that despite Pasteur's impressive achievements using a liquid medium, pure cultures were not guaranteed and that misleading results were being produced. Another of Pasteur's techniques, breeding a pure culture through 20 or more generations of mice, or other experimental animal, had the disadvantage of slowness.

On analyzing in 1880 the ideal conditions for culturing bacteria, Koch saw that the medium ought to be sterile or sterilizable, transparent in order to show the area of growth more clearly, and above all solid, so that mixing of bacterial species could not occur. Koch's first solution, in 1881, was the humble potato, followed by a meat extract in gelatine which could be made to set on sterile glass plates, or as a drop on a microscope slide. These were inoculated in a "noughts and crosses" pattern and cultured at a uniform temperature under a bell jar. If more than one species of bacteria developed they appeared as distinct colonies before merging into one another. By sampling and re-inoculating a clean plate before merging occurred, Koch obtained a pure culture of an organism.

This culturing technique was demonstrated by Koch at an international medical congress in London in 1881 and rapidly became a routine procedure. The following year, the wife of one of his assistants suggested that agar, an extract of seaweed should be used instead of gelatine (which many bacteria eat). The familiar petri dish, with its compact lid, was devised by an assistant, Richard Petri, in 1887. This technology of culture set the germ theory on a firm foundation.

After nearly 18 months study, and a great deal of confusion, Pasteur realized that the silkmoth larvae were infested with parasites picked up in the debris of the nurseries. As Pasteur also slowly realized, he was dealing with two distinct diseases, both of which depended upon the temperature, humidity and ventilation of the silkworm nurseries for their successful invasion. By adopting Pasteur's recommendation that the worms, moths and their eggs should be continuously sampled for signs of disease by microscopic examination, the French silkworm industry was slowly restored to profitability.

Pasteur's study of silkworm diseases came at a difficult period in his life, coinciding with the death of his father and a two-year-old daughter. The strain showed in 1868 when Pasteur suffered a stroke which left him paralyzed on his left side for the remainder of his life. From then on he had to rely upon assistants to perform his experiments for him, though his ability to use the microscope, for which his extreme shortsightedness appears to have been an advantage, remained unimpaired. In the same year, Pasteur was awarded an honorary degree by the University of Bonn; but he returned it in anger in 1871 as a result of the 1870-1871 Franco-Prussian War.

Pasteur's intense nationalism was also responsible for his interest in brewing. Why were German beers superior to French ones? Could he make French beers as good, so that France would never want to import beer again from her enemy? Again, as had been the case with wine, investigation revealed that beers were commonly spoiled through the presence of unwanted microorganisms. These studies led J.C. Jacobsen, founder of the Carlsberg brewery in Copenhagen, to establish a laboratory for the study of fermentation, where much important research on the biochemistry of proteins and the physical conditions of chemical reactions was carried out after 1876.

"When we see beer and wine undergoing profound changes because these liquids have furnished a refuge for microscopic organisms", wrote Pasteur in 1876, "how can we avoid the thought that phenomena of the same order can and must be found sometimes in the case of men and animals." With this pronouncement, at the age of 55 Pasteur began the study of infectious diseases, starting with cattle anthrax. By 1850 it had been shown that anthrax was probably spread by contagion. In 1863, the French physician Casimir Davaine (1812-1882) found microscopic, thread-like bodies in the blood of diseased cattle, a discovery that was confirmed by Robert Koch in Germany in 1876. Koch showed that the organism formed spores and that anthrax could be induced in healthy animals by inoculating them with anthrax blood. This "proof" that the thread-like bodies were the cause of anthrax was open to the objection that other possible causal factors in the diseased blood had not been eliminated. Pasteur's contribution was the demonstration in 1876 that by nurturing the bacillus in a liquid medium, inoculating a drop into a fresh medium, and repeating the process a hundred times, still produced a lethal disease in a healthy animal. In another, dramatic experiment, he showed that hens, which were normally immune to cattle anthrax, could catch the disease if their body temperature was lowered, enabling the bacillus to live.

Within the next decade, bacteriological isolation and confirmatory inoculation techniques became successfully applied to the study of tuberculosis, cholera, diphtheria and tetanus. In the lead in this field was Robert Koch, who introduced a more reliable method of culturing bacteria on a solid gelatine medium in petri dishes. Pasteur and the French school turned instead to the question of therapy.

"Fringe" Medicine

Fringe medicine

In the 19th century, funds for medical research institutes, such as the Pasteur Institute (1888), the Robert Koch Institute (1889 – the dates reflect the Franco-German rivalry of the times), the Lister Institute (1891) in Britain and the American Rockefeller Institute (1901), were all the result of private philanthropy or private subscription. These institutes dated from the period following the establishment of the laboratory-based germ theory with its promises of cures for disease by vaccines and drugs. In the absence of a state medical insurance system, the poor either operated their own small medical insurance societies or they relied upon a variety of cheaper "alternative" medical systems which would now be regarded by many doctors as unorthodox or part of "fringe" medicine.

Today, the fundamental difference between so-called "orthodox" and "unorthodox" medicine is that while the former relies mainly on restoring health through synthetic chemical drugs and through surgery, the latter (while it may use natural remedies, such as plant drugs) claims to awaken the patient's own healing powers – what ancient Greek doctors (the Hippocratics) called "vis medicatrix naturae", the healing power of nature. Most Western medicine is based upon "allopathy", another ancient Greek idea that disease is a disequilibrium and that something (an opposite of the symptoms) must either be given, or taken away, to restore the balance of health. For example, the humoral theory which was accepted right up until the late 18th century (♦ page 64) provided the rationale for the bleeding, leeching and purging which were the sole practical signs of the physician's authority. The germ theory (♦ page 178) rationalized allopathy by claiming that germs caused a patient's unease and that health could be restored by the physician engaging in chemical warfare against specific "enemy agents".

Samuel Hahnemann and homeopathy

Before this, however, the German physician Samuel Hahnemann (1755-1843) had argued in 1810 that "like cures like". He supposed that disease was really the body's own attempt to throw off poisonous matter which was undermining the body's natural vitality, rather than the specific effect of the poisonous matter itself. Hence any drug which replicated or produced similar symptoms to the disease could be used in minute amounts to heal by reinforcing the body's natural defenses. Hahnemann called his system homeopathy (Greek for "like therapy"). (He also invented the term allopathy.) Practical tests on healthy patients seemed to indicate that the process worked if infinitesimal amounts of simple drugs were used. For example, since poisoning by belladona (deadly nightshade berries) produces a flush like scarlet fever, belladona was used against this illness.

Allopathic medicine recognizes no mechanism by which homeopathy can succeed and explains its success as mumbo jumbo disguising a placebo effect. Hahnemann's system was strenuously opposed by pharmacists, since it involved no mixing of drugs and was less profitable for them.

▲ **Little Girl: "Please, sir, I want the hundred-thousandth part of a grain of magnesia." Young Chemist: "Very sorry, miss, but we don't sell in such large quantities." Although the subject of jokes, as in this "Punch" cartoon, homeopaths succeeded in gaining recognition, as in Britain in 1858. Interest in homeopathy remains strong, but there are fewer practitioners today than in the 1860s. Developed by the German doctor, Samuel Hahnemann, homeopathy was probably inspired by the rhetoric of 18th-century physicians who extolled the medicinal merits of drinking sea water or the mineral-rich water of inland spa towns.**

Herbalism

For example, herbalism had always been a traditional folk medicine. This medical botany received a considerable boost in 1840 when the gravely-named American Albert Isaiah Coffin (c.1800-1866), brought his system to Britain. In tune with industrialized society, Coffin likened the body to a steam engine. Disease was caused by a want of heat in the boiler. Hence he advocated herbs such as lobelia and cayenne pepper, which have a burning, irritating effect on the tongue and nostrils. Coffinism was a tremendous popular success in the north of Britain despite several fatalities from lobelia poisoning, which allopathic practitioners were quick to exploit critically. Partly because Coffinism, like other forms of herbalism which still survive, was not recognized by the 1858 Act, its advocates became extremely antiestablishment. They were prominent in agitation against the compulsory vaccination of infants against smallpox in the 1870s, which they saw as an invasion of personal freedom as well as a dangerous procedure. There was indeed some truth in the latter fear, given the use of unsterilized needles and the possibility of cross-infection.

Proprietary cures and "natural methods"

If allopathy ultimately triumphed through state registration of doctors, popular medicine remained little affected. The sales of proprietary medicines, many of which relied on a placebo effect, increased after 1860. Many Victorian entrepreneurs, like Thomas Holloway and Thomas Beecham, made their fortunes from "cure everything" pills whose innocuous ingredients included lard, wax, turpentine, aloes, soap and ginger! Although allopathic doctors were scathing of homeopaths, herbalists and naturopaths, ironically they were prepared to exploit "natural" methods where they were powerless to offer cures and where richer patients still commanded power over the doctor. Tuberculosis was the greatest killer in the 19th century, and even though the cause was identified by Koch in 1881, no vaccine was found.

In this situation, 19th- and early 20th-century allopathic doctors were fully prepared to recommend naturopathic environmental or climatic remedies such as sending consumptive patients to the warmer climate of southern France or Egypt and, from the 1870s, to the clearer atmosphere of clinics amid the alpine peaks of Switzerland. Doctors remained fairly helpless until the discovery of streptomycin in 1940 by American biochemist Selman Waksman (1888-1973), though improved hygiene had by then greatly reduced the rate of mortalities due to tuberculosis.

◄ **Quack medicine was particularly prevalent in 19th-century America, where there was a "cure for everything". Because the Shakers (a Quaker sect) were renowned for honesty, advertizers often sold their wares as "Shaker remedies". Many American patent medicines were outlawed by a Food and Drug Act in 1907.**

► **A popular form of 18th- and 19th-century medicine was drinking mineral waters at inland resorts (right, one of the spectacular Pump Rooms at Cheltenham, England). Such resorts also developed "balneotherapy" in which patients were immersed in baths for long periods.**

"Chance favors the prepared mind," said Pasteur, dismissing the idea that his attenuated cultures were an accidental discovery

The idea of vaccine therapy

In many societies the incidence of smallpox had been reduced by artificially inducing a mild attack through inoculation of the matter from a pustule on a sick person. This often (though not always) reduced the likelihood of catching the disease during an epidemic. At the end of the 18th century, the English physician Edward Jenner (1749-1823) had introduced the practice of vaccination, the introduction of matter from cowpox pustules, it having been noticed that dairymaids who caught cowpox seemed less likely to catch smallpox. Once Pasteur had identified the bacterial nature of several animal diseases, it was natural that he should wonder whether, like smallpox, they could be weakened (attenuated) sufficiently to stimulate some unknown body defense mechanism which would be able to destroy the more virulent form of the microorganism.

At the end of 1878 Pasteur chose to experiment with the microorganism which others had found to cause fowl cholera, a disease in which the bird weakens, sleeps and dies. On cultivating the bacillus, he found that whereas it had little effect on guinea pigs, it was lethal to rabbits. During the summer of 1879 research was abandoned while Pasteur and his staff took their annual holidays. On return, Pasteur found – not unexpectedly – that most of the cultures of fowl cholera had died off. Those that had not, he transferred to fresh culture infusions and tested for potency by inoculation into healthy fowls. To Pasteur's surprise the chickens showed no disease symptoms. Presumably the old cultures were useless? Nevertheless, to overlook no other possibilities, Pasteur injected the same fowls with fresh cholera. To everyone's surprise and delight they survived, while fowls which had not received what was now clearly an attenuated culture died.

It took Pasteur but a few months to discover that the secret of successful attenuation lay in exposing cultures to air for a long period at an even temperature of about 37°C. (If air was excluded, the culture retained its lethal power.) Pasteur again dismissed the notion that this discovery, which was to transform the lives of human beings as much as animals, had occurred by chance. He had been seeking an example of attenuation; his mind had, therefore, been prepared for any sign. Further proof of the validity of the technique came with the preparation of attenuated anthrax in 1881. (These attenuated forms, following the smallpox vaccination model, became known as vaccines.) The

▲ The statue outside the Pasteur Institute commemorates the Alsatian shepherd boy, Joseph Meister, who was savagely mauled by a rabid dog then treated by Pasteur in July 1885. By 1882, Pasteur had shown that rabies was an infection of the nervous system and that its virulence could be weakened by passing it through the spinal chords of some 25 rabbits. This attenuated virus conferred immunity upon healthy dogs and offered a way of vaccination.

► Following Pasteur's success in saving the life of Joseph Meister, rabies victims flocked to Paris, where Pasteur opened a clinic in the École Normale. Here Pasteur summons patients for vaccination.

▼ Worldwide donations from people and governments grateful for rabies treatment led to the establishment of the Pasteur Institute in 1888 as a rabies clinic (patients of different nationalities wait outside, below) and microbiology laboratory.

preparation of anthrax vaccine was complicated by the need to prevent it turning into spores by attenuating it at 42-44°C. The results were publicly demonstrated before the Agricultural Society at Pouilly-le-Fort in 1881. From then on, Pasteur's laboratory began to "manufacture" vaccines for sale to farmers and veterinary surgeons.

It was while this work was being completed that Pasteur turned his attention to rabies, a disease which can be passed from dogs and other animals to man. This, together with childhood memories of rabies victims, seems to be the reason why Pasteur chose a statistically unimportant human disease for investigation. All attempts to find the bacterial cause of rabies proved abortive. On the other hand, since it was clearly a disease of the central nervous system, Pasteur found that the brain tissues of dogs and rabbits provided an ideal culture medium for the disease. By leaving inoculated brains in dry air for a fortnight, attenuation occurred. The vaccine proved harmless to healthy dogs and rabbits and conferred immunity.

In July 1885, Joseph Meister, a boy of nine, arrived in Paris from Alsace after having been savagely mauled by a rabid dog. That he was brought to Pasteur is an indication of how widespread knowledge of Pasteur's work on animal rabies was at this time. All the indications were that the child had a month to live before the dreadful and fatal symptoms of hydrophobia would develop. With nothing to lose by the boy's death, Pasteur began a course of vaccine therapy. It was soon clear the symptoms of rabies were not going to appear.

► An electron micrograph of rabies virus (red) budding off cells of the infected host (green).

Scientific Medicine

Reform of hospital design

In the influential view of the English nurse and hospital reformer Florence Nightingale (1820-1910), huge multistory hospitals were insanitary places which actually bred disease in the air because of overcrowding, poor ventilation and bad drains. They were "gateways to death". Nightingale was a convinced believer in the miasmatic theory of disease, and her nursing experiences during the Crimean War in 1854 encouraged her to take a sanitarian view of hospital design. Wards had to be light and airy and built upon a pavillion principle, not more than two stories high so that outside air could circulate easily between the building blocks. Tall, narrow windows reaching from floor to ceiling encouraged the cross-circulation of air. Furnishings were to be kept to a minimum to avoid trapping miasma and to free nurses, who were to be scrupulously clean, from domestic duties for the fulltime care of patients.

Antisepsis, asepsis and anesthetics

A similar campaign for cleanliness in hospitals came from the "contagionist" side. In Vienna, the Hungarian obstetrician Ignaz Semmelweis (1818-1865) showed that puerperal fever, which led to the death of mothers, was due to surgeons cross-infecting patients, usually after postmortem examinations. In 1865 the Scottish surgeon Joseph Lister (1827-1912), learning of Pasteur's work, began to dress surgical wounds with cotton wool and bandages sprayed with carbolic acid to prevent microorganisms from entering the wound. He also sprayed the entire operating theatre.

Although this method of antisepsis enjoyed considerable success it was of limited value until the Nightingale sanitary principle of cleanliness (asepsis) was fully accepted. Indeed, like other surgeons, Lister conducted his antiseptic operations wearing a blood-soaked apron over his outdoor clothes. The first British surgeon to emphasize the importance of cleanliness and who also understood the significance of Pasteur's germ theory, was Thomas Spencer Wells, who used cold boiled water to wash his hands and instruments.

▲ *Florence Nightingale in the "hospital" at Therapia, Crimea.*

Once it was realized that germs on the skin and instruments were far more dangerous than those in the air, steam sterilization of instruments (introduced in 1886 by the Berlin surgeon Ernst von Bergmann, 1836-1907), rubber gloves and face masks began to replace the carbolic spray. Experience during World War I showed that antiseptics actually prevented healing and the practice of asepsis became universal.

The advent of aseptic, well-ventilated hospitals gave patients a greater chance to recuperate from operations. The introduction of anesthesia after 1846 not only freed patients from the pain of surgery, but allowed surgeons the time to do operations that were much more complicated and constructive than the lightning amputations to which they had been mainly restricted before. On the other hand, apart from vaccines developed to fight smallpox, rabies and diphtheria, doctors had no way of fighting infection until the 1930s.

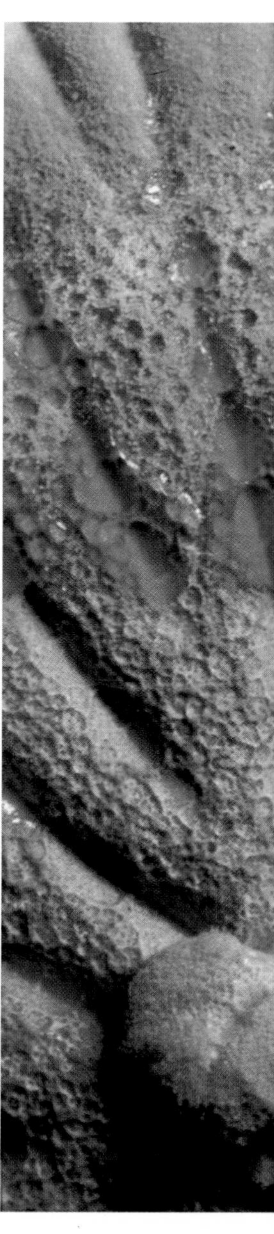

◄► *In the 19th-century antiseptic operation (left) the chloroformed patient is sprayed with carbolic acid from a steam aspirator devised by Lord Lister and the surgeons wear their everyday clothes. However, by 1914 in the operating theater in King's College, London (right), aseptic conditions prevail. Lister got the idea of using carbolic acid (phenol) as a disinfectant when he read of its success as a deodorant of raw sewage.*

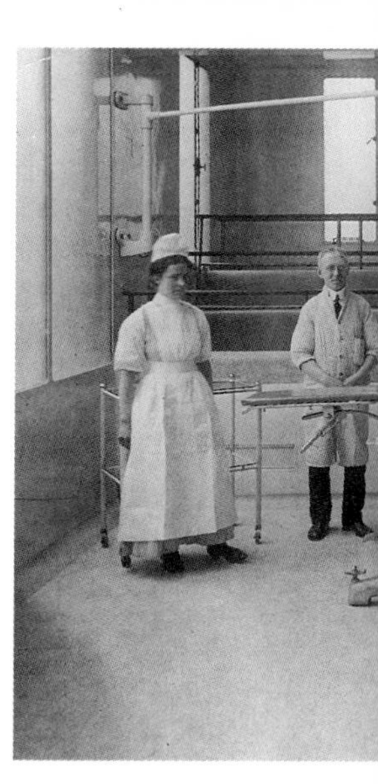

◄▲ *In the last quarter of the 19th century, German university and industrial chemists searched for pharmacologically useful chemicals. In 1886 Ludwig Knorr prepared 1-phenyl-2 : 3-dimethyl-5-pyrazolone, which had antipyretic (fever-reducing) properties. This was marketed by the Hoechst Company as "antipyrin" (above) rivaling the Bayer Company's "aspirin". Successful chemotherapy came only in the late 1930s. By culturing the mold Penicillium notatum (left) Florey and Chain produced the antibiotic "penicillin".*

The development of drugs

One consequence of the increased sophistication of organic chemistry in the 1880s was the chemist's ability to synthesize new substances of use in industry, such as dyestuffs and in pharmacology, as with the creation of "aspirin". Interest in the mechanism by which a dye is permanently "fixed" onto a colorless fabric led the German bacteriologist Paul Ehrlich (1854-1915) to speculate on the possibility of fixing a poisonous chemical onto target germs which had invaded the blood stream. This idea of a "magic bullet", or chemotherapy (as it is known), had its first success between 1909 and 1911 when Ehrlich tested an organic arsenic compound on patients suffering from syphilis. Used successfully to fight the appalling increase in venereal disease during World War I, this drug also freed from workhouses and lunatic asylums many inmates who suffered from the general paralysis which eventually overcame those suffering from tertiary syphilis.

Unfortunately, Ehrlich's "bullet" remained unique for many years while bacteriologists continued to try to fight disease with vaccine therapy. However, in 1935 the German chemist Gerhard Domagk (1895-1964) found that a derivative of the yellow dye chrysoidin cured mice infected with bacteria. Marketed rapidly as "prontosil R", it was soon discovered that the active chemical component was a sulfanilamide, from which a complete range of other bactericides (antibiotics) could be developed. It was in this context that two Britons, the Australian-born pathologist Howard Florey (1898-1968) and German-born biochemist Ernst Chain (1906-1979) developed an industrial process for the manufacture of penicillin – a mold whose anti-bacterial properties had in 1928 been noted, but not exploited, by Alexander Fleming (1881-1955). Although Fleming, Florey and Chain shared the Nobel Prize for Medicine in 1945, ironically it was Fleming who received all the popular acclaim.

Pasteur's laboratory was besieged by requests for rabies vaccine and hundreds of victims were brought to him from all over Europe

Pasteur's genius

The rabies treatment, the swansong of his career, was Pasteur's greatest triumph. Gifts of money enabled him in 1888 to erect his own private research laboratory, the Pasteur Institute, where he remained director until his death. Pasteur worked daily at his bench until September 1895, when he suffered a second stroke; he died on 22 September at Villeneuve-l'Étang, near Paris. He was given a state funeral and buried in a magnificent tomb within the Pasteur Institute.

Pasteur's genius lay less in great originality of mind than in a combination of virtuoso experimental skills, bulldog purposiveness, obstinacy, authoritarianism, extraordinary literary and polemical skills and a sense of theater which kept his name in the public eye. Enormously ambitious, both for himself and for French science, he delighted in solving practical problems and was astute enough to choose research areas which had a close bearing on industrial and agricultural processes and, ultimately, on human health and happiness. Logically, his work on optical isomerism should have led him to explore molecular theories of structure; instead he moved into fermentation research because of its industrial implications and financial rewards. The state rewarded him well with honors, finance and laboratory facilities that became the envy of British, and even the well supported German, scientists. Despite his overbearing manner and now unfashionable patriotism, Pasteur inspired an intense loyalty in his pupils and collaborators who saw him as a French national hero and "benefactor of humanity". This, quite rightly, has ensured that Pasteur's name has been immortalized in medical research centers all round the world and in large numbers of biographies, films and popular images of the scientist.

▲ Pasteur was buried in an elaborately-designed tomb within the Pasteur Institute. Among the mosaic tile decorations are two commemorating his research on rabies (a shepherd muzzling a dog) and (above) anthrax (a contented flock of sheep).

▼ Pasteur's death in September 1895 was followed by a host of heroic tributes throughout the world's press. They include this allegorical portrait from the "Supplément Illustré" in which the angel's ribbon carries the dedication "from a grateful universe".

Marie Curie 1867-1934

Growing up in Poland...A student in Paris...Partnership with Pierre Curie...Röntgen and "uranic rays"...The Curies discover polonium...and radium...Radioactivity investigated...Two Nobel prizes...X-radiography and radium treatment...PERSPECTIVE...Difficulties facing women scientists...Science in Eastern Europe...Alfred Nobel and his prizes...Radiation and medicine

"Often and often then, her weary spirit
Returns beneath the roofs to the corner
Where silent labor dwelled
And where a world of memory has rested."

So wrote Marie Curie (née Marya Sklodowska), recalling to mind the lonely, "obscure and happy" student years in the 1890s, spent poring over books in a Paris garret. Perseverance was the trademark of her skill as a scientist. After Einstein, Curie was the most famous scientist of her time, but she always retained much of the simplicity and self-discipline of the impoverished student, "like a child set before natural phenomena, which impress him like a fairy tale".

Marya Sklodowska in Poland

The youngest of five children, Marya Sklodowska was born in 1867 in the house where her mother had her school. Marya's father Wladimir Sklodowski had a good teaching post in a secondary school, but it is probable that his career was frustrated by the Russian authorities' determination to ensure that top posts were in the hands of Russians, or at least of Poles loyal to their regime. Although Wladimir had studied in St Petersburg, and was a teacher of science rather than literature, he must have been somewhat suspect. No doubt this only fuelled the patriotism of his family; the Sklodowskis shared to the full the longing for national redemption then so strong in the Poland which the Tsars had renamed "Vistulaland", and where all advanced education had to be in Russian.

When Marya was eight, her eldest sister died; her mother died of tuberculosis two years after. These tragedies overcast Marya's childhood more than the economic difficulties of her family. By the time she left school with many honors, at the age of 15, she had lost the Catholic faith in which she had been brought up. Formal higher education was not then open to women in Poland. Instead she was drawn into self-education circles. She was one of many. Everywhere in Europe, enthusiasm for new science and technology was booming. Physics and chemistry were striding ahead; electricity seemed about to transform industry and communications; radical ideas were fermenting, perhaps even more in central and eastern Europe than in the West, where modern science and industry had first been established. Education, especially in scientific subjects, was the key to this exciting progress. Marya joined a "flying University", a group of friends who tried to teach each other science, technology and modern political theories – this last subject always dangerous under Tsarist rule, and hence the others sometimes also, by association.

At the age of 18 she gave up this relatively carefree life to work as a governess, eventually hoping to save enough to study abroad, probably in Paris, where her elder sister Bronya was a medical student.

▲ Marie Curie at the height of her fame, 1910.

Physics and chemistry transformed

By the last years of the 19th century it was possible to suppose that physics and chemistry were in essentials complete. Scientists discovered several new elements which filled gaps in the periodic table of elements devised by Dmitri Mendeleev. In this period, too, physicists including William Thomson (Lord Kelvin), Hermann von Helmholtz and James Clerk Maxwell had worked out the laws of thermodynamics, electricity and the manifold exchanges of energy.

The discovery of X-rays in 1895 by Wilhelm Röntgen and of "uranic rays" by Henri Becquerel in 1896, transformed the situation. It was Marie Curie who took up Becquerel's work, and by her discovery of the elements radium and polonium in 1898 revealed that the immense strength of radioactivity within the atom exceeded that of any known chemical or mechanical forces. Research soon made it clear that atoms had an internal structure, described in the models of Rutherford and Niels Bohr. Investigation of the atom's central nucleus, the source of radioactive discharge, led within a few years to the discovery that the nucleus could be split – and so to nuclear weapons and nuclear power.

Curie also became involved with X-radiography, and subsequently with the medical potential of radium in diagnosis and cure; the work of her Radium Institute at Paris was seminal in research on the nucleus and in radiotherapy.

Marie Curie – Her Life, Work and Times

From governess to research scientist

Marya's life as a governess was an exacting one. A governess might well feel herself the intellectual superior and social equal of employers who veered between treating her as one of the family, and as a servant. When the son of one employer fell in love with her and sought to marry her, his family promptly reminded him that a wealthy young man does not marry a governess. Marya's correspondence with her friends and siblings during these four years after she left school convey an attractive picture of her adolescence – sensitive, idealistic, imaginative and studious. But four years as a governess were bound to reinforce a sense of frustration. As she wrote to her brother later:

"Life is not easy for any of us. But what of that? We must have perseverance and above all confidence in ourselves. We must believe that we are gifted for something, and that this thing, at whatever cost, must be attained."

On her return to Warsaw she was able to add some experimental training to her reading of physics and mathematics at a modest Museum of Industry and Agriculture, enough in fact to gain entry to the University of Paris in 1891. In Paris she stayed at first with her sister, now married to a Polish doctor who was a political exile. With them she could relax, but although she enjoyed the cosy atmosphere of the exile intellectual community, she realized that starting her university course at 24 she would need to give herself up wholeheartedly to her studies in order to catch up. Among many students who lived in chilly attics on a pittance, she was cut off by her sex and her nationality from the comradeship that she most enjoyed. Free from all distraction, she was first in her year, obtaining her degree in physics in 1893, and second, earning her mathematics degree in 1894. Following these achievements, she received a small grant for a research project on magnetism. Once it was complete she intended to go back to Poland, to teach science at a school and look after her father.

Marie Sklodowska, as she now called herself, was advised that she would get help from a bright young man then teaching at the Paris Municipal School for Industrial Physics and Chemistry. Pierre Curie (1859-1906) already had quite a reputation, partly for his work on magnetism, partly for the discovery and exploitation (with his brother Jacques) of piezoelectricity, the stress effect produced in certain crystals by electricity, and conversely the electrical activity produced by such pressure stresses. His school was neither prestigious nor did it pay well; it was one of many set up after the Franco-Prussian War of 1870-1871 to help France regain the ground her industry had lost to Germany, supposedly through the lack of scientific training which everybody agreed had made German industry so powerful and efficient. But the school was not a university.

Pierre Curie was not really interested in "industrial science". For him the intellectual joy of the chase was all that mattered; research should be undertaken for the love of knowledge, and that absorbed him fully. Such a man was particularly attractive to Marie, and despite a little hesitation on her part, they were wed the next year, 1895. He sought to persuade her that an individual could do little to achieve the humanitarian and national objectives she had at heart – she was not going to overthrow the Tsarist regime. But she could contribute as an individual to science. That ideal they shared. They were certainly very much in love. This close relationship expressed itself also in an equal partnership in the laboratory. Their sole relaxation was cycle rides into the remoter corners of the French countryside.

Historical Background

1863 January Revolution in Russian Poland crushed; systematic Russianization of "Vistula Land" began

1889 Universal Exhibition, marked centenary of F Revolution and importar scientific ind

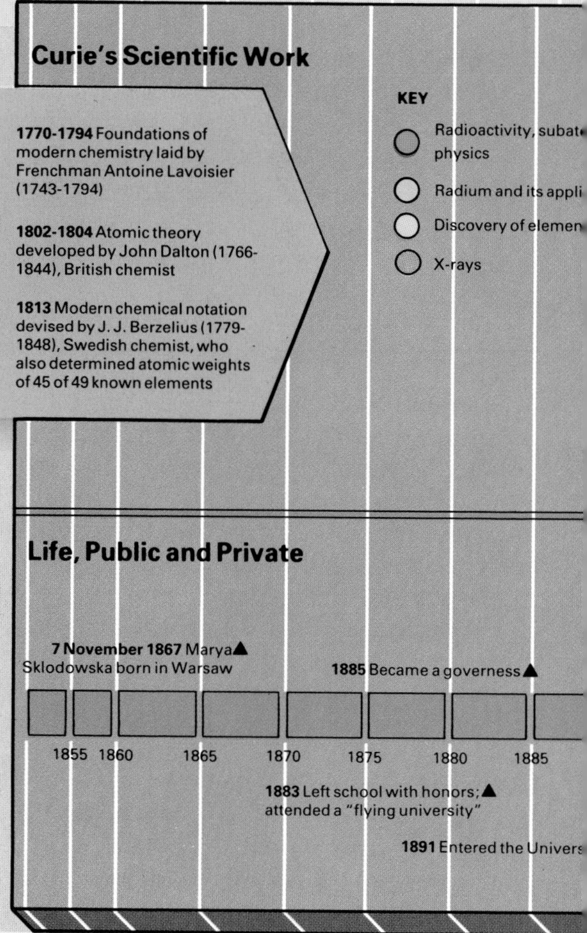

Curie's Scientific Work
KEY
Radioactivity, subat physics
Radium and its appli
Discovery of elemen
X-rays
1770-1794 Foundations of modern chemistry laid by Frenchman Antoine Lavoisier (1743-1794)
1802-1804 Atomic theory developed by John Dalton (1766-1844), British chemist
1813 Modern chemical notation devised by J. J. Berzelius (1779-1848), Swedish chemist, who also determined atomic weights of 45 of 49 known elements
Life, Public and Private
7 November 1867 Marya Sklodowska born in Warsaw
1885 Became a governess
1855 1860 1865 1870 1875 1880 1885
1883 Left school with honors; attended a "flying university"
1891 Entered the Univers

Scientific Background

1880 Pierre Curie (1859-1906) and brother Jacques discovered piezoelectricity, produced by pressure; invented piezoelectric quartz, or electrometer

1888 Pasteur Institute inaugurated with Louis Pasteur (1822-1895) as director

1876 William Crookes (1832-1919), British physicist, invented his evacuated glass "Crookes tube" which produced cathode rays; used later to produce X-rays

1858 Julius Plücker (1801-1868), German physicist and mathematician, first studied cathode rays

1869 Periodic table of elements published by Russian chemist Dmitri Mendeleev (1834-1907)

1854-1874 The elements cesium and rubidium discovered by spectrum analysis by German scientists R. W. Bunsen (1811-1899) and G. R. Kirchhoff (1824-1887)

1850 1855 1860 1865 1870 1875 1880 1885

-1906 France polarized by
yfus affair

1898 The Spanish-American War and British expedition to Sudan both saw X-ray units in service

1905 Revolution in Russia

1914-1918 World War I

1918 Independence of Poland proclaimed

1917 Revolution in Russia

1921 After war with Russia, Poland secured most of its territorial claims

1939 World War II precipitated by German invasion of Poland, which was partitioned between Germany and Russia

1897 Completed first scientific paper

1896 Began study of "uranic rays"; using Curie brothers' electrometer, discovered that radiation is a property of the atom, not chemical reaction

Early research covered
netic properties of tempered

April 1898 Announced that thorium, like uranium, emits rays (but discovery made by Gerhard Schmidt 2 months earlier)

July 1898 Announced discovery (with Pierre) of the "radioactive" element, polonium

December 1898 Announced discovery of radium

March 1902 After 3 years' hard labor, established atomic weight of radium

1904 Published *Researches on Radioactive Substances*

1910 With André Debierne isolated metallic radium by electrolysis

1910 Published *Treatise on Radioactivity*

1910 Provided pure radium chloride which in 1911 established the International Radium Standard

1914-1919 Organized over 200 mobile X-ray units for French army; trained French and US operators

1897 Daughter Irène born

1900 Started teaching at École Normale, Sèvres

Became research scientist;
ried Pierre Curie

1904 Daughter Ève born

1906 Pierre killed in traffic accident; Marie succeeded him as professor of physics at University of Paris

1914 Appointed head of radioactivity section, Paris Radium Institute

1914-1919 Director of the Red Cross Radiology Service

1920s-1930s Active in League of Nations International Committee on Intellectual Co-operation

4 July 1934 Died at Sancellemoz, Savoy, France

1900 1905 1910 1915 1920 1925 1930

)btained degree in physics,
g top of year

btained degree in
matics, passing 2nd in

1903 Received doctorate; shared Nobel Prize for Physics with Pierre Curie and Henri Becquerel

1911 Failed in election to French Academy of Sciences; awarded Nobel Prize for Chemistry

1921 First of 2 tours of United States; President Warren Harding presented Curie with 1g of radium

1905 Special relativity theory published by Albert Einstein (1879-1956), German physicist

1915 General relativity theory published by Einstein

1897 Existence of electron established by J. J. Thomson (1856-1940), British physicist

1934 Artificial radioactivity discovered by Irène (1897-1956) and Frédéric (1900-1958) Joliot-Curie, French physicists

1899 Alpha and beta particles discovered by Ernest Rutherford (1871-1937), New Zealand-born physicist

1911 Rutherford proposed nuclear theory of the atom

1919 Rutherford announced first artificial disintegration of an atom

1913 Neils Bohr (1885-1962), Danish physicist, proposed his model of the atom

1895 X-rays discovered by Wilhelm Röntgen (1845-1923), German physicist

1903 Laws of radioactive decay and transformation established by Rutherford and Frederick Soddy (1877-1956), British chemist

1910-1912 Soddy proposed concept of isotope to explain why radioelements exist with more than one atomic weight

1932 Neutron discovered by James Chadwick (1891-1974), British physicist, working at Rutherford's laboratory in Cambridge

1900 Gamma rays discovered by Paul Villard (1860-1934), French physicist

1895 Argon, first of inert gas elements, discovered by William Ramsay (1856-1916), British chemist

1896 "Uranic rays" discovered by Antoine Henri Becquerel (1852-1908), French physicist

1938 Otto Hahn (1879-1968), German chemist, and Austrian physicist Lise Meitner (1878-1968) and her nephew Otto Frisch (1904-1979) discovered nuclear fission

1942 First sustained nuclear reaction produced by Enrico Fermi (1901-1954), Italian physicist

1899 Radioelement actinium discovered by André Debierne (1874-1949), French chemist

1900 1905 1910

Marie Curie was attracted to "uranic rays" because little had been written about them, so she could plunge straight into original research

▲ **Marie and Pierre Curie were married in 1895. Their one major relaxation from work was the popular new sport of cycling. Marie's cycling activities were curtailed by the birth of their first daughter, Irène, in 1897.**

X-rays and "uranic rays"

When Marie Curie decided to work for a doctorate, she was one of the first women in France to attempt it. It was a bold step, all the more so since Marie was pregnant – her daughter Irène was born late in 1897. In 1895 the German physicist Wilhelm Röntgen (1845-1923) made a discovery in Würzburg that resounded through the laboratories of the world, and indeed outside them. Like many other physicists of the 1890s, he was working on cathode rays. Such rays or particles (now known to be electron beams) cause splashes of luminescence to appear on the walls of a largely exhausted, or near-vacuum, glass tube when electric current is discharged through the gas that remains in the tube. (These "Crookes tubes" were the invention of the English physical chemist William Crookes, 1832-1919.) Hopes of establishing the connections between light and electricity inspired these researches. Although Röntgen had covered his tube with a black paper, he suddenly noticed across the room a screen coated with barium platinocyanide gently glowing. On investigation he found out that this mysterious form of light would penetrate many substances opaque to ordinary light, and would pass through wood or rubber or even human flesh, so that only the bones of his hand cast a strong shadow on the screen. Better still, it would expose a photographic plate so that coins could be photographed inside a box, white lead showed up in the paint on the other side of a door, metal objects were revealed buried in human flesh, or fractures in the bones made clearly visible. The mysterious light he called an X-ray. Not since 1609, when Galileo announced the discovery of four new satellites around Jupiter, had there been so much public excitement concerning a scientific discovery (◆ page 40).

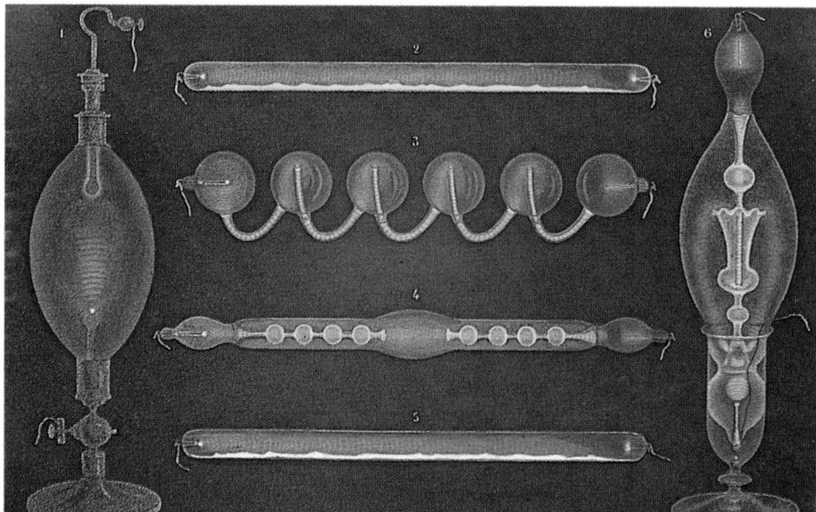

◄ **The discharge of electricity through tubes containing rarified gas was studied by Faraday, but became a particularly popular research topic later in the 19th century. The English physicist William Crookes (left) followed up work by German scientists to demonstrate that these discharges produced a new kind of "ray". Further research on these rays led to the discovery both of X-rays and of electrons.**

▲ **It was a time-consuming task to prepare tubes for the study of electrical discharges through gases. A glass tube with appropriate wires inserted had to be constructed. Then most of the air was extracted from the tube and the glass immediately sealed off. Heinrich Geissler (1814-1879) invented a new form of pump to remove the air, and was able to produce a variety of successful discharge tubes.**

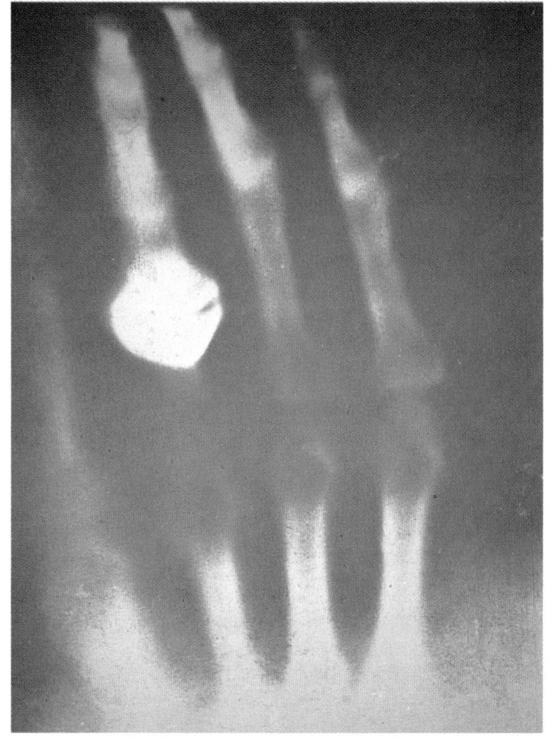

▲ ***The first X-ray: an image of his wife's hand, showing the ring she was wearing, made by Wilhelm Röntgen.***

▼ ***X-rays quickly caught the popular imagination: a German postcard of an "X-ray picture" on a beach.***

A young research student at Cambridge, England, at the time, the New Zealand-born physicist Ernest Rutherford (1871-1937) wrote to his fiancée that his professor, J.J. Thomson (1856-1940), "of course is trying to find out the real cause and nature of the waves, and the great object is to find the theory of the matter before anyone else, for nearly every Professor in Europe is now on the warpath."

Also on the warpath was the French physicist Antoine Henri Becquerel (1852-1908), then at the Paris Polytechnic. Becquerel had been in the audience when a fellow physicist showed the Academy of Sciences X-ray photographs of the skeletons of a fish and a frog. Becquerel had been working on problems of fluorescence and phosphorescence and believed that X-rays might be produced by fluorescing substances. One of these is potassium uranyl sulfate. He placed some crystals on metal foil covering photographic plates wrapped in dark cloth. Since he wished to observe the effect of sunlight on the crystals and the weather was overcast, Becquerel put them away in a drawer. Three days later he took out and developed the plates and was amazed to find that the silhouettes of the crystals "appeared with great intensity". This effect continued, and Becquerel brought out a number of papers on his "uranic rays" as he called them, since uranium compounds emitted them whether fluorescing or not. By contrast with X-rays, he could arouse only slight interest. At the session of the Academy to which Becquerel read his paper there were six papers on X-rays, and six more at the next meeting – but no one went on the warpath to learn more about uranic rays. Through 1896 literally thousands of articles were written about X-rays (*Nature* magazine had published over 150 by the end of April).

Notebooks of the Curies were still dangerously radioactive 75 years after they had entered their meticulous daily records

The discovery of polonium

Becquerel was almost alone in his investigation of "uranic rays". In 1897 he found that they made air electrically conductive, which enabled measurements to be made. Perhaps his silhouettes were too fuzzy and uninformative compared to X-rays. Even when scientists agreed that this was an effect quite different from fluoresence or phosphorescence, most assumed that Becquerel had just discovered some slightly unusual form of X-ray or possibly ultraviolet light.

As so little had been done on the subject, Marie Curie decided it would be perfect for her doctoral thesis. First she investigated whether any other substances might be the source of the "radiation". Since uranium was the heaviest element known, she tried thorium, the next heaviest. That worked. Unknown to Curie, Gerhard Schmidt at Erlangen in Germany was just two months ahead of her. However, he had not much interest in the topic and, after his publication of this discovery, soon turned to other problems. Curie now tried many other materials, but in vain. At this point the Curie brothers' discovery of piezoelectricity came in handy; they had devised an instrument of great sensitivity which enabled her to exploit Becquerel's discovery of the electric effect, and measure a degree of activity too feeble for electrometers previously in use.

One curious problem struck her. If the "uranic" rays were proper to uranium and thorium, then once the uranium was extracted from its ore, pitchblende, the residue ought to lose its radiation. In fact, the pitchblende without uranium often revealed greater activity than the pure uranium. Curie came to the conclusion that there must be something else in the pitchblende, though only tiny quantities of whatever it was: initial chemical analysis showed only known substances, which by themselves did not give off any uranic rays. Pierre now gave up his own work to join her in the search. In a small glassed-in workshop at the back of the School of Physics and Chemistry the Curies embarked on the work of endless filtering, crushing, boiling, stirring, skimming of pitchblende and its components. Marie became more and more convinced that she had found a new chemical element. By July 1898 she and Pierre were sure, and named it polonium, "from the name of the country of origin of one of us".

▲ **_A fragment of the uranium ore, autunite, fluorescing in ultraviolet light._**

Barriers to women in science

Until the last years of the 19th century such women as had engaged in scientific research did so within the framework of the family, normally in partnership with husbands, fathers or brothers. Before about 1830 most research was still home industry, funded by the scholars involved. When public institutions sprang up for the advancement, or diffusion, of science, the organizers gradually came to accept that women could reasonably form part of the audience. But institutions of higher prestige, and those which granted qualifications or recognition, as a rule continued to resist the admission of women, often quite bitterly. Neither the Royal Society of London nor the Academy of Sciences of Paris elected female members until well on into the 20th century. Universities could claim that they had begun as monastic institutions, and the older English universities stayed celibate until the late 19th century. Secular subjects like medicine and law were just as all-male as was theology. It was in supposedly conservative Italy alone that there were in the 18th century women professors in the universities and women members of academies. As the universities became the center for science teaching and research, and provided the essential qualifications to enter any learned profession, so women began to seek entry to them on the same terms as men. Excluded from laboratories and professional societies, women would hardly be able to take part in research, except in such field studies as were open to them: effectively only botany was held to be suitable. As mathematics demands few public facilities there were some prominent women mathematicians in the 19th century, such as Sophie Germain (1778-1831) in France and most notably Sonya Kovalevskaya (1850-1891) in Russia. But they too found recognition hard.

◄ **The Scot Mary Somerville received widespread recognition for popularizing science, rather than for carrying out original research. Marie Curie was one of the first women to receive proper recognition as a researcher. Even so, women scientists in the 20th century still receive less recognition than their work deserves.**

► **Sonya Kovalevskaya, unable to study mathematics in her native Russia, married primarily in order to go to Germany to study. There she was forced to seek private lessons because enrolment in classes was disallowed for women. Her Göttingen PhD was for work on differential equations. In 1884 she became the first woman professor at Stockholm.**

Ideology and practice of prejudice...

The biological theories of the 19th century were pressed into service to hold this pressure back; professors claimed that women's brains are smaller than men's, so they must be weak-minded and incapable of rational thought; evolution had developed home-making skills and attitudes among them; too much science would render women unfeminine and unsuited to be the mothers of future generations.

Women were admitted to the Paris medical school from 1868, but entry to other leading medical centers proved more difficult. It took a long struggle before women were allowed into London and Edinburgh universities. In a number of universities even after women were accepted they could not receive the degrees they had earned. There might be some encouragement, but only from a minority of male friends and family, and most had to devote far more effort to study than men before they were permitted to compete.

When women did get into universities they often had to put up with the patronizing attitude of their male professors. The British physicist J.J. Thomson wrote in 1886 (to a woman friend) that the women physics students "take notes in the most praiseworthy and painstaking fashion". But, he added, "the most extraordinary thing is that I have one at my advanced lecture. I am afraid she does not understand a word...and must be attending my lectures on the supposition that they are on Divinity...." When one of Thomson's first research students, Rose Paget, married him in 1890, it seems to have been assumed mutually that she would give up further laboratory work. That was certainly the normal expectation.

Even if the wife did collaborate with her husband, it was suggested that she was just a helpmate, as happened to Marie Curie. Her friend Hertha Ayrton (1854-1923) was apparently the first woman to be proposed for fellowship in the Royal Society (she was turned down on the grounds that their charter did not permit the election of married women). Ayrton wrote of a friend's work in astronomy, "she has not had a bit of recognition for it, just because no one will believe that if a man and a woman do a bit of work together the woman really does anything".

▼ **The Austrian physicist Lise Meitner (1878-1968), codiscoverer of nuclear fission. She was the second woman to gain a doctorate at Vienna University (1905), and found great difficulty in obtaining a post as a research scientist. She is perhaps the most outstanding physicist not to have obtained a Nobel prize. It was Meitner's colleague Otto Hahn (1879-1968) who received a Nobel prize for work on fission.**

...and their continuation

Over the years charters and regulations that appeared to forbid the admission of women were set aside, and since the end of World War I most of the formal barriers have been broken. Nevertheless prejudice remained, and probably remains still. Women are markedly under-represented in physics, chemistry and the earth sciences. Although the position is somewhat better in biology, there are far fewer women in the upper levels there too. Indeed, the under-representation of women becomes worse the higher you go from student through academic teaching to the national academies and the top awards. If the Nobel laureates are the top of the tree, then it must be said that up to the late 1980s only seven women had received science prizes – four of them the wives of Nobel laureates. Marie Curie and Irène (♦ pages 204, 208) are the only case of mother and daughter both winning science prizes, whereas there have been several father and son pairs.

Is this due to some difference in the female as against the male personality, whether physiological and inherited, or socially established and confirmed? Are women put off science at school; are they strongly encouraged to think that the scientific stress on toughness of mind and objectivity is not for them? Does scientific logic repel women? Some have argued that the ideology of science is too concerned with domination, and that its proclaimed objectivity conceals emotional presumptions, whereby women scientists are particularly resented by those who perceive them as soft and subjective.

Origins of East European Science

The historical legacy

Marie Curie was the most outstanding scientist to come out of Poland, since, in 1543, Copernicus launched the modern scientific endeavor with his theory that the Earth is a satellite of the Sun. The eastern European nations have played a less active part in the creation of science than those of western Europe. In the lands of Orthodox Christianity none of the intellectual and religious movements that shook the West had any impact. There was no Reformation, nor a Renaissance to revive a more secular outlook. Instead an authoritarian monarchy was buttressed by a mystical and authoritarian church. In the Balkans, Turkish rule further isolated the population, in which higher education was confined to the priesthood. Of the Russians and Balkan Slavs, very few had attended the universities of the West (they had none of their own) and they had not experienced the social upheavals and economic development of the 16th century. It is hardly surprising that they were barely touched by the encroachments of the new science. The works of Copernicus, Galileo and Harvey were not read in places where Aristotle had scarcely penetrated.

Poland, Hungary and Czechoslovakia were indeed linked to the West by the Catholic Church, and the Reformation affected all three. But Poland and Hungary were frontier lands. The huge Polish kingdom, which then included most of what is now Lithuania, Byelorussia and the Ukraine, stayed much more feudal in structure than western countries and stood apart from what was going on in the West. The great nobles absorbed the latest knowledge and the outlook of the western enlightenment, but when the last king of independent Poland, Stanislaus Augustus (1732-1798), tried to bring the latest principles and practice to his country, his aristocrats, who opposed increasing domination of Poland by Russia, brought the kingdom down.

"Westernization" of Russia

Tsar Peter the Great of Russia (ruled 1682-1725) had earlier seen the new way of life, and the new scientific discoveries that were welling up in Britain and the Netherlands, and he wanted the same for his country. Above all he saw how commerce and power depended on the sea and built himself a new capital, St Petersburg (now Leningrad), to be a great port. It was to be a thoroughly western capital, with an Academy of Sciences (founded 1724) based on that of France. Still, despite a few remarkable individuals such as Mikhail Lomonosov (1711-1765), for decades the Academy was manned by foreigners, mostly German.

As in Poland the nobility adopted a western way of life and took a keen interest in new scientific thought, but there was little independent local development, so engineers and doctors were often imported like the scientists. The armed forces were the first to assimilate modern ways – as in so many countries since. Under Tsar Alexander I new universities were set up, to extend higher education and open up new regions to south and east at Kazan (1804) and Kharkov (1805), and at St Petersburg (1804).

Isolation persists

Naturally the Poles and Baltic peoples resented the tsars' determination to "Russianize" their subjects. Marie Curie recollected the bitterness of her girlhood with the strain of speaking one language while she was officially obliged to study in another, the tongue of the oppressor. There were outstanding Russian scientists, such as Ivan Pavlov (▸ page 221) and Dmitri Mendeleev (1834-1907), who invented the periodic table. But on the whole political and language problems kept eastern Europe isolated from the West. For all their efforts to bring a backward school system up to date, reactionary governments could not truly open their doors to new ideas. Those who wanted to study science often had to travel to the West to do so, particularly if they were women, Jews, or politically suspect, as so many young intellectuals were.

▲ The Russian Dmitri Mendeleev made a major contribution to 19th-century chemistry. His recognition that chemical elements could be grouped by their properties into families was only finally explained theoretically with the development of quantum mechanics in the 20th century. His systematic listing of the elements showed that some still remained to be discovered. Polonium and radium, discovered by the Curies in 1898, fitted into Mendeleev's periodic table.

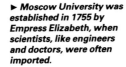

► Moscow University was established in 1755 by Empress Elizabeth, when scientists, like engineers and doctors, were often imported.

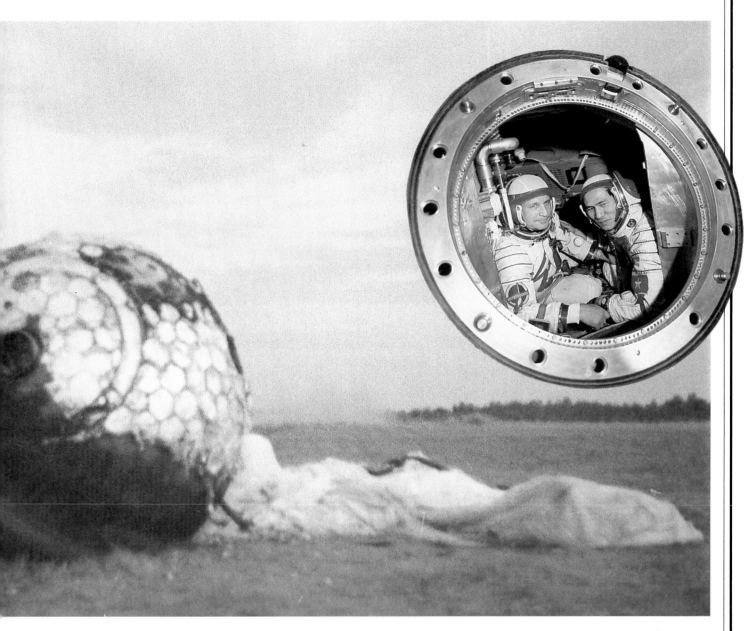

Modern successes and a failure

The fall of the empires of the Tsar and the Austro-Hungarian Kaiser after World War I ended these restrictions. But civil war, dictatorship, and nationalism all too frequently succeeded them. In the interwar years, racial and political persecution were rampant in eastern Europe. The tsars had been replaced by what was intended to be a state based on scientific principles, one in which science was to receive the greatest encouragement. And yet, in the Lysenko affair (▸ page 236), theories of inheritance which apparently did not match the state's presuppositions could be thrust aside. The last 30 years have seen many triumphs for Russian and east European science, not least the Soviet space program.

But the Lysenko episode delayed Soviet acceptance of the ideas (symbolized by discovery of the DNA double helix in 1952-1953) of modern molecular biology. The adverse effect of this on biomedical science is still apparent today.

▲ The landing craft of Yuri Gagarin after the first manned space flight (1961) and (inset) astronauts (one of them a Vietnamese) on a 1981 space mission. In the 19th century, many of the eminent scientists in Russia came from abroad. After the Revolution in 1917, the Soviet Union depended primarily on native scientists; but the growth of space research after World War II was assisted by captured German scientists and engineers. It greatly surprised the rest of the world when, in 1957, the Soviet Union successfully launched an artificial Earth satellite before the United States. This led to the "space race" between the two powers.

Curie named polonium to honor her homeland, but she became famous primarily as the "Frenchwoman" who discovered radium

The discovery of radium

Curie believed that, since three elements produced "uranic rays", some new force of nature was involved. All the chemical battering, all the changes in temperature to which a substance could be subjected, did not affect its capacity to give off these peculiar rays. Unlike the processes that normally generate heat and light around us, this radiation was not the consequence of chemical changes. There was apparently no molecular dissociation or combination involved. So this had to be a phenomenon of the atom itself. To this power of uranium, thorium and polonium, Curie gave the name of radioactivity (ray activity). Soon she realized that the radioactivity of polonium was not enough to explain the level of radioactivity in pitchblende: somewhere in those piles of ore lay minute amounts of something even more powerful, perhaps as much as 4,000 times as radioactive as uranium. Eventually the Curies pinned it down to something contained in part of the barium they had extracted. As its radioactivity had to be enormous they named it radium. The discovery was published by the Academy of Sciences in 1898. It made Marie Curie famous.

She did not find acclaim overnight. However exciting it might be for a chemical explorer to discover a new element, announcements of new elements, which were not infrequent, were met with understandable suspicion. Since the invention of spectroscopy in the 1860s by the German physicist Gustav Kirchhoff (1824-1887) and chemist Robert Bunsen (1811-1899), at least a dozen new elements had been identified, mostly rare earths and inert gases. But more than one "discovery" turned out to be a mistake. The spectral lines of two elements might combine to give the impression of a third, new one. In the same year of the Curies' discovery, 1898, William Crookes, who had discovered the element thallium in 1861, thought he had found a new element. This he named victorium, in honor of his queen's jubilee. But the "element" proved to be just an illusion.

▲ ***The Curies' laboratory in which they discovered polonium and radium, about 1900. Originally it was a storeroom attached to the École de Physique et Chimie in Paris. Apparatus for measuring radioactive decay is scattered about: most of it was designed to detect the ionizing effects of radioactive particles.***

► ***A gold-leaf electroscope used by the Curies to measure the minute quantities of radium they were able to process. The emission from the radium ionized the air in the instrument. This caused the inclined metal strip in the glass case gradually to fall. How long this took was an indication of the amount of radium present.***

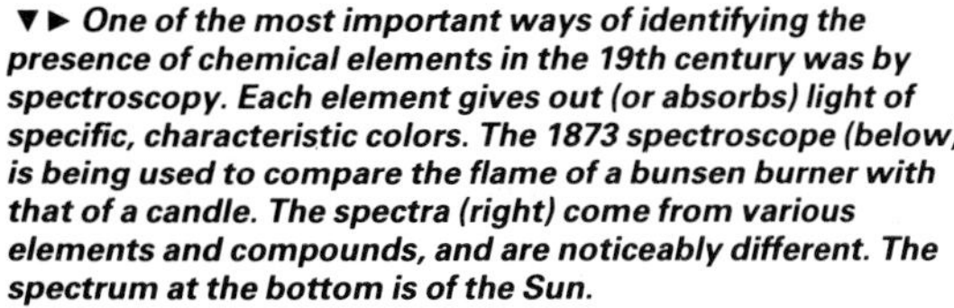

▼► ***One of the most important ways of identifying the presence of chemical elements in the 19th century was by spectroscopy. Each element gives out (or absorbs) light of specific, characteristic colors. The 1873 spectroscope (below) is being used to compare the flame of a bunsen burner with that of a candle. The spectra (right) come from various elements and compounds, and are noticeably different. The spectrum at the bottom is of the Sun.***

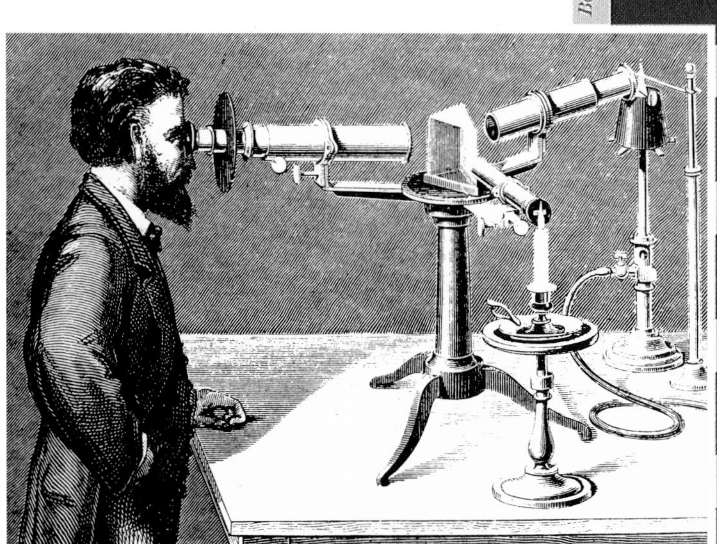

► *Pierre and Marie Curie in their laboratory examining a tube containing radium. Apart from suggesting the confused idea of "rays" that then existed, this drawing also shows that Pierre was initially considered the more important researcher of the two.*

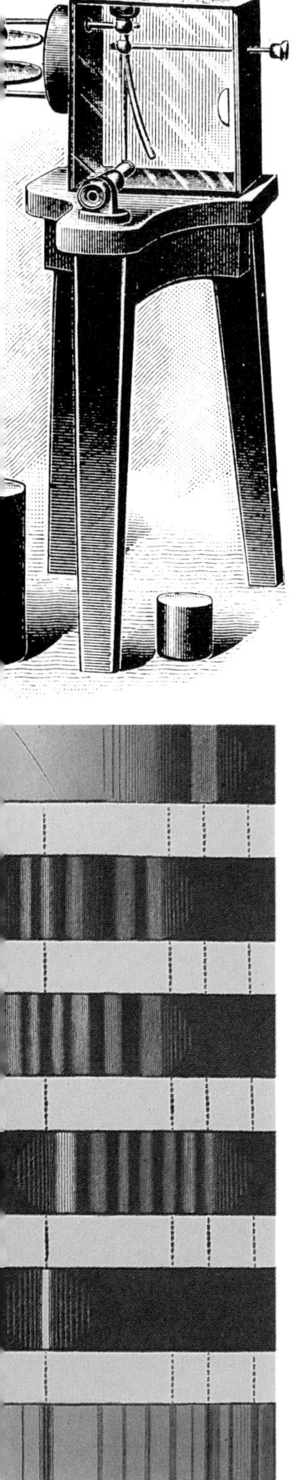

"The best sprinters in this road of investigation are Becquerel and the Curies in Paris" – Ernest Rutherford

▼ The Curies portrayed on the cover of a popular French magazine. It appeared in 1904, when the Curies had just gained worldwide fame for their work on radium.

► The Curies and others soon realized that the rays given off by radioactive materials were not all identical. Some were electrically charged particles, while others behaved like light waves. The charged particles could be detected by passing them between the poles of a magnet, which deflected them to one side. The Curies used the apparatus shown here to study electrons emitted in radioactive decay.

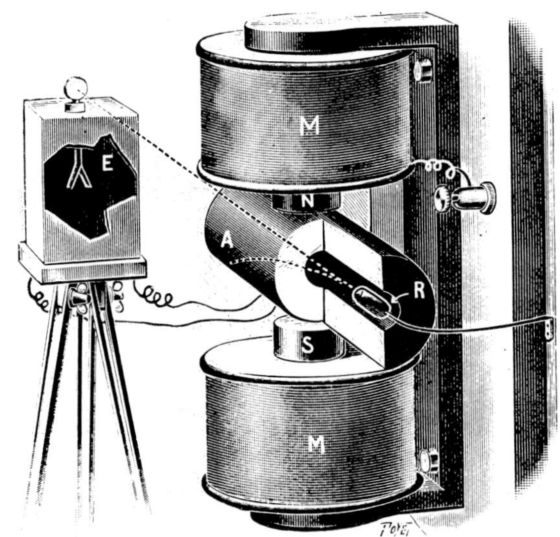

To confirm irrefutably that radium was indeed elementary, the Curies had to obtain a sample large enough and pure enough to reveal its own atomic weight. This required much more purification, and a bigger sample. Through a contact of Pierre's in Vienna they were able to procure what they wanted from the piles of waste, from which uranium had already been removed, dumped in the pinewoods near the pitchblende mines in Bohemia. These residues were delivered in sackfuls to the municipal school in Paris, where the Curies began the fractional separation of the different constituents. A hanger at the back of the school was offered for a laboratory, for they were now carrying out a laboratory job on an industrial scale, in terms of the quantities that were being treated. Furthermore, the "laboratory" was not very suitable: it had been a temporary dissecting-room, and a leading German chemist described it as "a cross between a stable and a potato-cellar". Apart from the discomfort, it was very difficult to keep the samples pure amid so many draughts. It took the Curies three more years, until March 1902, to extract from the tonnes of material one decigram pure enough to assign to atomic weight 225.93 (it is now calculated at 226.025), immediately below uranium and thorium.

The analysis of radioactivity

During the last two years of the century, radioactivity itself was analyzed. Two forms of ray were distinguished according to how far they could penetrate barriers set before them. The radioelements could be identified by the two types, named by Ernest Rutherford

▲► Ernest Rutherford and his laboratory in Cambridge. It proved very difficult to determine just what happened when radioactive materials disintegrated. Rutherford was a major figure in bringing some kind of order to the understanding of radioactivity. He showed that radioactive materials could give off two types of radiation, labeled "alpha-rays" and "beta-rays" (later joined by "gamma-rays"). He subsequently demonstrated that, in radioactive decay, atoms underwent successive transformations into different elements. These studies ultimately led him to the discovery of the atomic nucleus.

Page Los Angeles Examiner

HOPE!

(Copyright, 1914, by the Star Company.)

RADIUM

WINSOR McCAY

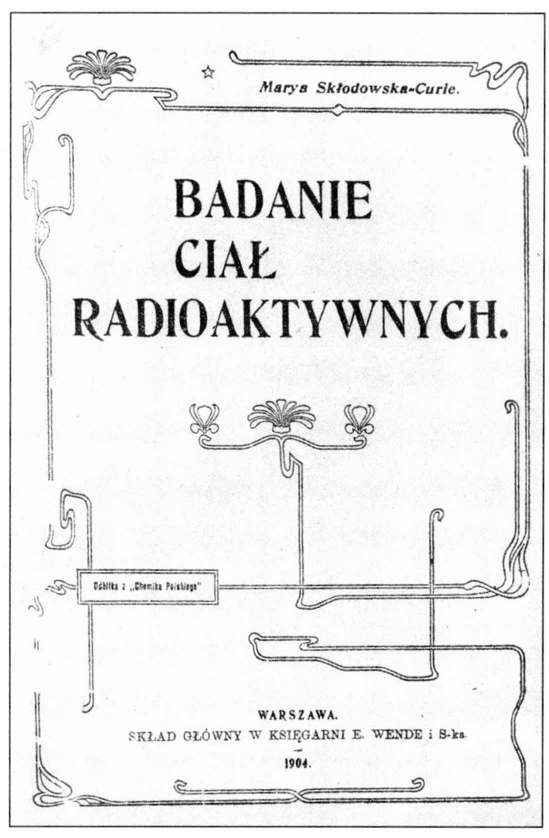

Marya Skłodowska-Curie.

BADANIE CIAŁ RADIOAKTYWNYCH.

Odbitka z „Chemika Polskiego"

WARSZAWA.
SKŁAD GŁÓWNY W KSIĘGARNI E. WENDE i S-ka.
1904.

◄ *Extravagant claims for radium made in the press in 1914 included energy sources and miraculous cures.*

▲ *The title page from a Polish edition of Marie Curie's book "Researches on Radioactive Substances" published in Warsaw, 1904.*

alpha and beta rays. A third type of "undeviable ray" was captured by the Curies' fellow-Parisian, Paul Villard, in 1900 (they were later called gamma rays). Changes in radioactivity led the way to more radioelements: André Debierne (1874-1949), a chemist who joined the Curies' team and became their closest friend, in 1899 discovered one to which he gave the name actinium (a kind of Greek equivalent of radium, the "ray element"). More followed, ionium, mesothorium (both of these later redefined as isotopes of thorium, and of radium or actinium respectively); polonium was rediscovered at least once; there was briefly a "Nipponium". Indeed there seemed to be too many new elements now, identified by their radioactivity, and their atomic weight, but sharing most of their chemical properties with others.

Radioactivity was problematic in other aspects too. Substances which had not been radioactive before easily caught the condition if left close to radioactive substances. Was it something they gave off, a gas or some other carrier? Rutherford, now in charge of his own department at McGill University at Montreal, cautiously termed it an "emanation". The Curies saw that this emanation "induced radioactivity". Attempts by Becquerel to isolate whatever was radioactive in uranium chlorate, and by Crookes to do the same with uranium nitrate, led to confusing results. Apparently it was possible to extract a mysterious "X" which was radioactive, leaving the uranium inactive; but this "X" then proceeded to lose its radioactivity, while the original uranium regained it. Ernest Rutherford obtained the same results employing the element thorium.

"A continuous snowstorm silently covering every available surface" was how Frederick Soddy described radioactivity

Radioactive decay and transformation

Working with the British chemist Frederick Soddy (1877-1956), Ernest Rutherford showed that the radioactivity induced by thorium was an effect of an emanation which was, as Soddy put it, "a continuous snowstorm silently covering every available surface with this invisible, unweighable but intensely radioactive deposit". By 1903 Rutherford and Soddy had worked out the law of radioactive decay and transformation, showing how such radioelements as their inert, gaseous "emanation" are the products of changes of one element into another. This gave the prospect of several years' research, tracing the pedigrees of radioelements, through the series which linked them, and establishing the particular radiation mix of each stage along the line.

Meanwhile Pierre Curie was looking into the properties of radioactivity. Not long after the discovery of radium, the Curies had realized that it was luminous, and glowed gently in the dark, "strongly enough to read by," wrote Marie, "using a little of the product for light". Radium gave off heat, enough to melt its own weight of ice in an hour, it was claimed. In an experiment which Pierre carried out in 1903, one end of a thermocouple (a device to measure temperature differences) was inserted in a bulb containing barium chloride which had proved by its radioactivity to contain radium, the other end into barium chloride free of it. The radioactive bulb was 1.5°C hotter than the other. At first the Curies thought that there might be as much as 50,000,000,000 calories in a gram of radium. Although this turned out to be an exaggeration, the truth was remarkable enough. As one popular science book of the time claimed, there would be enough there to raise 10,000 tonnes a mile high!

The longlasting power of radioelements was not the least intriguing thing about them. As early as 1899, Becquerel had noted that the radiation of his uranium went on "undiminished after imprisonment for three years in a wooden box encased with lead". Although some scientists, including for a time the Curies themselves, wondered if radium was not defying the very laws of energy conservation, their original conviction that radioactivity must be some disturbance within the atom provided a truer explanation: these "heavy" atoms were not permanent bodies, but unstable. Rutherford showed that there were shortlived elements and devised the "half-life" as a measure, taken from the time in which a substance would lose half its radioactivity.

Evidently uranium itself, and probably radium since it was so powerful, would have very long "half-lives". As the English comic paper *Punch* put it:

"Radium very expensive, the source of perpetual motion,
Take but a pinch of the same you'll find it according to experts,
Equal for luminous ends to a couple of million candles,
Equal for heat to a furnace of heaven knows how many horsepower."

1903: a doctorate and a Nobel prize

In the meantime the Curies were finding new posts to enable them to live more comfortably, at the price of a tougher teaching load, Pierre at the Polytechnic and ultimately at the University of Paris, where he finally became professor of physics in 1904, Marie as his chief of laboratory. Since 1900, she had been teaching physics at the highest women's college in France, at Sèvres. In 1903 she received her doctorate, basically a summary of her work so far. The same year, rather unusually for someone who had just achieved a doctorate, she and Pierre shared with Becquerel the Nobel Prize for Physics.

"The wealthiest vagabond in Europe"
Alfred Nobel (1833-1896) was one of the most imaginative of 19th-century inventors. A restless, melancholy man, Nobel built up a fortune out of his explosives and other patents. But he had no children to whom he could leave it.

His father Immanuel, an unsuccessful architect in his native Sweden, migrated to Russia where in 1842 he designed the mines that were later to be used in the Crimean War to defend Sevastopol by land and sea. That launched the family as armaments manufacturers. In the 1860s, Immanuel and Alfred Nobel developed manageable combinations of the powerful but unstable explosive nitroglycerine with other materials that would stabilize it; the final version Alfred patented as dynamite in 1867. This and other techniques for the better ignition of explosives and his "blasting gelatine" revolutionized civil engineering; they made it possible to cut through ridges of rock for canals and excavate the foundations for the abutments of bridges on a scale which marks the late 19th century. Nobel then turned his attention to military explosives, where the main requirement was for a smokeless powder, and achieved success with his combination of nitroglycerine with guncotton, known as ballistite (1888).

Alfred expanded his interests and eventually he became an international figure, for his companies had branches in a score of countries. After his childhood he only lived in Sweden for a few years, in the 1860s, to set up nitroglycerine manufacture, and again toward the end of his life when he took over the Bofors ironworks, for conversion to a munitions factory. Otherwise, he lived in France, in Italy, and traveled constantly – the wealthiest vagabond in Europe, he called himself. In the end he came to dislike those very forces of military competition that had made him rich. Money had never brought him happiness, nor release from loneliness and distrust.

◀ **Toward the end of his life, even as he pursued chemical experiments in search of still deadlier weapons, Alfred Nobel was inspired by his one-time secretary Baroness Bertha von Sutter, to take an interest in the organization of peace congresses, to solve the world's problems by rational discussion instead of by war. Nobel did feel that his explosives would also contribute to world peace, as wars must now be so destructive that no state would dare engage in them. He did not tell the baroness, or anyone else, about the will he drew up a year before he died in December 1896. He left his fortune to a fund for various prizes.**

▼ **Nobel intended to honor by his prizes work done in the previous year, but that turned out not to be feasible. US geneticist Barbara McClintock waited 40 years after publishing her major results before receiving her prize (below) on the 150th anniversary of Nobel's birth.**

The prizes

The prizes Nobel established in his will included ones for literature "of an idealistic tendency" and for work "for the fraternity of nations, for the reduction of standing armies and for the holding of peace congresses". With these went three prizes for scientific achievement in physics, in chemistry and in physiology or medicine.

While the awards for peace and literature have often been the subject of criticism, the science prizes have seldom been controversial. Few have received a prize who were not agreed by nearly all those working in that science to have earned it. And only a few who might be held to have deserved a prize did not get one, at least eventually.

The idea of science prizes was already well established. Yet the Nobel prizes almost immediately caught the attention of the world. Perhaps that was due to the large sums of money involved. The Nobel laureates were henceforth identified as the very peak of their profession, agreed to be among the best in the world.

The physics and chemistry prizes are awarded by committees chosen from the Swedish Academy of Sciences, the physiology prize by the Carolinian Institute, its medical equivalent. Originally these committees just asked the professors of the relevant subjects in Scandinavian universities for their nominations, but now they have widened their scope, and ask for opinions from professors in at least six universities further afield. Past Nobel laureates are also invited to propose possible successors.

The Nobel prizewinners

Not surprisingly, then, a few countries have dominated in the Nobel stakes. Germany was first, and won the lion's share of science prizes until the 1930s, but the onset of Nazism and the flight of refugees to the United States brought about a clear American lead that has continued ever since. But even after World War II many North American laureates have been immigrants. Up to 1986 United States citizens had won 47 physics prizes, 57 medicine and 29 chemistry prizes. Only in chemistry did Germany now come anywhere near, with 24, while in physics and medicine/physiology Britain came a poor second with 20 in each, and Germany third with 15 and 11 respectively.

*Before World War II many Nobel laureates in the sciences, no less than in literature, became household names – like Einstein and Curie. Their views on any subject were demanded and heard with great respect. Perhaps that was not always wise; at least two physics laureates – Philipp Lenard (**▶ page 236**) and Johannes Stark – became ardent Nazis. In smaller countries laureates were respected sages who cast luster on their compatriots – that happened to Curie in Poland and Bohr in Denmark, for example. Since 1945, however, the scientists honored have been mostly the heads of large laboratories, and the prizes are accepted as celebrating the work of the whole team, not so much for individual genius. As a result the announcements cause less stir than once they did, and the prizewinners are little known to the general public.*

Radiography and Radiotherapy

The early use of X-rays

From the very first weeks after Röntgen published his observations of X-rays on 28 December 1895 (♦ page 192), doctors saw that they would be invaluable. Hospitals had their X-ray apparatus at work by the summer of 1896. In 1898 the army of the United States was using X-rays to look for bullets in the limbs of soldiers wounded in the Spanish-American war, and the British army sent out an X-ray unit with its expedition to the Sudan. However, it was not until the heavy casualties of World War I that all armies decided it would be necessary to put this new technique of diagnosis into the hands of military surgeons.

At first the new device was often used too enthusiastically, as it came only gradually to be realized that overexposure would be dangerous. A reddening of the skin might be noticed first, or a loss of hair. When these injuries healed, normal skin grew over the damaged area, so that a part which had previously been infected or diseased was now found to be clear. Even before 1900 attempts were made with some success to treat lupus, ulcers and skin cancer by exposing affected areas to X-rays. Radiotherapy was even used for cosmetic purposes, to remove excess hair. Once it came to be realized that X-radiation is not all identical, only the shorter-wave "soft" rays were used against these superficial ailments while "hard" X-rays were beamed to attack tumors deep in the body.

Radiography continued primarily as an investigative tool. Techniques evolved to look for damage to the lungs, and the swallowing of substances opaque to X-rays made it feasible to trace activities and abnormalities in the soft organs of the body. Through the interwar period the dangers of too high a dose of radiation were not fully appreciated. Certainly in the 1940s fluoroscopes designed to enable a doctor to examine a patient for minutes at a time were quite common. Indeed, every respectable shoe shop had one to show the fit of a new pair of shoes.

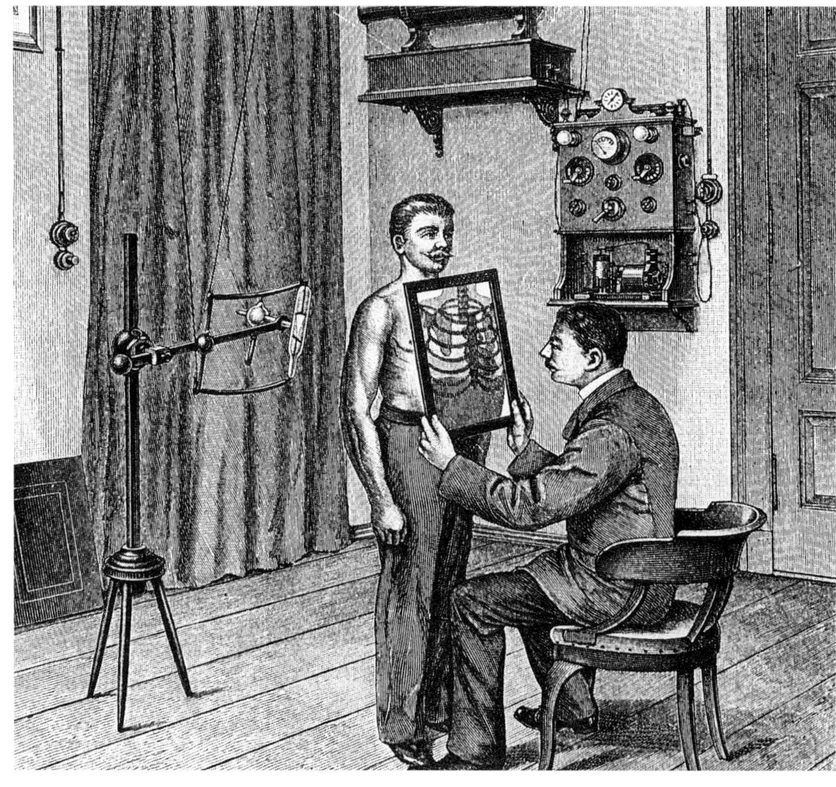

▲ A 1903 print showing a doctor in his surgery with a patient. An X-ray tube is on the stand at the left. X-rays from the tube are stopped by the patient's bones, so the doctor sees a picture of them on the screen. Both patient and doctor are being exposed to a hazardous amount of radiation. Marie Curie and Irène, as well as many other pioneers, died of radioactivity-induced illnesses.

▼ The importance of X-rays for medical diagnosis was quickly recognized. Primitive X-ray apparatus, such as this used in the South African (Boer) War (1899-1902), was developed for use with patients. This equipment has three components: a machine (in the background) to produce high-voltage electricity; an X-ray tube (in front); and a fluorescent screen (on the left) to receive the X-ray picture.

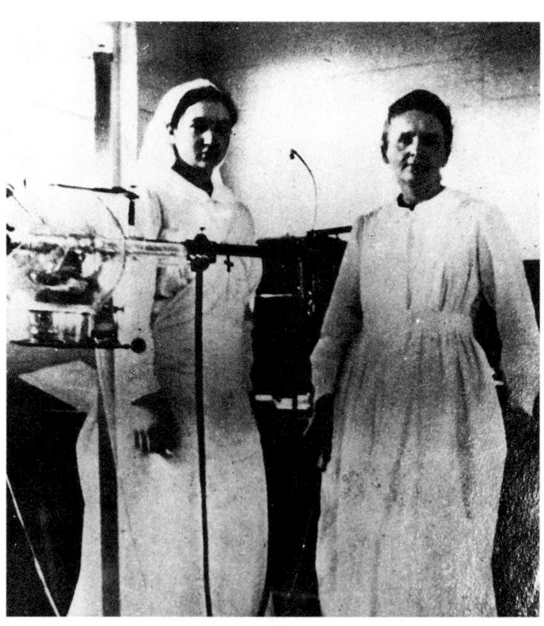

▼ X-rays proved of particular value in World War I when they were used to detect not only broken bones, but also the position of bullets and fragments of shrapnel. Marie Curie became the director of the military radiology service in France and pioneered the development of mobile X-ray units. All nations had similar units and Marie helped train American operatives. She is seen here with her daughter, Irène.

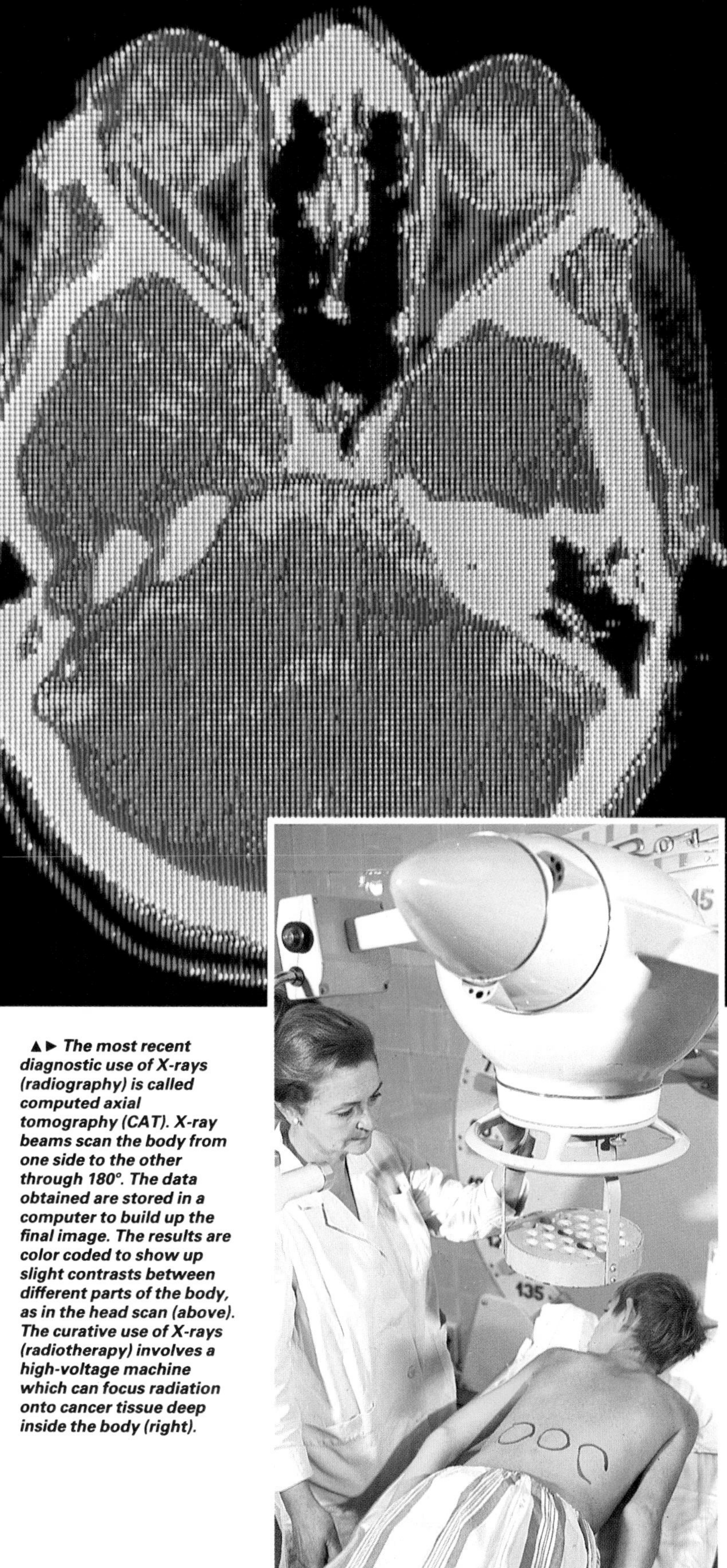

▲► **The most recent diagnostic use of X-rays (radiography) is called computed axial tomography (CAT). X-ray beams scan the body from one side to the other through 180°. The data obtained are stored in a computer to build up the final image. The results are color coded to show up slight contrasts between different parts of the body, as in the head scan (above). The curative use of X-rays (radiotherapy) involves a high-voltage machine which can focus radiation onto cancer tissue deep inside the body (right).**

Radium therapy

Radium also had quickly won a reputation as some kind of magic substance, the wonderful gift of scientific genius to relieve suffering humanity (◆ page 201). A Dr London claimed he had cured blindness in mice by radium, which defied physiology as it appparently defied physics. The English chemist Frederick Soddy proposed that radium emanation (now called radon) could be employed against tuberculosis – patients would breathe in over a cup containing some radium compound (◆ page 206). Minute quantities of radon were inserted into gold grains placed beneath the skin so that their radiation could affect the surrounding area. But as with X-rays, the destructive side of radium was soon apparent. The Curies, Becquerel, and Crookes began to notice sore spots where they had carried radioactive samples unprotected.

After some years radiographers began to appreciate that it was only the highly penetrating gamma radiation that produced these effects, and that these rays would strike through to a wider area unless just the right dosage was applied. Not until the 1930s, however, was it possible to agree on satisfactory units of measurement and ways to measure radiation, so that correct doses could be established. Then gamma rays could be aimed at precisely the area the surgeon wanted to hit, and in just such quantities that cancerous or unhealthy growth would be reduced to a minimum. Seldom, however, could it be wiped away altogether. For more superficial illness radium needles, grains or pads could be inserted where required so that radon and radiation would spread out from these sites. However, even today it is difficult to ensure that all cancerous cells have been destroyed, that none have seeded themselves elsewhere.

Modern radiography

The phenomenon of artificial radioactivity, discovered by Frédéric and Irène Joliot-Curie (◆ page 208), has led to the replacement of "natural" radium and radon by radioactive isotopes of elements like cesium and cobalt. Experiments have been made since World War II to avoid some of the problems of gamma radiation, by using beams of electrons or neutrons instead. Safer and more effective techniques of scanning the body have multiplied, extending the principles of the differential penetration or reflection of radiation, to see what is normally invisible.

Modern techniques include CAT (computed axial tomography) and NMR (nuclear magnetic resonance), in which the area to be examined is placed in a magnetic field, within which the nuclei of the body's atoms align themselves. When the field is cut off, the atoms return to their original disarray and as they do so emit radio signals, but as every element does so slightly differently, the change reveals the pattern of the elements, and so their respective behavior. In another development, PETSCAN (positron emission tomography scan), some very short-lived radiochemical is fed into the bloodstream, making it possible to observe which parts of the brain become more active in particular functions, because they then take up more blood.

The Curies both suffered radiation sickness, but Pierre's last paper was about how radium could heal rather than hurt

Trials and successes

While the money from the 1903 Nobel prize came in useful for the Curies' research (they always refused to patent their techniques for the extraction of radium and polonium, which might have made them very rich), they disliked the high focus of journalistic interviews and publicity. Both shy at heart, they suffered more than most from suddenly finding themselves in the limelight. Moreover, later that summer Marie gave birth to a stillborn baby, after another wearisome pregnancy. However, the next year she had her second daughter, Eve.

Then in April 1906, Pierre died when his head was crushed by a wagon wheel in a traffic accident. Marie was overwhelmed and took a long time to recover. Later the same year, she was appointed to succeed him, the first woman science professor in France. Pierre's notorious absentmindedness may have contributed to the accident, and it has been suggested that all the radon he had been breathing in for so many years, all the radiations in his antiquated laboratory, may have damaged his brain, at least so far as to affect his alertness. He had himself demonstrated in an experiment on himself in 1900 that radioactivity produced an unpleasant burn. Many visitors noticed that Marie's fingers were sore and reddened from constantly handling radioactive material: Becquerel and others also reported burns. An American who tried tasting a radium compound found hs heart and kidneys affected badly, and suffered from hallucinations. However, Pierre's last paper dealt with the opposite possibility, that radium could heal rather than hurt. Quite early on, it was used to attack ulcers, lupus and skin cancers, and there were hopes that it could be used against rheumatism and diabetes, even, according to Soddy, tuberculosis. The theory went that diseased tissue would be destroyed; the new tissue that replaced it would be healthy.

Such hopes inspired ever greater fascination with radium, and to a lesser extent other radioelements. Marie could now devote herself to the development of techniques for their isolation. She was the obvious person to be charged with the task of establishing a pure radium standard, which was achieved in 1911. That year was however marred for her by the failure to be elected to the Academy of Sciences (as Pierre too had failed, at his first attempt) and by a silly scandal about her relationship with the physicist Paul Langevin (1872-1946), a former student of Pierre who had succeeded him at the Municipal

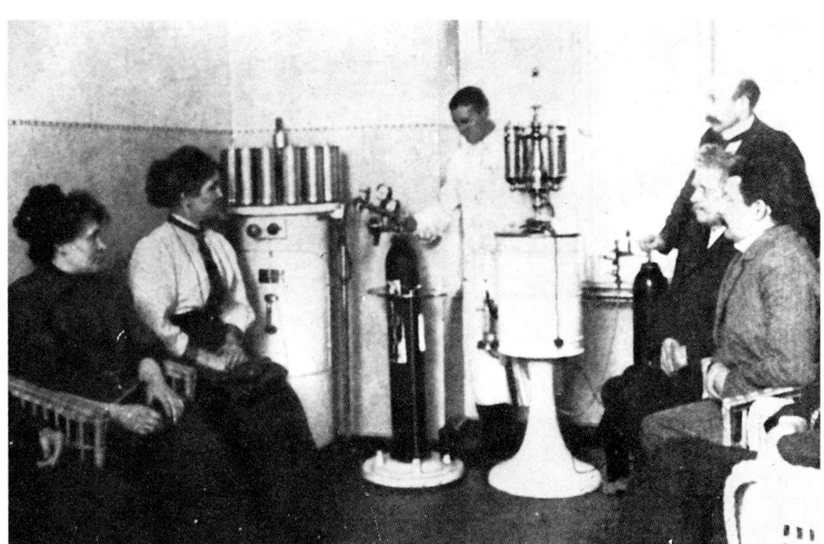

◄ **Patients receiving radium therapy in 1934 in a clinic at the Radium Institute in Berlin. Within a few years of the discovery of radium, special institutes for handling radium had begun to appear. They were concerned with research into radium and its applications.**

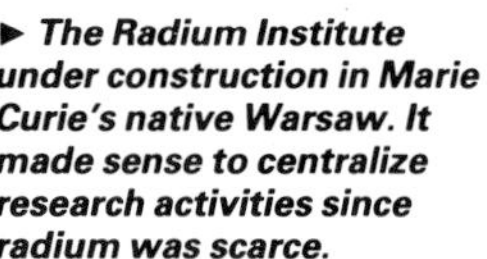

► **The Radium Institute under construction in Marie Curie's native Warsaw. It made sense to centralize research activities since radium was scarce.**

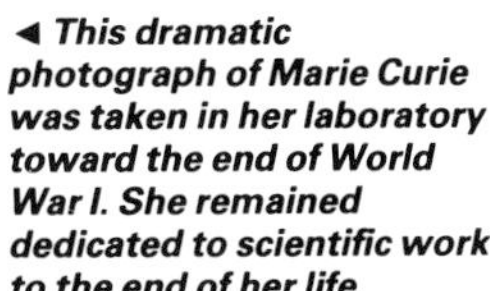

◄ *This dramatic photograph of Marie Curie was taken in her laboratory toward the end of World War I. She remained dedicated to scientific work to the end of her life.*

► *Marie Curie received worldwide recognition for her work. In 1913, during her honorary doctorate ceremony at Birmingham University in England, she is seen seated with the President of the University, Sir Oliver Lodge (left) and Sir Gilbert Barling (Vice-chancellor). Standing are her fellow-graduands, from the left: R.W. Wood (the US physicist), H.A. Lorentz (the Dutch physicist) and S.A. Arrhenius (the Swedish chemist). The last two were Nobel prizewinners.*

▲ *Marie Curie died in 1934 of leukemia, almost certainly a result of her lifelong work with radioactive materials and X-rays. The celebrations of the centenary of her birth in 1967 included the production of this commemorative medal.*

School, and then took over Marie's job at Sèvres. She may perhaps have been too naive to be discreet in her attempts to help a friend whose brilliance she recognized. Langevin needed encouragement in his scientific work and in his private troubles, and Marie felt, as she wrote to him, "a tender and maternal affection". The whole thing was blown out of proportion by enemies she made, unwittingly, at the time of her candidature for the Academy.

Perhaps in consolation, the Swedish Academy instead presented her with a second Nobel prize, this time in chemistry, for the discovery of radium itself. This remains the only time the same person has won two Nobel prizes for science. She turned down offers to go home to Poland to teach, but went to Warsaw to open a laboratory for research into radioactivity. In Paris the university and the Pasteur Institute (◀ page 186) agreed to launch a joint "Radium Institute" with two sections, one to investigate the medical potential of radioelements, the other to pursue their physical and chemical properties, and further improve methods of extraction. The latter was named for Pierre Curie, and Marie was to be its director. The building was completed in July 1914, a few days before World War I broke out.

Wartime work with Irène

The authorities eventually recognized that their brightest scientific brains were better employed in research than at the front line. Marie Curie learnt that, while radium therapy might not be useful, X-radiography would be invaluable for locating bullets or pieces of shrapnel. Although already practiced for a dozen years in civil hospitals, it was barely appreciated by the military. She therefore spent the war years organizing a mobile X-ray service for the French army. Her daughter Irène, not quite 18 when the war began, soon joined her. Marie herself in the early days took a very active part in these examinations. Given the lack of awareness of the dangers of prolonged exposure, this must have injured her health still more. Later on Marie and Irène gave courses to instruct nurses in the use of X-ray equipment, at first for France, then for other armed services.

Irène and Frédéric Joliot-Curie's discovery of artificial radioactivity brought the Curie family its third Nobel prize, in 1935

Director of the Radium Institute

After the war Marie Curie returned to her institute. The director of a modern research laboratory has perforce to be more an administrator, the inspirer of successful research in others, than an originator. This role only really began in the last quarter of the 19th century, as before that laboratories were too small to require it; perhaps the German Hermann von Helmholtz (1821-1894) at Berlin was the first. By modern standards the laboratories of the 1920s were still modest, but by comparison with earlier days, sizeable sums were now being invested, and the solitary worker was being replaced with the team.

Among the workers at the Paris Radium Institute was Irène, who married Marie's research assistant Frédéric Joliot (1900-1958) in 1926. During these years Marie's reputation spread; she made two visits to the United States to raise funds for radium research, for her own Radium Institute and for the one in Poland. She became the center of adulation such as no scientist had ever received. Perhaps her war work added to this fame, among people who might have been vague about the differences between X-radiography and radiotherapy using radium. Fame bewildered her, although she saw its advantages in the promotion of what she wanted to remain pure science, unadulterated by the need to find industrial application for knowledge. Medical application, if not strictly pure, was certainly more laudable, so the exaggerated hopes for radiotherapy, which she shared, undoubtedly kept her going on these punishing tours. She was frequently ill, and in the summer of 1934 died of leukemia, the disease presumably being linked to her excessive intake of radiation.

▲ Irène Curie learned much of her science working alongside her mother. In 1926 she married Jean-Frédéric Joliot. Working together, they narrowly missed discovering the neutron, but were awarded a Nobel prize in 1935 for their discovery of artificial radioisotopes. During World War II France bought up Europe's entire stock of heavy water (deuterium oxide) on the initiative of the Joliot-Curies, then transferred it to Britain. (Heavy water was the key to building an atomic bomb or a nuclear reactor, where it acts as a moderator.) Active in the French Resistance, the Joliot-Curies later made major contributions to construction of Europe's first atomic pile at Châtillon, Paris, where Irène was photographed on a visit in 1948 (top). They were Communist Party members and were subsequently removed from their positions. Above: irradiated fuel in a modern steam-generating heavy-water reactor.

Sigmund Freud 1856-1939

Freud's family...Dabbling in many subjects...Brentano, von Brücke and Charcot...Free association...Dreams and the unconscious...The pleasure principle...The ego, id and superego...Electra and Oedipus complexes...The Vienna Psychoanalytical Society...Growth and schisms of psychoanalysis...PERSPECTIVE...What is science?... The brain as a physical object...Social attitude to mental disorder...Impact on the arts

Sigismund Freud was born in Freiberg in Moravian Germany (now Příbor, Czechoslovakia) on 6 May 1856. After 1875, Freud was to shorten his first name, which could be loosely translated as "leading mouth" or "leader" and was also commonly used in antisemitic jokes at the time, to Sigmund – another version of Siegmund, father of Siegfried in the 13th-century German poem, *Nibelungenlied*, which Richard Wagner drew upon for his symbolic music drama, *The Ring*. In 1856 Freud's father, a poor Jewish wool merchant, was 40, with two grown-up sons from a previous marriage, and his 21-year-old mother was younger than her stepsons. Today, when divorce and remarriage are so common, few people would find much to comment on about this family situation. Freud, however, was to view childhood and relationships with parents as a key to adult mental stability. He therefore later saw his family history as an important factor in his own psychic development.

A revolution in ideas

In the 16th century, Nicolaus Copernicus revolutionized our understanding of our place in the Universe by showing that the Earth was not at its center (♦ page 40). Three hundred years later Charles Darwin destroyed the view that man was a unique creation (♦ page 167). It remained only for Sigmund Freud to deliver the final revolutionary blow to "human megalomania" and "the naive self-love of men", by suggesting that humans were not rational beings and that the difference between normality and abnormality rested upon a knife edge. (The parallel with Copernicus and Darwin was drawn by Freud himself.)

Some scientists and critics would deny Freud the status of "great scientist", or even that he was a scientist at all. No one, however, could seriously deny that his writings and the system of psychoanalysis (Freud's own term for analytical psychology) he created have had the most profound effect upon 20th-century men and women's understanding of themselves as individuals, as members of families, and as social, political and artistic beings. No other 20th-century scientific figure, not even Einstein, has had such an extraordinary and direct effect upon our lives.

Whether we still accept everything that Freud claimed as "true" or not, our attitudes toward sexuality, authority and religion are profoundly different from those of our 19th-century great-grandparents. We live in the post-Freudian age.

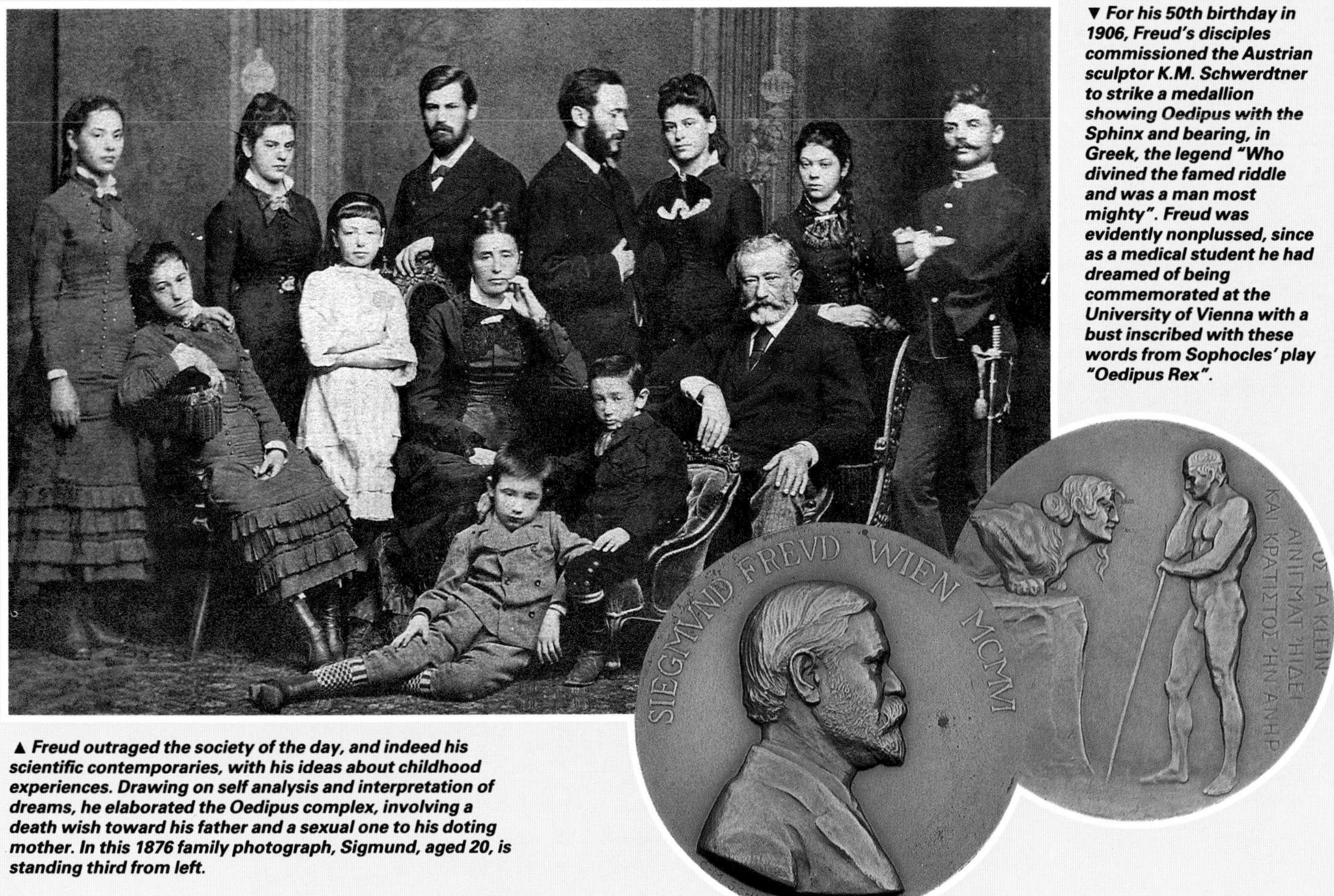

▼ For his 50th birthday in 1906, Freud's disciples commissioned the Austrian sculptor K.M. Schwerdtner to strike a medallion showing Oedipus with the Sphinx and bearing, in Greek, the legend "Who divined the famed riddle and was a man most mighty". Freud was evidently nonplussed, since as a medical student he had dreamed of being commemorated at the University of Vienna with a bust inscribed with these words from Sophocles' play "Oedipus Rex".

▲ Freud outraged the society of the day, and indeed his scientific contemporaries, with his ideas about childhood experiences. Drawing on self analysis and interpretation of dreams, he elaborated the Oedipus complex, involving a death wish toward his father and a sexual one to his doting mother. In this 1876 family photograph, Sigmund, aged 20, is standing third from left.

Sigmund Freud – His Life, Work and Times

Freud the biologist

When Freud was four, his family moved to Vienna, where he was to spend his life until 1938, when Hitler's antisemitic policies forced him to leave, and settle in London. Although Freud was never an orthodox Jew, and became agnostic in his religious opinions, his Jewishness was important to him, mainly because his career developed at the same time that European Jews were leaving their ghettos and becoming, with difficulty, assimilated into society. Although he rapidly abandoned the idea, Freud had initially decided to be a lawyer because this was one of the few careers open to Austrian Jews.

However, he was inspired by Goethe's example to interest himself in science. When he entered the University of Vienna in 1873, he decided to take a medical degree before embarking upon scientific research. Belonging, as he said of himself, to the class of sedentary humans "who can be found for the largest part of the day between two pieces of furniture, one formed vertically, the chair, and one extending horizontally, the table", Freud took eight, instead of the usual five, years to graduate. This was not through laziness, but because he dabbled in so many subjects, making him eventually into one of Europe's most cultivated men. For example, although it was not required of medical students, he elected to study with the German philosopher Franz Brentano (1838-1917), whose book *Psychology from the Empirical Standpoint* (1874) introduced Freud to the idea of "the unconscious". (This idea, which Brentano actually dismissed, had been invoked by generations of earlier philosophers to explain how ideas and forgotten memories could sometimes rise to the surface as if floating up from a hidden reservoir.)

Another influential teacher was the German physiologist Ernst von Brücke (1819-1892), who, like his close friend Hermann von Helmholtz (1821-1894), professor of physics at Berlin, was determined to explain biological phenomena in terms of chemistry and mechanics. From 1876 to 1882, Freud spent much time working in Brücke's physiological institute, where a team of biologists was investigating the fine anatomical structure of nerves with a view to discovering how they triggered off both physiological and mental phenomena. Here Freud "found rest and satisfaction", he wrote, "and men whom I could respect and take as my models." One of these friends was the Austrian physician Josef Breuer (1842-1925), soon to be the "John the Baptist" of psychoanalysis, and who was already famous for showing that respiration was controlled by a reflex action.

Freud's first scientific paper, in 1877, was the result of dissecting some 400 mature eels in order to confirm that they possessed testes! Brücke then asked him to investigate the large nerve cells in the spinal chords of larval lampreys – another eel-like creature – and the nerve systems of crayfish and crabs. From these zoological investigations Freud concluded that there was ample evidence that the nervous systems of lower animals had developed into the more sophisticated brains of higher species. This work taught Freud that since biological effects always had causes, mental events triggered by the excitation of nerves must also ultimately receive a physical explanation.

In the mid-1870s, therefore, Freud seemed poised for a distinguished, though not very exciting, career as a research zoologist. However, in those days there were no research grants and consequently, unlike Brücke's colleagues who were either men of independent means or occupying the few existing university assistantships, Freud was completely dependent upon his aging father for support.

Historical Background

■ 1848 Revolutions shook Austria (centered on Vienna), Germany, France, Italy, Hungary

■ 1866 Austria ceded leadership in Germa[ny] to Prussia

■ 1867 Austro-Hunga[rian] monarchy establis[hed]

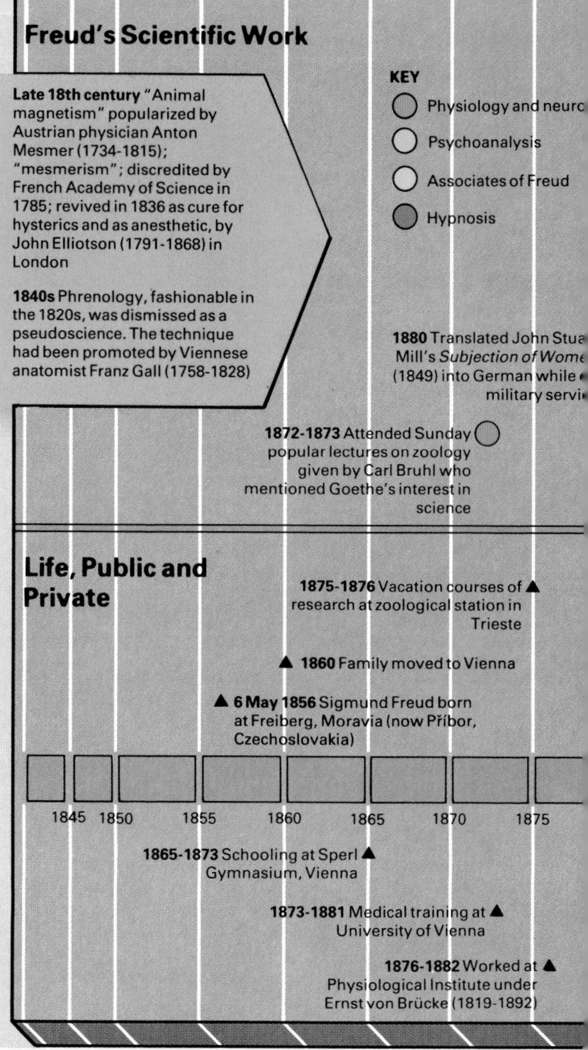
Freud's Scientific Work
KEY
Physiology and neuro
Psychoanalysis
Associates of Freud
Hypnosis
Late 18th century "Animal magnetism" popularized by Austrian physician Anton Mesmer (1734-1815); "mesmerism"; discredited by French Academy of Science in 1785; revived in 1836 as cure for hysterics and as anesthetic, by John Elliotson (1791-1868) in London
1840s Phrenology, fashionable in the 1820s, was dismissed as a pseudoscience. The technique had been promoted by Viennese anatomist Franz Gall (1758-1828)
1880 Translated John Stua Mill's Subjection of Wome (1849) into German while military servi
1872-1873 Attended Sunday popular lectures on zoology given by Carl Bruhl who mentioned Goethe's interest in science
Life, Public and Private
1875-1876 Vacation courses of research at zoological station in Trieste
1860 Family moved to Vienna
6 May 1856 Sigmund Freud born at Freiberg, Moravia (now Příbor, Czechoslovakia)
1845 1850 1855 1860 1865 1870 1875
1865-1873 Schooling at Sperl Gymnasium, Vienna
1873-1881 Medical training at University of Vienna
1876-1882 Worked at Physiological Institute under Ernst von Brücke (1819-1892)

Scientific Background

1868 Joseph Breuer (1842-1925) identified feedback mechanism involved in respiration

1862 Jean-Martin Charcot (1825-1893) began neurological department at La Salpêtrière in Paris; began his work on hysterical patients in 1872

1843 Mesmerism renamed "hypnotism" by Manchester surgeon James Braid (1795-1860)

Mid-19th century In Germany, experimental psychology developed by G. H. Fechner (1801-1887), H. von Helmholtz (1821-1894) and E. H. Weber (1795-1878)

1860s Russian physiologist Ivan Sechenov (1828-1905) studied automatic reflexes

1840 1845 1850 1855 1860 1865 1870 1875

iermany and Austria-
ıry formed dual alliance

1914-1918 World War I, precipitated by assassination of Austrian Archduke Franz Ferdinand

1920s In much reduced Austria, political unrest increased, e.g. between "Red" Vienna and "Black" rural factions

1933 Nazi dictatorship set up in Germany; Freud's works burned

1938 German troops occupied Austria, Nazis seized power, Austria part of Third Reich

1939-1945 World War II

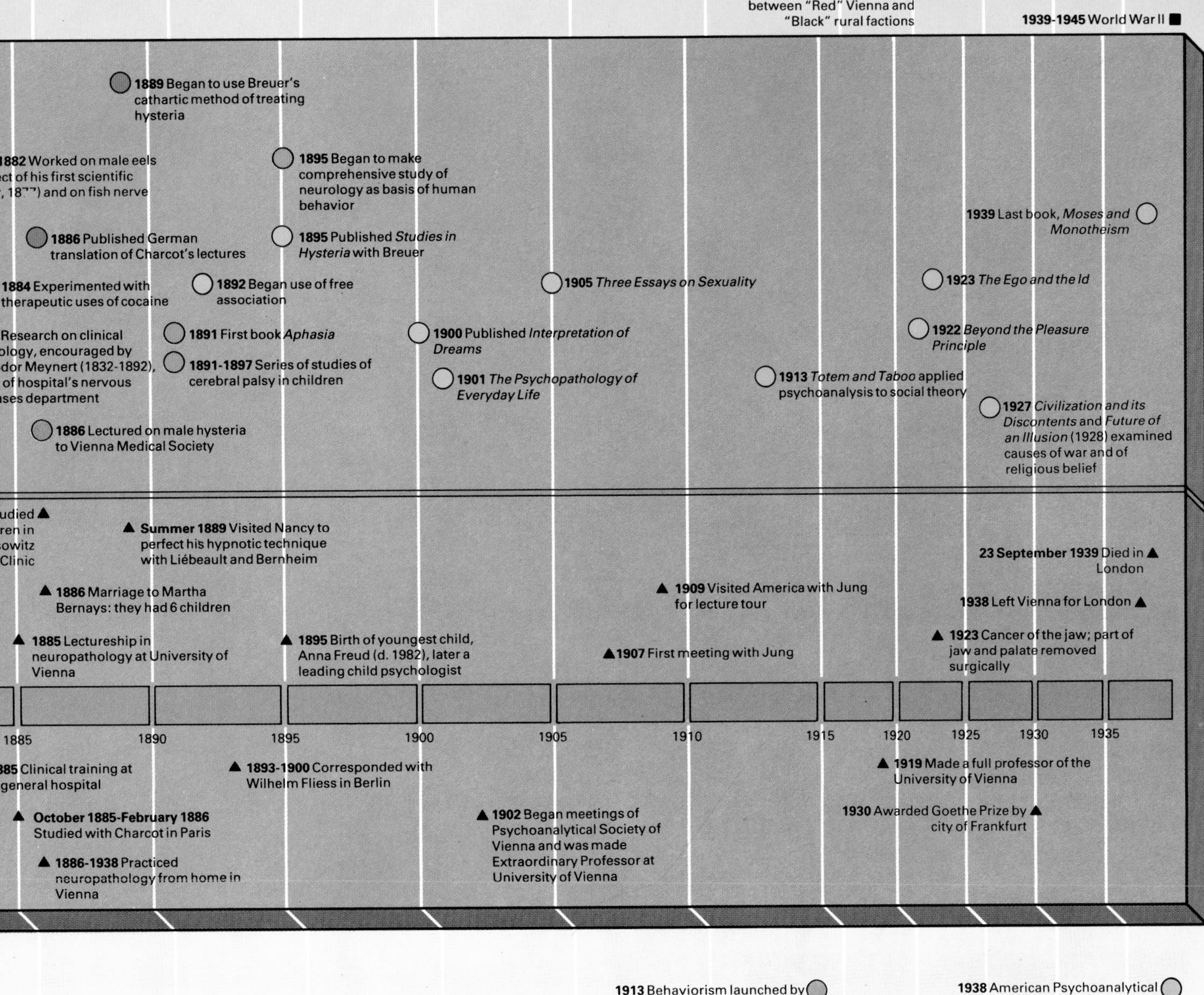

1889 Began to use Breuer's cathartic method of treating hysteria
5-1882 Worked on male eels
ıject of his first scientific
er, 18⁷⁷) and on fish nerve
s
1895 Began to make comprehensive study of neurology as basis of human behavior
1886 Published German translation of Charcot's lectures
1895 Published Studies in Hysteria with Breuer
1884 Experimented with therapeutic uses of cocaine
1892 Began use of free association
1905 Three Essays on Sexuality
1939 Last book, Moses and Monotheism
1923 The Ego and the Id
82 Research on clinical
urology, encouraged by
eodor Meynert (1832-1892),
ad of hospital's nervous
eases department
1891 First book Aphasia
1891-1897 Series of studies of cerebral palsy in children
1900 Published Interpretation of Dreams
1901 The Psychopathology of Everyday Life
1922 Beyond the Pleasure Principle
1913 Totem and Taboo applied psychoanalysis to social theory
1927 Civilization and its Discontents and Future of an Illusion (1928) examined causes of war and of religious belief
1886 Lectured on male hysteria to Vienna Medical Society
Studied
ldren in
ssowitz
ic Clinic
Summer 1889 Visited Nancy to perfect his hypnotic technique with Liébeault and Bernheim
23 September 1939 Died in London
1886 Marriage to Martha Bernays: they had 6 children
1909 Visited America with Jung for lecture tour
1938 Left Vienna for London
1885 Lectureship in neuropathology at University of Vienna
1895 Birth of youngest child, Anna Freud (d. 1982), later a leading child psychologist
1907 First meeting with Jung
1923 Cancer of the jaw; part of jaw and palate removed surgically
1885
1890
1895
1900
1905
1910
1915
1920
1925
1930
1935
-1885 Clinical training at
ıa general hospital
1893-1900 Corresponded with Wilhelm Fliess in Berlin
1919 Made a full professor of the University of Vienna
October 1885-February 1886 Studied with Charcot in Paris
1902 Began meetings of Psychoanalytical Society of Vienna and was made Extraordinary Professor at University of Vienna
1930 Awarded Goethe Prize by city of Frankfurt
1886-1938 Practiced neuropathology from home in Vienna

1880-1882 Breuer treated "Anna O" for hysteria using hypnosis

1880s Hippolyte Bernheim (1837-1919) began serious study of hypnotism at Nancy, France, with Auguste Liébeault (1823-1904)

1884 Carl Koller (1857-1944) in Vienna showed cocaine has anesthetic effects on the eye, a discovery arising from experimentation suggested by Freud

1890 American William James (1842-1910) published *Principles of Psychology* presenting the approach of dynamic psychology that stemmed from Darwin's principle of evolution

Late 19th and early 20th centuries Vienna flourished as cultural and scientific center

Early 1900s Importance of synapses established by British physiologist Charles Sherrington (1861-1952); later he stressed importance of integration of reflexes

1897-1901 Russian physiologist Ivan Pavlov (1829-1905) developed the concept of the conditioned reflex

1913 Behaviorism launched by American James B. Watson (1878-1958)

1908 First psychoanalytical periodical founded

1910 International Psychoanalytical Association founded at Nuremberg

1911 Viennese Alfred Adler (1870-1937) left Freud to found own school of individual psychology

1914 The Swiss Carl Jung (1875-1961), founder of analytical psychology, left Freud's school

1938 American Psychoanalytical Association broke with International Psychoanalytical Association over need for medical qualification of psychoanalysts

1924 Austrian-born US psychoanalyst Otto Rank (1884-1939) broke with Freud

From 1920s Gestalt psychology established by Max Wertheimer (1880-1943), Kurt Koffka (1886-1941) and Wolfgang Köhler (1887-1967). Köhler also studied animal psychology

From 1950s Behaviorist ideas placed on firm footing in publications of Burrhus Skinner (b. 1904), US psychologist

1885 1890 1895 1900 1905

Determination to make his mark on the world was always a driving force in Freud

From biology to neuropathology

As a Jew, Freud had slight chance of a financially-rewarding career in science, or of rapid promotion to a university post. If he practiced as a physician and tried to carve out for himself a medical specialism, the chances of fame and fortune were much higher. In 1882 he fell in love with Martha Bernays, the daughter of a cultured Jewish family in Vienna. Although he could not afford to marry her until 1886, their engagement, coupled with the guilty feelings he experienced concerning his parents' financial sacrifices on his behalf, encouraged him to move from biology into neuropathology. They had six children, one of whom, Anna (1895-1982), became an influential child psychologist. A grandson, Lucien (born 1922), has become an important British painter.

The study of hysteria

In July 1882, Freud joined the staff of the Vienna general hospital and slowly worked his way through the various clinical sections before settling into the department of nervous diseases under the direction of Theodor Meynert (1832-1892), the world's foremost expert on human brain anatomy. Meynert encouraged Freud to continue with his anatomical work tracing nerve tracts in human brains while at the same time working on clinical neurology in the hospital itself and in an adjacent clinic for subnormal children. This clinical work was to lead to Freud's first book, *Aphasia* (1891), in which he showed that, contrary to established opinion, loss of memory or speech or certain other brain functions, was not due to damage (lesions) to specific areas of the brain supposedly associated with these functions, but to a general deterioration of less circumscribed areas. Although the book established his reputation as an expert neurophysiologist, it brought him neither the fame nor the fortune which he was desperate to find during the early years of his marriage. For example, he also dabbled with histo-chemistry (the science of staining microscopic tissues) in

► At his clinic in Paris, where Freud studied in 1885, Jean-Martin Charcot attempted to discover an underlying type ("grand") representing an illness in its most complete form; lesser forms were "petit". After studying epilepsy and polio, Charcot investigated hysteria, whose "grand hystérie" form is illustrated in this caricature. His use of hypnosis in studying hysteria greatly influenced Freud, who translated Charcot into German.

▼ Shaking palsy was a neurological disorder first described by Englishman James Parkinson, in 1817. Charcot was the first clinician to differentiate its principal symptoms of tremor and peculiar gait from multiple sclerosis. Supposing that such symptoms were reduced when patients went on long bumpy road and train journeys, Charcot devised a vibrating hat driven by an electric motor. Parkinson's disease, caused by a failure of nerve signals, can now be treated with levodopa.

◄ "Anna O", Freud's and Breuer's most celebrated patient. Her real name was Bertha Pappenheim. The daughter of a wealthy Viennese merchant, she later became an advocate of women's rights and did much social work for European Jews.

► Charcot mistakenly divided hysteria into four stages: epilepsy, large movements, hallucinatory actions and delirium. Few patients showed all of these stages, with the exception of some women like Blanche Wittmann, the "Queen of Hysterics". For Charcot, hypnotizability was a major symptom and, not surprisingly, he found that all four stages were displayed under hypnosis. Despite gullibility and the theatricality of his lecturing (captured in this painting of 1889 which Freud owned), Charcot brought hysteria and hypnosis into medical research.

the hope of making a breakthrough in bacteriology, a new science which had gained Louis Pasteur (◀ page 178) and Robert Koch (◀ page 180) world renown in the 1880s. This determination to make his mark on the world also explains Freud's flirtations with the properties of cocaine. In Germany recently, exhausted soldiers had been fed on the drug to give them extra energy. Using the drug on himself in 1884, Freud found it acted as a stimulant and as an anesthetic (he narrowly missed its application in eye operations). Unfortunately, he also published the claim that it could be used to wean drug addicts off morphia. For years Freud was remembered in Europe and America as the doctor who had released a fourth addictive scourge on the human race to join tobacco, alcohol and morphine. This reputation was to cloud his attempts to launch psychoanalysis in Vienna.

A large number of the mainly female patients Freud had to deal with in Meynert's clinic were diagnosed as suffering from "hysteria" – a loose term for a variety of physical symptoms and behavior (such as blindness, numbness, lameness, tics) which had no apparent physical cause in neurological damage. (The word hysteria literally means "womb"; the ancient Greeks had associated such symptoms in women with a malfunction of the uterus.) One of Breuer's patients at this time, Bertha Pappenheim, the daughter of a family friend, had come to him with distressing hysterical symptoms which he had had some success in removing, one by one, under hypnosis. Breuer found that if he got Bertha to remember some past emotional crisis under hypnosis – usually events associated with the nursing of her dying father – and then told her hypnotically that the symptom would disappear, relief was obtained. Breuer freely discussed the case (known anonymously as that of "Anna O") with Freud and called the treatment the "cathartic method", that is, the purging treatment.

In 1885, convinced that hypnosis was a key to understanding and curing hysteria, Freud studied in Paris with the charismatic Jean-Martin Charcot at La Salpêtrière lunatic asylum in Paris.

At La Salpêtrière, Jean-Martin Charcot (1825-1893) had gained a European reputation for his kind and sympathetic treatment of lunatics and for showing that hypnotized people could be made to show hysterical symptoms. According to Charcot, hypnotizability was merely another symptom of hysteria (he could claim this because not everyone can be hypnotized), and he showed that men were as much prone to hysteria as women. Finally, he attributed the cause of hysteria to an inherited derangement of the brain: the sins of past generations were visited upon the present.

Although Charcot showed that under hypnosis hysterical patients could be made to lose some of their symptoms, at least temporarily, he was convinced of its incurable nature. (His critics showed later that an elaborate theory of hysteria, which he developed from hypnotic experiments, was a figment of the responses made by patients to Charcot's suggestions.) Privately, like Freud, he also seems to have been aware that the causes of most hysterias were sexual in origin.

Freud learned a good deal from Charcot, whose work he translated into German. What impressed him most, however, was that although patients claimed no memory of what Charcot had made them do under hypnosis, the entire experience was recalled when Charcot placed his hand on their foreheads and said authoritatively, "now you remember!" Freud reasoned that if the "pressure" technique overcame a hypnotic subject's amnesia it might also unlock an hysterical patient's memories previously blocked by nervous degeneration. (He was also encouraged to try the technique because he proved to be a poor hypnotist.) Freud's technique was to place patients on a couch with their eyes closed and ask them to recall when their physically disabling symptoms had first begun. If reminiscences stopped, or seemed to be blocked, he applied pressure, frequently finding that this stimulated new memories or, more puzzlingly, seemingly irrelevant dreamlike images. More bizarrely, by talking freely to Freud in this way, and without the hypnotic suggestions used by Breuer, many patients obtained relief from their symptoms. By trial and error Freud discovered that hypnosis, pressure or electrical stimulation (which he also tried) were not necessary: therapeutic benefits followed simply, if slowly, from "free association" as he called this process which bears a close similarity to the Roman Catholic ritual of confession. Patients still had the choice of releasing their repressed memories or not. As Freud put it, he could give people a rail ticket to normality, but it was up to them whether they made the journey.

Although Breuer and Freud collaborated on a discussion of their treatment in *Studies in Hysteria* (1895), Breuer, who was by then a well-established physician with a rich clientele, was unwilling to explore the nature of the catharsis in more detail – mainly because it rapidly became clear that the emotional crises which precipitated hysterical symptoms in later life were of an explicitly sexual nature. "Anna O", for example, as we now know (it was censored in the *Studies*), developed a phantom pregnancy which she attributed to Breuer. He had stumbled upon what Freud was later to call "transference", a redirection of childhood feelings and desires whereby patients became erotically involved with their therapists. (By a modern, alternative interpretation, the whole case of "Anna O" could be attributed to her frustration at living in male-dominated Vienna.)

Breuer therefore left Freud free to explore the nature of hysterias and neuroses – territory which could gain a physician notoriety and ostracism unless extreme care was exercised.

Why are astronomy, chemistry, physics and biology accepted as orthodox sciences, but astrology, alchemy and phrenology labeled pseudo or deviant sciences? Why is treatment by the administration of conventional medicines made by pharmaceutical companies (allopathy) considered to be respectable, but naturopathy, herbalism, homeopathy, osteopathy, acupuncture or faith healing considered to be unorthodox and labeled "fringe medicine" (♦ page 182)? Well into the 20th century, medical care for the vast majority of ordinary people relied on self-care and the use of many of the techniques and beliefs of so-called fringe medicine. The answer to the medical question is largely to be found in a social change – the rise to dominance of professional doctors whose practice of allopathy ousted other forms of therapy. Hypnosis and phrenology are good illustrations of this social construction of science.

"Mesmerism" ridiculed

Hypnosis first came to widespread attention toward the end of the 18th century, when the Austrian physician Anton Mesmer (1734-1815), inspired by contemporary ideas concerning a rarified, all-pervading ether, argued that a person's good health depended upon an efficient flow of ether through the body. Some people had the power to influence this flow by moving their hands in a pattern over the body, just as moving a magnet over iron filings alters their configuration. Although dismissed in Paris by a team of scientists as a mass delusion (it was causing disturbances during the revolutionary period), Mesmer's system of "animal magnetism", or mesmerism, proved extremely popular. After much controversy, it was taken up again in 1836 by John Elliotson (1791-1868), a friend of Charles Dickens, who was professor of medicine at the University of London. Elliotson, who had been the first British doctor to use the stethoscope, claimed that mesmerism could be used both to cure hysterical patients and as an anesthetic in surgical operations – this was a decade before the introduction of the volatile, flammable liquid ether and of chloroform as anesthetics. However, when he claimed that certain "sensitive" women, when hypnotized, were able to diagnose disease (by clairvoyance), he was sacked.

The slow acceptance of hypnotism

Mesmerism clearly challenged contemporary theories of disease and, more seriously, threatened the livelihoods of the general practitioners who were slowly gaining in social prestige among the middle and upper classes. However, because mesmerism had been taken up by itinerant showmen, it was comparatively easy to dismiss it as undignified, disreputable and "pseudoscience".

Although James Braid (1795-1860), a Manchester surgeon who renamed mesmerism "hypnotism" in 1843, explained it in terms of the hypnotist's suggestions acting on the imagination of the subject, British doctors refused to take it seriously. In 1878, however, Charcot, who was working with hysterical patients in Paris, argued that the hypnotic state was due to brain derangement. Other French investigators of

hypnotism, such as August Liébeault (1823-1904) and Hippolyte Bernheim (1837-1919) at Nancy, preferred Braid's view. Because this French dispute involved medical men only, it persuaded the British Medical Association (BMA) in 1891 to review the credibility of hypnotism. By then, the social and professional status of British doctors was secure, and in 1893 the BMA accepted hypnotism as a genuine phenomenon which might be used, with caution, in therapy.

The decline of phrenology

Phrenology (▶ page 216) moved in the opposite direction – from being welcomed as a useful ideology for medical practitioners to adopt in the 1820s, when their authority in society was weak, to being dismissed as a pseudoscience in the 1840s.

▼ In this print of 1794, by which date Mesmer had fled from France to Switzerland, the "mesmerist" passes lines of force from his hands toward the female patient, who enters a state of "crisis". Hypnotism remained medically unorthodox until the 1890s.

► Mesmer's hypnotic therapy, with its magnets and "celestial influences", was exposed by the French Academy of Sciences in 1785. In this cartoon, Benjamin Franklin holds the commission's report, Mesmer is shown as an ass, while gullible patients fly off on broomsticks.

Ironically, although the grounds for rejecting phrenology were empirical – that the shape of the skull did not follow the contours of the brain – by the end of the 19th century, neurophysiologists had demonstrated beyond doubt that various motor and sensory powers were localized within the cortex.

Popper's falsification criterion

In the 20th century, philosophers of science, such as the Anglo-Austrian Karl Popper (born 1902), have argued that there is a criterion for distinguishing between science and pseudoscience. As a young man in Vienna after World War I, Popper was impressed by the way a difficult and, on the face of it, implausible, theory such as Einstein's general theory of relativity could produce a confirmable prediction about the apparent displacement of stars seen during an eclipse of the Sun (▶ page 241). By contrast, Freud's (and Adler's) psychoanalysis, equally as difficult to accept as Einstein's, offered no predictions upon which it could be tested. For Popper, psychoanalysis is a pseudoscience because it cannot be refuted; it is always flexible enough to explain away any difficulty.

In fact, however, there are many scientific theories which do not satisfy the falsification criterion. Popper has claimed, for example, that Darwin's theory of natural selection is untestable, though most Darwinians feel that subsequent studies of the genetic structure of animal populations have provided one of its greatest predictive triumphs. What matters in the end is that falsification usually leads to a theory's adjustment rather than to its dismissal; and this adjustment is a matter of social agreement between scientists who hold powerful professional positions and who control the purse strings of scientific funding.

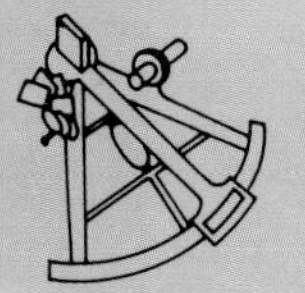

Study of the Brain

The seat of the rational soul

Early philosophers had regarded the brain as a repository for the rational soul which marked the difference between people and animals and plants. Commentators during the Middle Ages located the intellectual functions of imagination, reason, common sense and memory in four chambers, or ventricles, within the brain. (These four ventricles are now known to contain cerebro-spinal fluid which filters, protects and cleans the brain cells.)

Although the Renaissance revival of dissection encouraged examination of the cerebral convolutions of the brain, elements of the medieval view were still used in the 17th century by the French scientist and philosopher René Descartes (1596-1650) in his mechanization of physiology. According to Descartes, the pineal body at the base of the brain was the seat of man's rational soul which, on receiving sensory messages, controlled the flow of animal spirits through hollow nerves to power muscular responses. The human body was like a garden automaton – a moving figure worked by hydraulic power at that time a feature of formal gardens. This primitive idea of reflex action had to wait two centuries before more detailed knowledge of nerves allowed it to be refined.

▲ *Phrenology was a gift to satirists. Franz Gall, the science's founder, is also given an interesting head in this cartoon of 1830. Phrenologists placed moral and religious attributes at the top of the skull, intellectual ones at the sides, and "animal" faculties, such as the desires for food and sex, at the base.*

▼ *Although phrenology had lost its scientific credit by the 1850s, it remained a popular art, as this chart published in 1923 shows.*

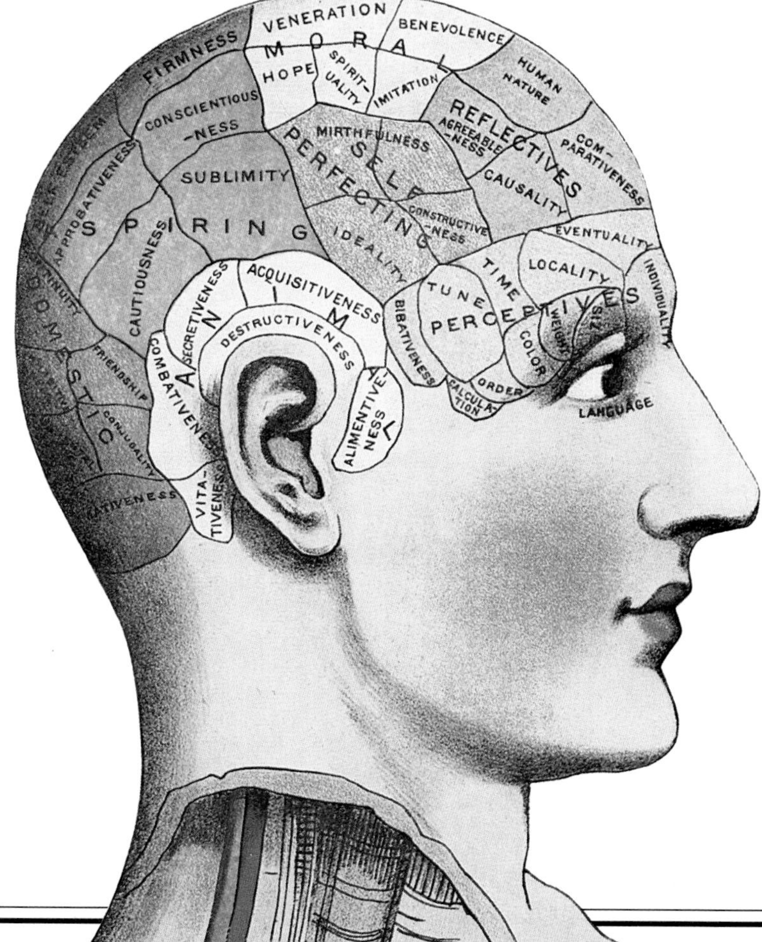

► *In 1879 Charles Sherrington identified nerve junctions which he called synapses. Nerve impulses pass across the synaptic gaps (red) through the release of chemicals such as norepinephrine which then trigger electric impulses in the adjacent neurone (nerve cell) or muscle. This electron micrograph shows the chemical transmitter as red and yellow spheres.*

▼ *Freud began his career as a neurophysiologist at the University of Vienna where illustrated lectures on the brain were popular.*

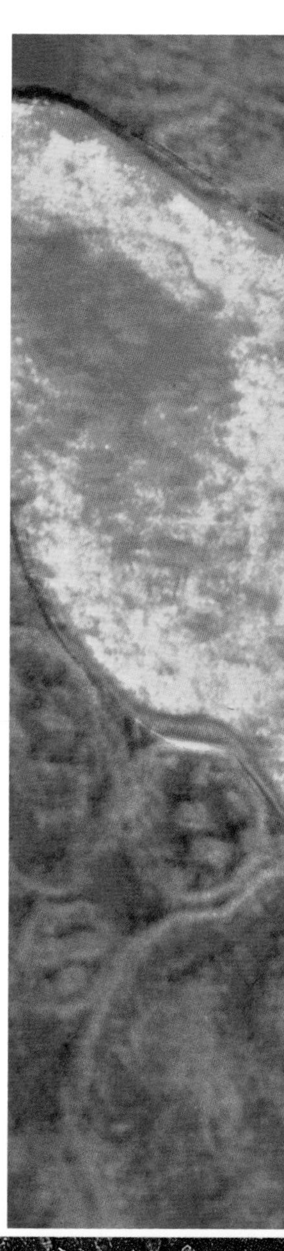

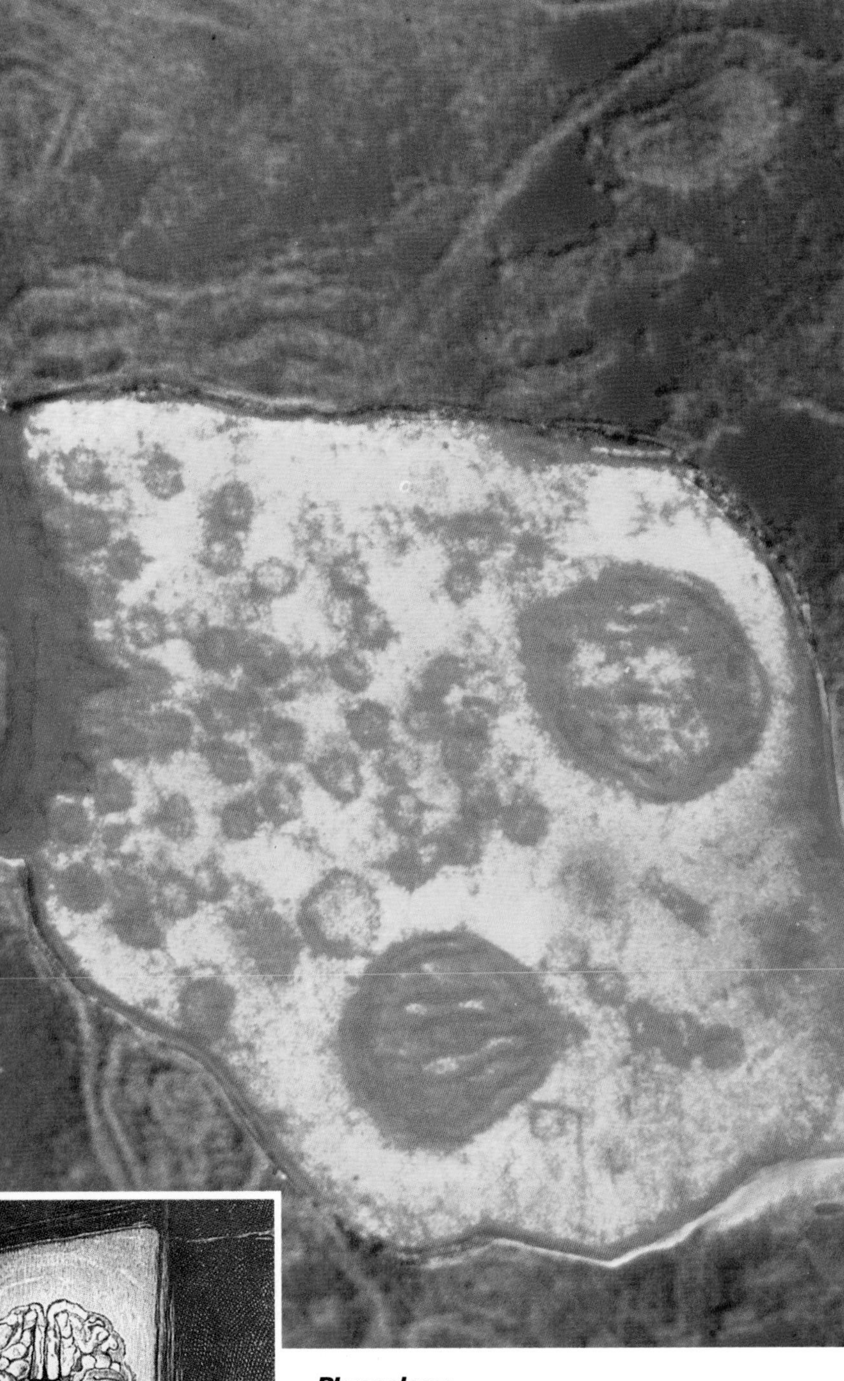

Phrenology
Descartes' contemporary, the English anatomist Thomas Willis (1621-1675), preferred to locate the brain's functions within the convolutions of the brain itself, rather than in the ventricles or the pineal, a view which influenced the development of phrenology at the end of the 18th century. Phrenology was "the science of the mind" which claimed that the brain could be mapped into 35 distinctive regions, each of which controlled a particular characteristic of human behavior, such as intelligence, sexuality, love of food or children, egotism, etc. The extent to which these traits of personality and behavior were developed (or not) in the individual depended upon the relative size of the appropriate area of the brain. This, again, was reflected by the shape of the skull, which could, accordingly, be "read" as a guide to character. This doctrine was enthusiastically expounded between 1815 to 1840 by the Viennese anatomists, Franz Gall (1758-1828) and Johann Spurzheim (1776-1832), and their Scottish disciples, George (1788-1858) and Andrew (1797-1847) Combe.

Treatments and experiments
Of obvious apparent practical value in choosing a mate or a servant, or avoiding pickpockets, phrenology also suggested that criminals and the insane might be reclaimed by encouraging the development of faculties which were less antisocial in content. For example, a person with an enlarged bump of self-regard who believed himself to be Napoleon might be cured by encouraging him to work with animals and, by this means, exercise a different, compensating, area of the mind.

This naive, materialistic phrenology had been discredited by experiments by the 1840s. Its treatments, when they worked at all in the asylum, probably did so because they reinforced the "moral method". Nevertheless, asylum doctors like the Scot David Ferrier (1843-1928) used electrical stimulation experiments later in the century to show the essential validity of the idea of brain localization. Much of the British work on cerebral localization was published in "Brain", a journal founded in 1878 by the British physician James Crichton-Browne (1840-1938).

Reflexes and nervous circuits
By the 1860s, the Russian physiologist Ivan Sechenov (1829-1905) had developed the idea that animal behavior depended upon automatic reflexes, whose actions could be intensified, or reduced, by the higher centers of consciousness in the brain. His pupil Ivan Pavlov (1849-1936) showed how reflexes could also be triggered by substitute stimuli. For example, the dog which salivated on seeing food could be trained to do so when hearing a bell ring before the food was presented. Pavlov supposed that new nervous circuits had been developed in the cortex to allow these "conditioned reflexes" to occur – an idea which was the basis for Freud's psychology. The work of the English physiologist Charles Sherrington (1861-1952) at the beginning of the 20th century differed from Pavlov's in that he traced the actual pathways taken by a signal from the skin to the spinal chord and brain, and back along nerves to the muscles. This work established the importance of the synapses – the junctions between nerves – and revealed the significance of inhibitory signals which allow one muscle in a pair to relax so that its partner can move. Sherrington's most important insight, for which he was awarded the Nobel Prize for Medicine in 1932, was that the secret of nervous coordination lay in the integration of reflexes.

The American James B. Watson (1878-1958) seized upon the work of Pavlov and Sherrington to develop a new experimental psychology based solely upon the observable, objective "facts" of behavior rather than on abstract concepts such as "mind" and "id".

Dreams and the unconscious

Freud began to apply the free-association method of therapy after 1892. He soon found that patients' memories continually harped back to childhood and to sexual experiences of an incestuous character. Charcot and other clinicians were clearly incorrect in considering neuroses (the collective term for hysteria, anxiety and obsession) to be hereditary. They were instead caused by the memory of "unpleasant" events within the individual's own life. By "repressing" these sexual memories into some unconscious part of the mind, Freud concluded, a poison was left "unconsciously" (subconsciously) which could emerge as mental and physical symptoms in later life. Relief was to be sought by purging and bringing the memory to the surface.

But had these childhood sexual experiences *really* happened? Freud intitially believed they had, and in 1896 went on record that the basis of female hysteria was actual, or attempted, incest by fathers of their daughters at the age of three or four. Such a conclusion was bound to shock and was embarrassing to Freud himself since many of his patients were daughters of family friends. As cases accumulated, the thesis seemed more and more preposterous and outrageous. Could thirty percent of Viennese parents really be seducing their children?

In 1897 Freud admitted his error – at least to himself, for he did not publicly abandon the theory until 1905 – and suggested instead that his patients' memories were fantasies. At least one historian has argued that patients' memories of child abuse were probably only too true and that, by changing his mind for fear of the immorality he had uncovered, Freud was to base psychoanalysis on fantasy rather than reality. However, if patients were fantasizing, it implied that pre-pubescent children were sexually active – a revolutionary idea which would offend traditional notions of morality, religion, education and child-rearing, as Freud made explicit in *Three Essays on the Theory of Sexuality* (1905).

A factor in Freud's abandonment of the seduction theory was his growing involvement in the analysis of the dreams which patients described under free association. On subjecting them to scrutiny, Freud saw that dreams worked on two levels. On the surface, the "manifest content" of disjointed visual experiences (i.e. what happened) was analogous to mysterious hysterical symptoms. But if these were caused by repressed emotions, dreams might also have a "latent" (hidden) content which, when analyzed, provided a "royal road to the unconscious". In a famous dream, Freud saw a medical colleague injecting a patient named Irma with trimethylamine, a simple organic substance found in fish oil, which has no medical use. On analyzing the dream by free association for its latent content, Freud's first thought was relief that the injector was a colleague, and not he himself, who would be blamed if the injection had serious consequences. Freud knew only too well that his close friend and confidant, Wilhelm Fliess (1858-1928), a nose-and-throat specialist, had recently operated upon one of Freud's female patients named Emma and had negligently left gauze in the wound, which had hemorrhaged after the operation. Emma, who was legally Freud's patient, had nearly died and Freud had hushed up the incident, the full details of which only became known in 1966. Although Freud bowdlerized the dream for publication in 1900 in order to save Fliess's feelings, he saw the dream as his unconscious mind criticizing Fliess's conduct. More importantly, the dream seemed to represent a *wish* that he had not been, even formally, associated with Fliess's incompetence.

▲ *Freud's confidant during the years he was developing psychoanalysis was the Berlin surgeon, Wilhelm Fliess, here photographed with Freud in the 1890s. In their massive and intimate correspondence, Fliess effectively acted as Freud's psychoanalyst. Fliess held some bizarre notions which Freud also toyed with: the relationship between the nose and the female genitals, the idea that men and women were bisexual, and an obsession that men have a 23-day energy rhythm comparable to the 28-day menstrual cycle in women. Their friendship lasted until 1903.*

◄ *In "The Interpretation of Dreams" (1900), Freud argued that dreams represent the fulfillment of subconscious wishes and anxieties. Since dreams were analogous to hysterical symptoms, it followed that hysterias were not memories of real physical seductions, but the repressed fantasy wishes that dreamers had experienced as children. By analyzing his own dreams and those of patients, Freud deduced that "the unconscious makes use of a particular symbolism, especially for representing sexual complexes". Candles, knives, ties or any long object, represented the penis, an oven or ship represented the female genitals, while climbing stairs indicated sexual intercourse. Such symbolism also underlay myths and fairy stories. The work of artists such as René Magritte, left in "The Secret Player" (1927), was inspired by this interpretation.*

▼ *Dreams have been a powerful theme in the world's literature and art, as in Rembrandt's biblical "Joseph Telling his Dream". Joseph is later cast naked into a pit by his jealous brothers for dreaming of his domination over them.*

Freud's first foray into the unconscious lives of people, in *The Interpretation of Dreams* (1900), was taken further in 1901 in his most popular book, *The Psychopathology of Everyday Life*, which explored the phenomenon of what soon became known as "the Freudian slip". Why do we sometimes forget the names of people, or make slips of the tongue or pen when speaking and writing? Or play about and fiddle with things as we speak? These small things, Freud said, were not trivial. "They always have meaning...and it turns out once again that they give expression to impulses and intentions which have to be kept back and hidden from one's own consciousness or that they are actually derived from the same repressed wishful impulses and complexes which we have already come to know as the creators of (hysterical) symptoms and the constructors of dreams". For Freud, then, everyday actions of an apparently random, undetermined kind had their roots in mental states that the conscious mind was unaware of. When, as Freud recalled, a friend generalized that men were more fortunate than women, who could only succeed if they were pretty, because men only needed "five straight limbs", he revealed the improper wishes of his subconscious mind! This could also be applied to humor. In *Jokes and their Relation to the Unconscious* (1905), Freud suggested that the basis of much humor was "allusion accompanied by omission", just as in dreams or slips of the tongue.

Unconscious sexual feelings for parents are a natural stage in a child's development, Freud believed

The id, ego and superego

The notion that energy is involved in maintaining repression is a reminder that Freud, as a trained man of science, did try to develop a theoretical explanation of psychoanalysis, based upon his acquaintance with neurophysiology. This was a quarter of a century before biologists became acquainted with the role of chemical hormones.

From his study of reflex action Freud knew that reflex contractions were proportional to the amount of stimulus along a nerve. (In fact more than one nerve channel is involved, and overstimulation leads to desensitization.) Like other biologists of the time he assumed that, for example, food-searching behavior was a response to an internal signal which set off reflex responses via the brain; learning consisted of the creation of nerve pathways in the brain (the cortex) during the individual's lifetime. Freud speculated further that the feeding baby was a model of this learning process. When internal signals of hunger were initially experienced, the baby kicked and screamed in an attempt to close down these painful signals. The baby learned that only when the mother suckled did the signals cease, because the hunger signals were now passing through nerves in the brain which served to turn the baby's head toward the breast for its eyes to see the nipple. Freud summed this process up as "the pleasure principle", the process whereby the mind finds ways of discharging nervous energy from the body or the outside world. The similarity with Pavlov's idea of a conditioned reflex is clear.

In the 1890s Freud called the interrelated tangle of brain pathways which were unconsciously created in any learned experience the Ego, a term already used in contemporary psychology to refer to personality. In order to explain the mechanism whereby the child consciously learned to suckle at its mother's breasts, Freud had to postulate an "inhibiting ego" which allowed variation in behavior depending on the way the breast was offered. For various reasons, in the 1920s, Freud redefined the conscious "inhibiting ego" as the *ego* and renamed the earlier concept of the instinctive, subconscious ego as the *id*. (In the original German, Freud used the words Ich (I) and Es (it), but his translators, anxious to provide psychoanalysis with a scientific, technical language, used their Latin equivalents, ego and id, instead.) To this dichotomy, which was a development of ideas proposed by a disciple named Georg Groddeck (1866-1934), Freud added a third concept, the *superego*. This last division of the mind – a subconscious part of the ego, was developed, Freud suggested, as a result of parental and societal criticisms and prohibitions and replaced the ethical concept of "conscience". Guilty feelings were thoughts, or actions, which were not in accord with the superego. The therapist's task was to bring the subconscious processes of the id under the control of the conscious ego, a treatment which involved exploring both the id and the superego.

Another way of looking at the mind was to see it beset by three conflicting demands: the body's biological urges or instincts (food, warmth, sex) which came from within; the stimuli which came from the body's environment and to which it had to respond (danger, sexual invitation); and moral demands which might easily conflict with instinct or external reality (for example, risking your life to save another's). It will be seen that these demands correspond to the id, the ego and the superego, and that it is the ego which has the heaviest and hardest task in compromising between conflicting demands. These compromises emerged as dreams, hysterical symptoms and defensive

▲ *Electra plotted with her brother Orestes to kill their mother, Clytemnestra, to avenge the murder of their father, Agamemnon. Freud, a classical scholar, saw this myth, and that of Oedipus, as symbols of unconscious desires between children and parents.*

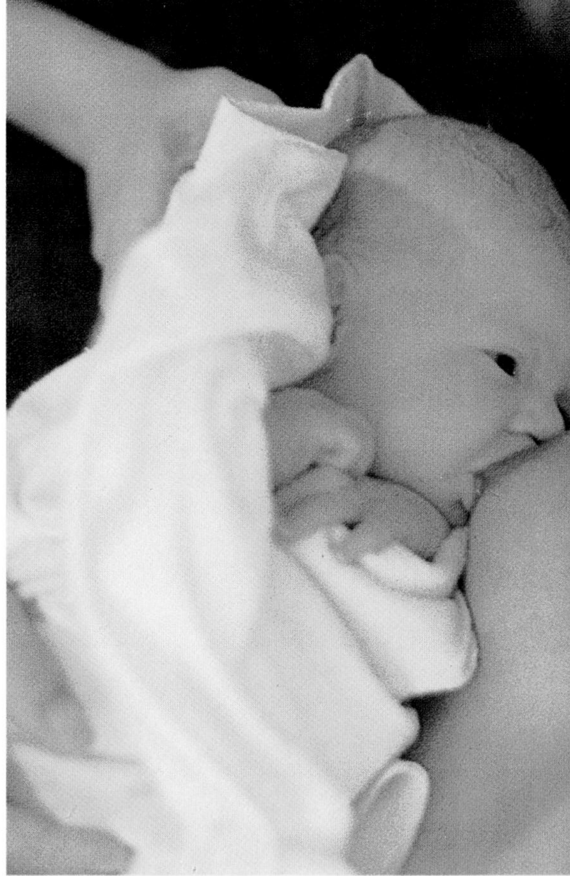

▼ *Freud used the feeding baby as a model of the learning process whereby nerve pathways were laid down in the brain. He came to distinguish the mind's instinctive desires (id) from the conscious self (ego) and its restraining power (superego).*

▼ *A skilled surgeon, the Russian psychologist Ivan Pavlov here directs the preparation of a gastric fistula in a dog whereby a channel is made from the stomach to the outside of the body. Secretions were collected, measured and analyzed in reponse to various stimuli, thus revealing the existence of conditioned reflexes.*

► *Finding Freud's authority as "father of psychoanalysis" questioned by Adler, Jung and others, the Freudians formed the "Committee" in 1912 to prevent further quarreling. This 1922 photograph shows (standing) O. Rank, K. Abraham, M. Eitingon, E. Jones, and (seated) Freud, S. Ferenczi and H. Sachs.*

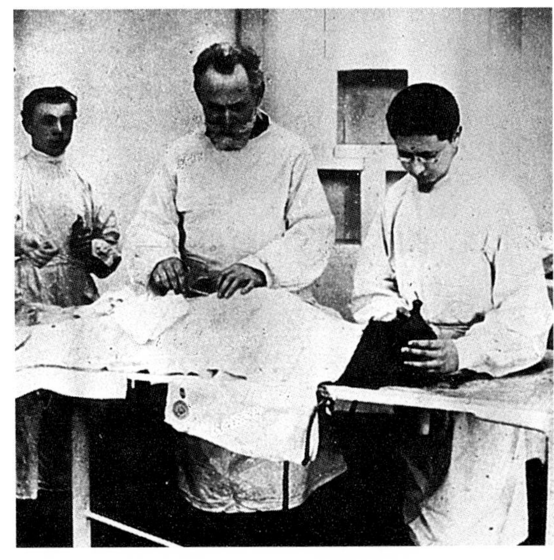

mechanisms such as "displacement" (as when a man gratifies the Oedipal urge by marrying a woman who resembles his mother), or "sublimation" (for example, the redirection of sexual or aggressive feelings into works of art, or scholarship).

Since both hysteria and normal dreams had sexual connotations, Freud's neurophysiological theory implied that cortical nerve pathways had been opened up in childhood. Just as they were developed by instinctive hunger pangs so were they opened up by sexual longings, or the "libido". In *Three Essays on Sexuality* Freud suggested that children went through three stages of sexual development: oral, anal and genital. All these were self-centered, pleasure-seeking activities: infants were autoerotic. If development from one phase to the next, or on to adult heterosexual feelings, was inhibited, retarded, or failed to run its full course, neuroses and hysteria would result. Freud believed that it was quite normal for children to have unconscious sexual feelings for their parents – boys for their mothers and girls for their fathers – and hence jealousy and feelings of rivalry between boys and their fathers and girls and their mothers. He described these feelings and the Oedipus and Electra complexes, referring to the Greek myths of King Oedipus, who murdered his father and married his mother, and of Electra, who avenged her father's death by arranging for the murder of her mother. Such complexes, which were suppressed through the superego's barrier to incest, were a normal stage in every boy or girl's development. (In self-analysis Freud recognized from his own dreams that as a child he had seen his father as a rival for the affections of his mother.) These complexes also led to children's interest in the differences between the sexes. The subconscious fears engendered by childish fantasies – penis envy in girls, castration fears in boys, the false impression that sexual intercourse necessarily involved violence, or that babies were born through the lower bowel – were, Freud believed, the major sources of adult neuroses. By repressing these fears into their unconscious (the id) children escaped tensions and anxieties which, in any case, were normally resolved by the arrival of heterosexual drive at puberty. Nevertheless, these repressed anxieties remained beneath the surface to emerge in adults as dreams and slips of the tongue, and occasionally as neuroses and hysteria.

Asylums and Treatment

Humane treatment or "casting out"?

Madness greatly exercised legislative, social and medical thought throughout the 19th century. Before then, most doctors believed that irrational behavior of any kind was a falling back to an animal state. Accordingly, there was every justification for treating the mad to the restraints, beatings and food that would have been given to a domestic animal. Restraints, nakedness, cold, poor food, bleeding and giving purges and emetics to induce vomiting and evacuation, were all methods employed to weaken and tranquilize them.

According to one view, more humane treatment of mental illness began toward the end of the 18th century, and was exemplified by the order of the French physician Philippe Pinel (1745-1826) to release lunatics from their chains in the Paris poorhouse in 1793. Its development can be related to the growth of a social conscience among evangelical members of the Church of England and the nonconformist sects, and to the development of a utilitarian political philosophy which argued that governments had a duty to legislate to insure the greatest happiness of the majority. It followed that the building of public lunatic asylums under government direction represented progress.

An alternative perspective suggests that the motives for reform were less altruistic. Given that European society was undergoing the drastic economic changes of urbanization and industrialization, it proved convenient to classify as deviant those unsociable people who posed a threat to its order and rationality by removing them from open society. This process of "casting out" coincided with the rise of a specialized breed of medical practitioners, the "alienists" (psychiatrists), who were able to advance their professional and social status by claiming over laymen a unique fitness to look after, treat and cure the mentally ill.

By the beginning of the 19th century, madness was seen as a physical disease of the brain and,

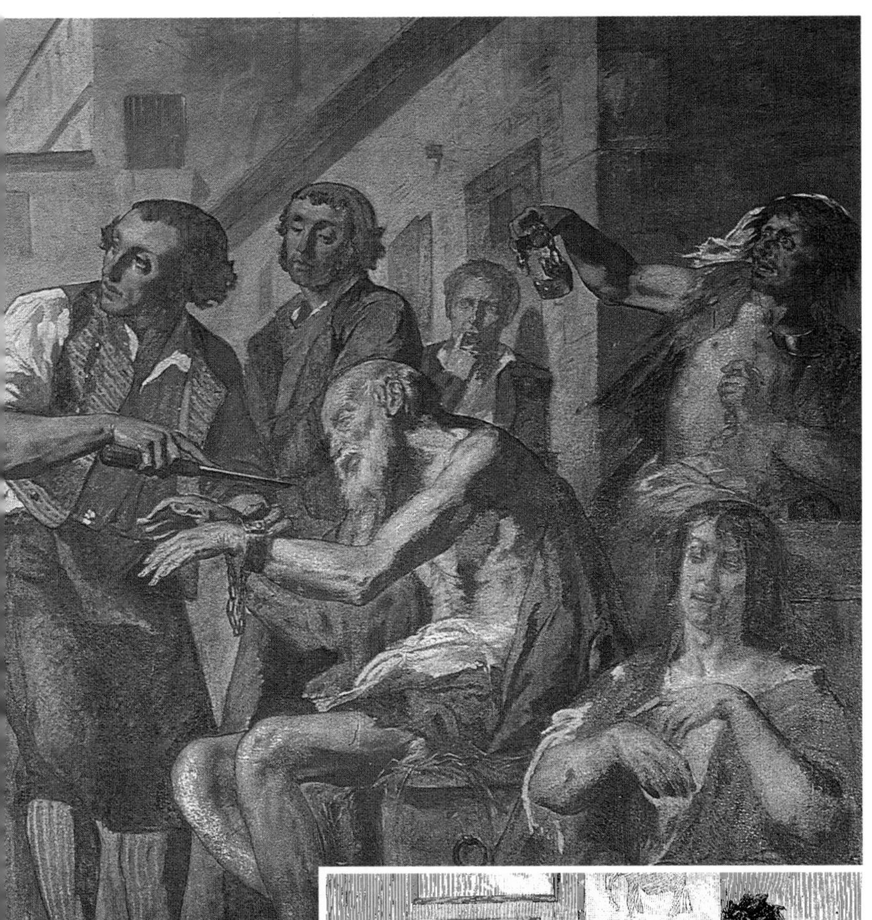

▲► Philippe Pinel (like Charcot, later) worked at La Salpêtrière in Paris. He had chains and other restraints removed from the insane in 1793 (see print above). By transforming the lunatic asylum from a prison into a hospital, he helped establish psychiatry as a profession and encourage educators like Léontine Nicolle (right) to teach subnormal children.

◄ The moral management of the insane encouraged asylum superintendents to hold dances, plays and festivities in which patients participated – as in this entertainment at the Middlesex County Lunatic Asylum, Colney Hatch, in 1853. By encouraging male patients to farm and garden or make and repair shoes, and females to launder, sew and make hats, asylums combined therapy with self-sufficiency.

hence, no different in principle from a physical ailment. Such mental diseases might be explained, it was believed, as due to a lack of moral willpower, or to inherited imperfections. Kindness, personal contact with the asylum superintendent, living in the country as opposed to the dangerous town, and a mild system of rewards and punishments, could be used to induce in a patient a personal sense of responsibility and an internal or "moral" desire on his, or her, part to return to reasonable behavior.

Although this kinder system of "moral management" abolished the grosser abuses of chains and purges, innovations such as the straitjacket (which allowed patients to walk about without injuring themselves, others, or property), the padded room, and the shower bath (for cooling the passions) could also be used for purposes of punishment – especially as inmates began to outnumber custodians.

The new asylums

Large lunatic asylums to house mad paupers (who could not afford the fees of private madhouse owners) began to be built in Great Britain from 1810 onward. They were made obligatory after 1845.

By the 1920s Sweden had adopted moral management and begun to build state asylums. In Italy and Germany, however, a variety of care systems persisted until the political unification of these countries occurred in 1860 and 1870 respectively. By then, Europeans and Americans had paralleled the British practice of identifying insanity as a medical problem to be dealt with by psychiatrists and supported by government legislation. Although such asylums had excellent records for discharging "cured" patients back into the community, they rapidly became 800-1,000-bed warehouses for incurables – especially the senile and those suffering from "general paralysis of the insane" (GPI). After 1912, GPI was recognized as tertiary syphilis and was partially cured by infecting patients with malaria, the consequent high fever destroying the syphilitic infection in the brain. (The disease is now controlled by penicillin.)

In 1841, British alienist doctors in charge of lunatic asylums formed an association, which became the Royal College of Psychiatrists in 1971. Through their "Journal of Mental Science", alienists reported their cases and the neurological research which increasingly began to occupy them. Research made by James Crichton-Browne in Yorkshire's West Riding lunatic asylum in the 1870s was particularly important for the development of diagnostic tools for distinguishing psychiatric (psychotic) from neurological diseases such as GPI and epilepsy. Both Browne and the London alienist Henry Maudsley (1835-1918) believed that much insanity was hereditary. Their fear that madness was increasing in late Victorian Britain was to encourage interest in eugenics and in intelligence testing during the early 20th century.

In 1935, the Portuguese neurologist Egas Moniz (1874-1955), learning that removal of the frontal lobes of people suffering from brain tumors caused major personality changes, began making prefrontal leucotomies (lobotomies) on psychotic patients. Between 1936 and 1949 (when Moniz was awarded the Nobel Prize for Medicine) over 10,000 leucotomies were performed in American psychiatric hospitals alone. Psychosurgery, which is now regarded as a shameful episode in the history of psychiatry, was replaced after 1950 by the use of tranquilizing drugs such as chlorpromazine (largactil). Electroplexy (electroconvulsive therapy, ECT) has also been widely used in controlling mental disease since the 1960s. Most western countries have to a greater or lesser extent closed their asylums and discharged their inmates into the community as outpatients. Only the criminally insane remain incarcerated.

▼ *In 1925 the great Austrian film director, G.W. Pabst, supported by Freud's disciples K. Abraham, H. Sachs and S. Bernfeld, made a film "Geheimnisse einer Seele" (Secrets of a Soul). Sachs wrote the book "Psychoanalyse" to accompany the film.*

► *Painting is a powerful method of psychotherapy. In this 1960s painting by a Canadian patient in the Maudsley Hospital, London, the skull is sawn open revealing a maze of obsessive memories of childhood, wood-cutting and psychiatrists.*

▲ *For Jung, dreams, fairy stories and religions arose from a shared "collective unconscious" common to all cultures (a kind of evolutionary memory) which complemented the individual consciousness. Neurosis, the "suffering of the soul which has not yet discovered its meaning" was treated by integrating the individual consciousness with the archetypal symbols of the collective unconscious.*

The growth of psychoanalysis

Freud's free-association therapy was a long process requiring months of regular meetings with a patient, who therefore tended to come from the wealthier parts of society. Therapy was continued until Freud was convinced that all the subconscious "complexes" had been brought to the surface. He usually judged the success of a treatment less by the decline of any physical symptoms than by the "transfer" on to the analyst of motives and attributes of the patient's parents, siblings or lovers, who had been the source of the suppressed feelings. Since Freud was aware that unscrupulous analysts could exploit both the long-term nature of the therapy financially and the intimate nature of the sessions sexually, he was determined to keep personal control of his system of analysis for as long as possible.

This was achieved initially through the meetings of clinical friends in Freud's apartments from 1902. This evolved into the Vienna Psychoanalytical Society in 1906. Systematic translation into English of Freud's writings had begun the year before, and this encouraged the interest of a Welshman, Ernest Jones (1879-1958), Freud's official biographer, and the first of Freud's American disciples, J.J. Putnam (1846-1918), a professor of neurology at the University of Harvard. At the same time, important support came from Switzerland, where Carl Jung (1875-1961) was working in the Zurich University psychiatric clinic. In 1908, Jung arranged for an international meeting of Freudian therapists in Salzburg, attended by 42 people.

In 1909, encouraged both by Putnam and by Stanley Hall (author of an enormous book, *Adolescence*, published in 1904) Freud visited America, where he gave five elegant lectures in English on his ideas and methods. When published, these did much to popularize psychoanalysis in North America. Although the image created by American films and novels that middle-class Americans are unable to make decisions without consulting their "shrinks" is a gross exaggeration, it is undoubtedly true that psychoanalysis has become institutionalized in American culture to an extent far beyond that in Europe. The reason for this seems to lie in the fact that psychoanalysis was taken up there by orthodox medical practitioners, who from 1925 barred unqualified colleagues from practicing. Moreover, it was a therapy which could be modified pragmatically to suit the patient. Since many Americans had great faith in their doctors, patients accepted Freudian therapy and its several variations with good grace. Freud himself was dismayed by this medical pragmatism and argued strongly in *The Question of Lay Analysis* (1926) that a psychoanalyst did not have to be medically qualified to practice his therapy.

Earlier, in 1910, when the second international meeting of psychoanalysts was held at Nuremberg, an International Psychological Association had been established with the explicit intention of cultivating and promoting "the psychoanalytic science as inaugurated by Freud both in its form as pure psychology and in its application to medicine and the humanities". However, mindful of antisemitism (➧ page 236) and the consequent danger of psychoanalysis being dismissed as a Jewish idea, Freud had Jung appointed as its permanent president. The Association continues to this day as the principal organization of Freudian psychoanalysis.

Unfortunately, as in the early Christian church, quarrels and rifts among members of the Association soon led to secessions of disciples who then founded rival schools. In 1911, Freud's Viennese pupil, Alfred Adler (1870-1937), who in 1907 had developed the idea that

▼ ***Crippled by rickets, childhood illnesses and accidents, Alfred Adler came to see a struggle for power by the weak and the inferior as the basis of human existence. He described the theory of the "inferiority complex" in 1907 and remained one of Freud's strongest allies until 1911, when he broke away to found a rival society for Free Psychological Research. Adler's success among American psychoanalysts proved galling to Freud, who remained bitter until Adler's death. The popularity of psychology and psychoanalysis in 1930s-America is shown in the writings and drawings of the humorist James Thurber. His cartoon "House and Woman" beautifully encapsulates the sex war and feelings of inferiority.***

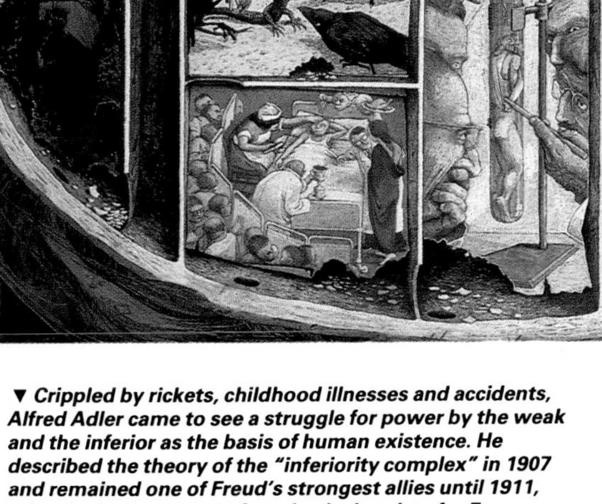

feelings of "inferiority" were the primary cause of adult neuroses, broke with Freud. Adler believed that aggressiveness and the struggle for power were more important than sexual longings or the Oedipus complex in explaining human personality. Adler's "individual psychology" was to prove particularly popular in America – to Freud's annoyance – because it suggested how parents could improve their children's chances in the power struggle of life. An equally bitter schism developed between Freud and his leading collaborator, Carl Jung, after 1914. The reasons for this rupture were due both to personality differences and to Jung's conviction that Freud had overplayed childhood sexuality. Jung went on to create his own striking and influential brand of "analytical psychology" according to which our mental lives are determined by a "collective unconscious" inherited from past generations and mappable from dreams, mythologies, alchemy and religions.

To prevent further schisms, in 1912 Freud, Jones, Otto Rank (1884-1939), and a few others formed a committee, or inner circle, whose task was to maintain and teach orthodox Freudianism. Members, who controlled the important publication *Internationale Zeitschrift für Psychoanalyse*, had to promise not to disagree publicly with any particular dogma without first discussing it with the committee. All remained steadfastly loyal except Rank, who broke away in 1929 to develop the influential idea that birth was the trauma underlying most of mankind's anxieties.

Freud and the Arts

Analyzing the work of artists

Questioned by a critic, the American poet Conrad Aiken confessed that he had been profoundly influenced by Freud from about 1912; but, he added, "so has everybody, whether they are aware of it or not." Freud's influence on writers and artists has been particularly strong: it was brought about in two ways. Firstly, he provided examples of the way in which psychoanalysis could be applied to past works of art; secondly, he directed artists' attention to the way in which psychology could be used in understanding contemporary people and their inner motives and struggles.

As a young man, Freud had been much struck by the way the Italian art historian Giovanni Morelli had been able to attribute unknown Renaissance paintings to particular famous artists by detecting some idiosyncratic quirk of the painter's style (such as the way earlobes or fingernails were depicted). Like Conan Doyle's detective, Sherlock Holmes, who belongs to the same period, both Morelli and Freud took up seemingly marginal clues from which they could construct a plausible case. This diagnosis by signs and symptoms was used by Freud to analyze Michelangelo's famous statue of Moses in San Pietro in Vincoli church in Rome. In an essay published in 1914, Freud argued from the particular way Michelangelo portrayed Moses that he had been trying to capture Moses struggling to repress his fury at the sight of the Israelites dancing around the Golden Calf. Earlier, in 1910, from a single sentence in the notebooks of Leonardo da Vinci, he had suggested that the artist had been over-mothered in childhood and turned into a homosexual. Unfortunately, in 1923 it was shown that Freud's analysis turned on a German translator's false rendering of the Italian word for a child's kite as a vulture. Nor were art historians convinced by Freud's essay on Michelangelo. Nevertheless, both essays were influential, as were Freud's studies of the motives of Shakespeare's characters such as Lady Macbeth, Hamlet and Shylock. Despite risible applications by uncritical analysts who overlook the elementary point that authors create characters for purposes of plot, or that historians must always look at counter-evidence to establish their case, psychoanalysis has become an established part of 20th-century literary criticism and historical biography.

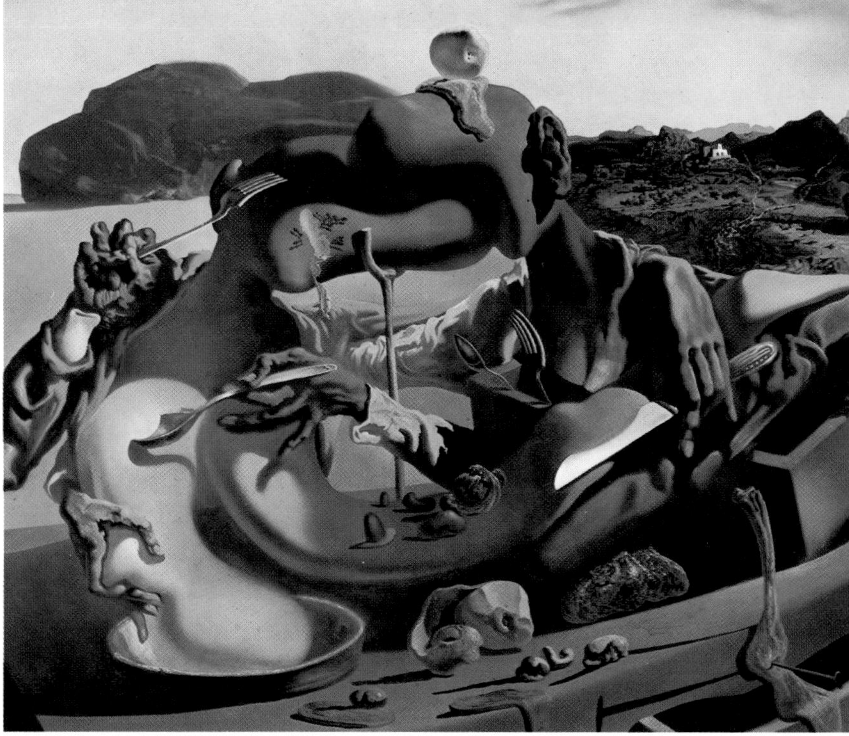

▲ In 1924 the French poet André Breton published a Surrealist Manifesto, urging writers and artists to expose the subconscious without the restraint of the superego. Painters like Salvador Dali were soon drawn to surrealism with its dream landscapes, hallucinations and symbols.

Art and the subconscious

Art and literature, and to some extent musical composition, were also transformed by Freud. In 1919 the French poet André Breton joined the Dadaist group of experimental writers whose aim

◄ An illustration from Thomas Mann's "Blood of the Walsungs". Mann acknowledged Freud's influence but as an artist felt "disquieted and reduced" by his ideas.

► "Standing Female Nude with Crossed Arms" by Egon Schiele (1910). The stark, angular nudes of the Austrian painter Egon Schiele, a student in Vienna in 1906, echo the suppressed sexuality which lay at the basis of Freud's exploration of the unconscious. In Germany, Schiele influenced the development of Expressionism, in which, hostile to illusion, the artist conveyed inner feelings.

◄ The "Bloomsbury set" writers Leonard Woolf and his wife Virginia Woolf (left) founded the avant-garde Hogarth Press in London in 1917. Despite Virginia Woolf's scepticism of Freud's teachings, the Press began to publish English translations of Freud's work by James Strachey from 1924. The Woolfs visited Freud in 1937 and found him a "screwed up shrunk very old man", but with an "old fire now flickering".

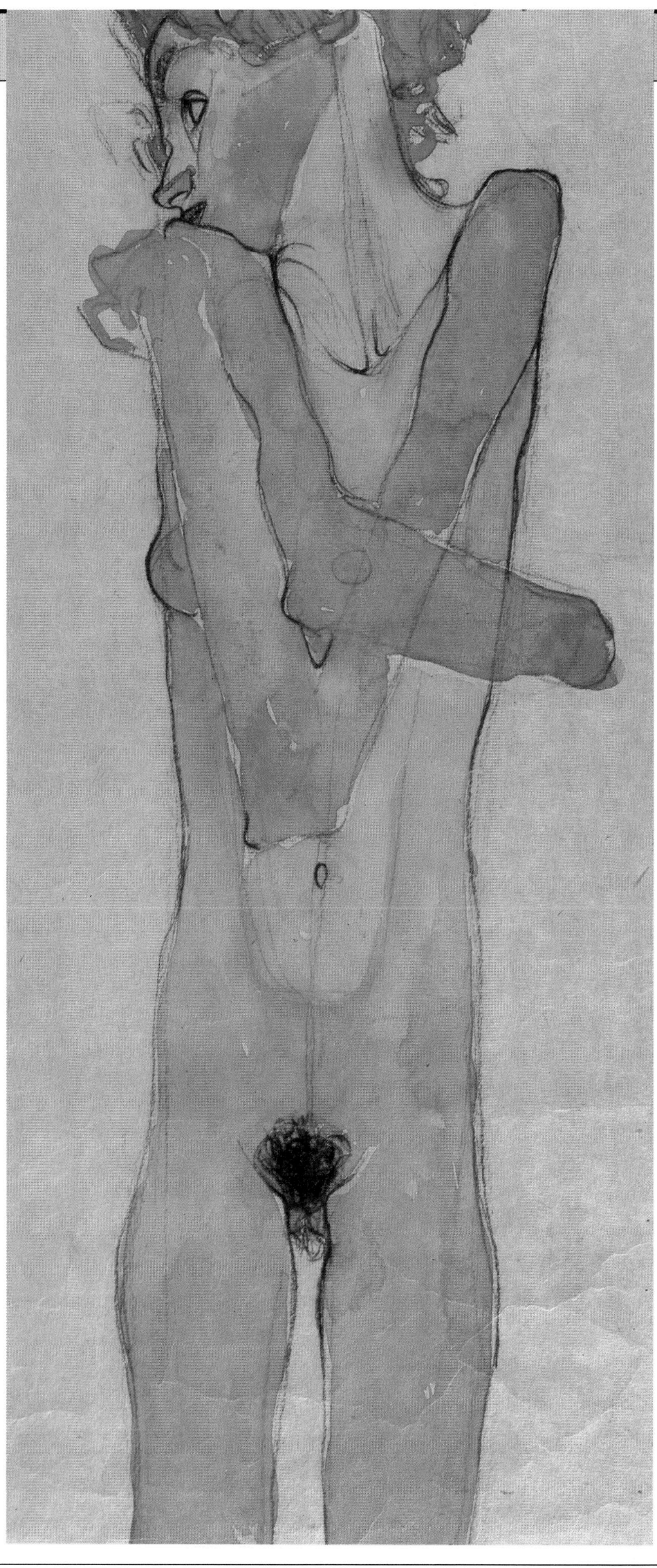

was to shock the middle classes and destroy their values by exploring such techniques as "automatic", or free-association writing. In 1924 Breton published a surrealist manifesto according to which writers and artists had a duty to express the subconscious without the restraint of the superego. Painters and sculptors such as Salvador Dali, Hans Arp, Joan Miró, Max Ernst and René Magritte were soon drawn to surrealism with its dream landscapes, hallucinations and symbols which, by association, evoked other images. Working on Freudian lines, all these artists strove to eliminate any consciously logical thought in their work – they claimed to work by pure, subconscious impulse, striving to express "the marvelous which is beautiful". Poets sought for similar inspiration. As Dylan Thomas put it, "poetry must drag further into the clear nakedness of light more of the hidden causes than Freud could realize."

The psychological novel also came to fruition in the 1920s and 1930s as writers attempted to explore the inner feelings of their characters and analyze behavior in terms of inner, hidden drives. For example, May Sinclair in "Mary Olivier: A Life" (1919) exploited Freud's Oedipus conflict and the love-hate relationship between a mother and her daughter. The novel was written in the third person and adopted what Sinclair called a "stream of consciousness" technique (the phrase had been used earlier by the American psychologist William James (1842-1910) to refer to the way any one idea is fringed with overtones of others) whereby key events, sensations and images are presented to illustrate Mary Olivier's development. The technique was further refined by Virginia Woolf in "To the Lighthouse" (1927) and "The Waves" (1931) in which narration was eliminated and replaced by sequential, poetical, moments of living.

The Irish novelist James Joyce, living in exile in Trieste and Zurich, was inevitably influenced by psychoanalysis in writing "Ulysses" (1922). Freud's dream theory actually determined the form of Joyce's final experimental novel "Finnegans Wake" (1939) which employs puns, wordplay and Freudian slips in a multitude of languages, ancient and modern. Other novelists, such as Henry Handel Richardson (the pen name of Ethel Richardson), Franz Kafka and Thomas Mann, used Freudian motifs to explore lovers' jealousies, neuroses, child-parent relations or the alienation of artists from society.

Inevitably, film makers were also influenced, as can be seen in the surrealist work of the Spanish director Luis Buñuel, or in the psychological thrillers of Englishman Alfred Hitchcock. However, in 1924, when Hollywood producers tried to get Freud to advise them on the psychological problems of love (presumably to make more realistic films), he politely declined. Freud's emphasis on childhood also inspired a few composers, like the Austrian Anton von Webern, and the American Charles Ives, to draw upon subconscious memories from their infancy, while in the strange and alienating "Pierrot Lunaire" (1912), the Viennese composer Arnold Schoenberg portrayed in speech-song the subconscious inner soul trying, and failing, to communicate.

By the 1930s psychoanalysis was applied to the study of language, religion, law and the origins of civilization itself

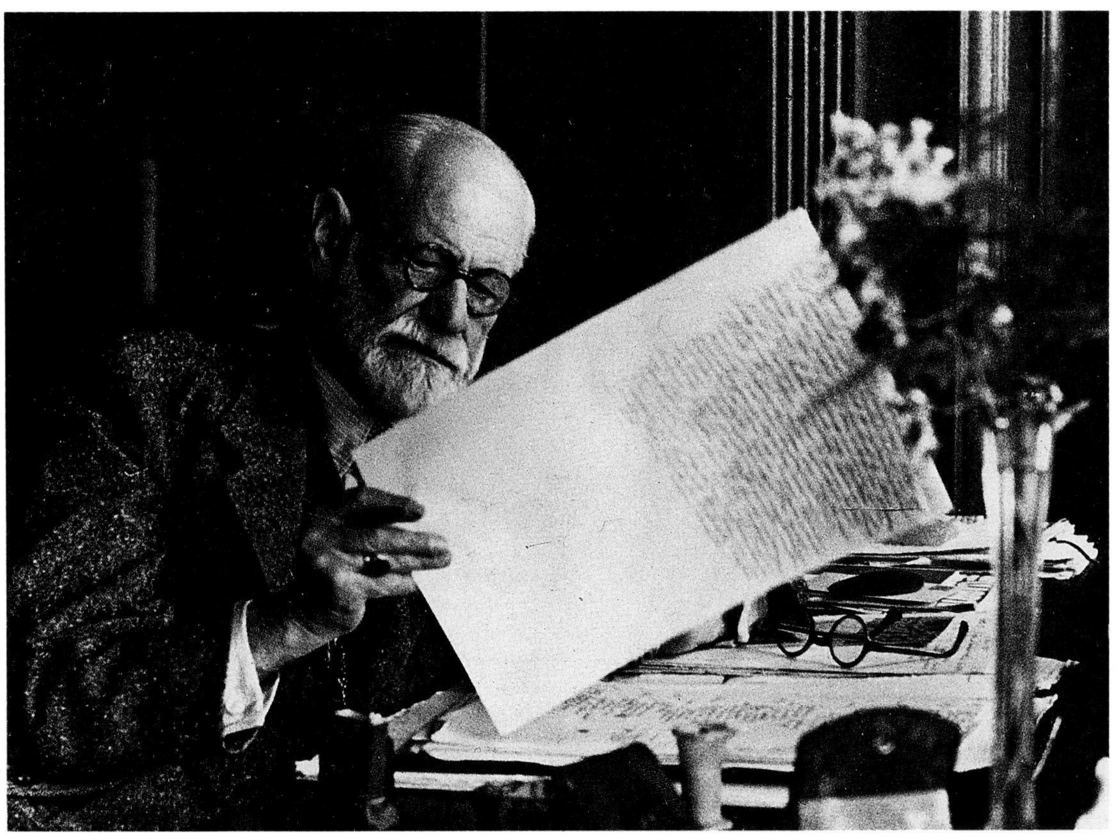

◀ **Exiled in London in 1938, Freud composed his last work, "An Outline of Psychoanalysis". Although warned by Fliess of the dangers of excessive cigar smoking, Freud's jaw cancer was not diagnosed until 1923 when he was already 67. A long course of radium therapy culminated in 1924 with the removal of Freud's right upper jaw and palate and the fitting of an uncomfortable metal roof to his mouth. Although speech was difficult he was able to work normally until 1937 when further operations were needed. In September 1938 he reminded his personal physician of his promise to administer euthanasia by morphine. Life, he said, is "nothing but torture and makes no sense any more".**

The appalling mental casualties of World War I trench warfare had provided fairly convincing proof that the subconscious mind was involved in producing physical symptoms and abnormal mental behavior. Both victors and vanquished came to see the value of psychotherapy in the treatment of shell-shocked victims and the standard neurology textbooks began to devote more attention to psychoanalysis. Mental patients in the state asylums began to benefit from psychiatrists' new confidence that they understood the unconscious workings of the human mind, and their awareness that the distinction between neuroses and madness was blurred. At a more popular level, too, and encouraged by Freud himself in such writings as *Totem and Taboo* (1913), *Civilization and Its Discontents* (1927) and *Moses and Monotheism* (1939), psychoanalysis began to be applied to the study of language, customs, religion, law and the origins of civilization itself. In consequence, it began to influence the way artists and writers painted and wrote about the human condition (◀ page 226).

In 1933 Hitler banned psychoanalysis because of its Jewish associations. Freud remained in Vienna until the Nazis seized power in Austria in 1938 and carried out searches of his house. With the help of Ernest Jones, Freud emigrated to London, where he died from cancer of the jaw, the disease he had fought since 1923, on 23 September 1939. After cremation, his ashes were placed in one of the many Greek painted vases he had loved to collect, and deposited in the crematorium at Golders Green in London.

Despite the diversifications and modifications of Freud's original teachings, and the disagreement that still exists over the best therapeutic techniques to be used, the Freudian inheritance remains intact. Psychoanalysis in the late 20th century is no longer to be equated with Freud, but Freud's influence remains paramount.

▲ **As Charles Slackman's drawing "Freud on Freud" (1970) suggests, the validity of Freudian psychoanalysis will always be controversial. Does it tell us more about Freud's own mind and culture than our own?**

Albert Einstein 1879-1955

"He will never make a success"...The Swiss Patent Office...Dispelling the ether...The 1905 papers on special relativity, molecular effects and radiation...The general theory of relativity...Public life and private research... PERSPECTIVE...Is there but "one science"?...Einstein's presence in literature...Technical advances based on atomic research...United States' science and immigration

The two great geniuses associated with the word "gravitation" – Newton and Einstein – though separated by over 200 years, shared a number of characteristics. For example, both began at an early age the work that led to their important discoveries and both became involved in politics later in their lives. But there are important differences, not least that Einstein's Germany underwent more drastic upheavals than Newton's England. The forces that obliged Einstein to take refuge in the United States have had a major impact on the practical application of 20th-century physics. More important, in the history of science, Einstein overturned fundamental ideas on the nature of time and space that had held sway since the time of Newton.

Early days

Albert Einstein was born in the small town of Ulm in south Germany on 14 March 1879, but his family soon moved to the nearby Bavarian capital, Munich. His father was in the electrical engineering business, so Einstein, from his early days, became acquainted with what was then the very young electrical industry. As a schoolboy, he was thought to have very little ability (it has been suggested, without much evidence, that he was dyslexic). When Einstein's father asked what career his son should follow, the headmaster replied: "It doesn't matter; he will never make a success of anything." Certainly, the young Einstein was introspective and rather isolated. His family was Jewish in an overwhelmingly Catholic city. At the same time, they took no active part in religious observances, and so were not integrated into the local Jewish community. During these years, Einstein began to discover topics that did interest him. Like Newton, he was attracted by the synthesis of classical geometry made about 300 BC by Euclid: he particularly liked its lucidity and certainty. This formed part of his growing interest in mathematics – one of the few good results he later saw in his schooling, for he quickly came to detest the regimentation of school life.

In 1894 his father's business failed, and the whole family apart from Albert moved to Italy, settling in Milan. Within six months Einstein quitted his school without taking the leaving certificate and trekked south to join his family. He found the relaxed atmosphere of Italy much more acceptable than the discipline of Germany; but his father's finances continued to be precarious, and Albert had to consider his future career. It was decided he should follow in his father's footsteps and train to become an electrical engineer. Without a school-leaving certificate, the openings for further education were limited. However, the Swiss Federal Polytechnic School in Zurich had a high reputation, and accepted students purely on the results of an entrance examination. Einstein sat the examination and failed. After a year's intensive schooling in Switzerland, he took it again and passed.

Einstein's revolution

Space and time are basic to all scientific observations, so understanding them has always been a concern of scientists. For centuries, scientists supposed that there was some kind of universal background to which relative measurements could ultimately be referred. In Aristotle's picture of the world everwhere outside the Earth was occupied by a solid transparent "aether", which could be used to describe motions. By the end of the 17th century, Newton had shown that, in order to fit the observations, any such all-pervasive ether would need to be very tenuous. Nevertheless, in the 19th century, this Newtonian ether was still widely accepted as the background against which measurements of time and space could be made.

Einstein boldly decided that the whole idea of some kind of universal framework, against which measurements could be made, was wrong – all observations are relative. He developed this idea to show that gravitation could be thought of as a distortion of the space around an object. As Faraday shifted the focus of attention from electric or magnetic bodies to the properties of the space round them, so Einstein did the same for gravitation. But Einstein was responsible for a still wider-ranging revolution. Along with Niels Bohr, he was a founder of quantum mechanics. His analysis of space, time, matter and radiation has provided the foundation of much of 20th-century physics.

▼ Einstein at the Swiss Patent Office in Berne.

Albert Einstein – His Life, Work and Times

Student life and after

Einstein's student years were marked again by his independent approach: he spent as much time reading the original works of the great 19th-century physicists as in attending lectures. Nevertheless, he graduated satisfactorily in 1900. By this time he was concerned with physics rather than electrical engineering, and his first hope was for a university post at Zurich. But no offer came his way – perhaps because of his independent attitude as an undergraduate. He survived on temporary teaching posts until midway through 1902, when, aided by a certain amount of influence behind the scenes, he was offered a post at the Swiss Patent Office in Berne. There he stayed, learning how to discern the basic ideas of patent applications and express them clearly, for the next seven years.

Einstein had enjoyed his student life in Switzerland. He contrasted the more democratic atmosphere there with his memories of Germany. He soon decided to transfer from German to Swiss nationality, and was so eager that he actually relinguished his German citizenship before being accepted by the Swiss. Consequently, Einstein entered the 20th century as a stateless person, and did not become Swiss until 1901.

A few months after starting work at the Patent Office, Einstein married a Hungarian, Mileva Marič, who had also been studying mathematics in Switzerland. Within a few years of graduating, he had become a married Swiss civil servant – not the most obvious background for an assault on the theoretical physics of his day. Einstein later emphasized how wrong this view was; the job provided him with the security and time he needed to come to grips with the basic problems of physics.

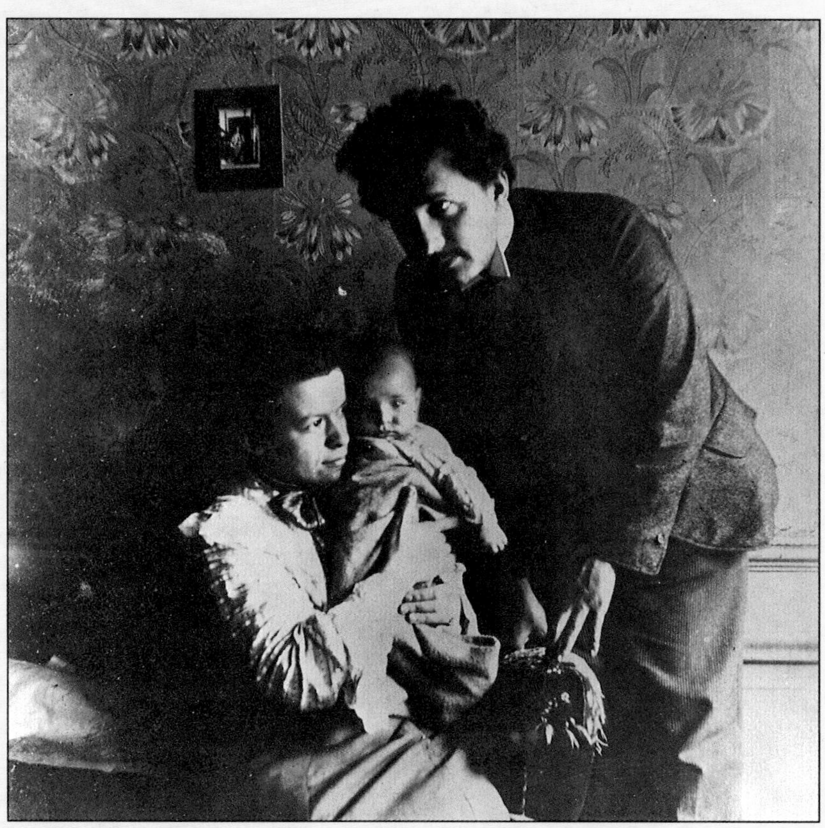

▲ *Einstein with his wife Mileva and their first son Hans Albert in 1904.*

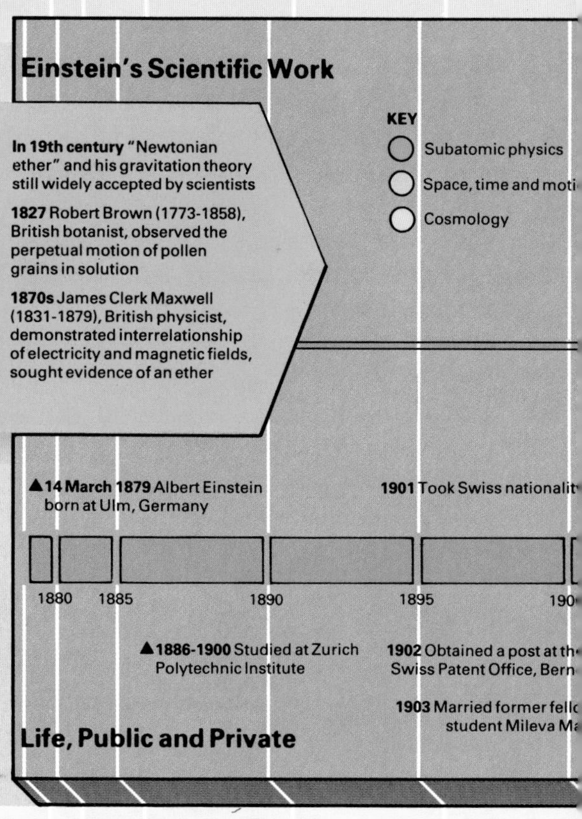

Historical Background
■1884 First conference of Zionist movement held in Prussia, after pogroms in Russia and Poland
Einstein's Scientific Work
In 19th century "Newtonian ether" and his gravitation theory still widely accepted by scientists
1827 Robert Brown (1773-1858), British botanist, observed the perpetual motion of pollen grains in solution
1870s James Clerk Maxwell (1831-1879), British physicist, demonstrated interrelationship of electricity and magnetic fields, sought evidence of an ether
KEY
Subatomic physics
Space, time and moti
Cosmology
▲14 March 1879 Albert Einstein born at Ulm, Germany
1901 Took Swiss nationalit
1880
1885
1890
1895
▲1886-1900 Studied at Zurich Polytechnic Institute
1902 Obtained a post at th Swiss Patent Office, Bern
1903 Married former fell student Mileva M
Life, Public and Private

Scientific Background

1900 John Rayleigh (1842-1919), British physicist, tried to explain how a hot body radiates its heat: applying then acceptable physics, he failed

1900 Max Planck (1858-1947), German physicist, solved Rayleigh's problem by assuming that heat is emitted from bodies in small bursts (quanta)

○ **1883** Ernst Mach (1838-1916), German physicist and philosopher, suggested that Newton's ideas on space and time must be replaced

○ **1887** Heinrich Hertz (1857-1894), German physicist, observed effect of light on electrical properties of metal surfaces (in c. 1886 discovered radio waves)

○ **1887** Albert Michelson (1852-1931), US physicist, and Edward Morley (1838-1923), US chemist, failed to detect ether experimentally

○ **1889** George Fitzgerald (1851-1901), Irish physicist, suggested that bodies contract along the direction in which they are moving

○ **1892** Hendrik Lorentz (1853-1928), Dutch physicist, also suggested contraction

1897 Existence of electron established by J. J. Thomson (1856-1940), British physicist ○

1899 Alpha and beta particles discovered by Ernest Rutherford (1871-1937), New Zealand-born physicist ○

1875 1885 1890 1895

n late 19th century Rivalry
ased between European
ers in economic, territorial
colonial spheres

1914-1918 World War I

1917 Revolution in Russia

1920 League of Nations inaugurated

1929 Wall Street Crash heralded increased economic depression worldwide

1933 Nazis came to power in Germany

1939-1945 World War II, "the physicists' war"

16 July 1945 First atomic blast, followed (August) by bombs on Hiroshima and Nagasaki

1945 United Nations formed

1948 State of Israel proclaimed

905 Paper on the quantum of
ght and the photoelectric effect
ublished in *Annalen der Physik*

905 Paper on Brownian motion

905 Paper on special relativity,
On the Electrodynamics of
Moving Bodies"

1911 Asserted equivalence of gravitation and inertia

1916 Completed work on general relativity

1916 Work on the absorption and emission of radiation, including stimulated emission, the basis of lasers

1918 Began work on unified field theory (gravitation and the quantum combined into a single theory); continued this for rest of his life

1924 Einstein took up Bose's work

1930 Produced model of the expanding Universe

1908-1914 Taught at universities in Berne, Zurich and Prague

1914 On outbreak of war, Einstein's attachment to pacifism increased

1919 Married cousin Elsa

1925 Joined Board of Governors of Hebrew University, Jerusalem

1933 Left Germany after Nazis came to power, took up post at Institute for Advanced Study, Princeton (held till 1945)

1952 Declined request to become president of Israel

18 April 1955 Died in Princeton, New Jersey

1910 1915 1920 1925 1930 1935 1940 1945 1950

1914-1933 Professor of physics and director of theoretical physics at Kaiser Wilhelm Institute

1921 First visit to United States

1921 Nobel Prize in Physics, primarily for work on photoelectric effect

1939 Wrote to President Roosevelt about the potential of atomic energy

1940 Became US citizen

02 Philipp Lenard (1862-1947),
rman physicist, showed that
ctrons are released from
tal by light, but in an
expected way

1905 Lenard won Nobel prize for work on cosmic rays; also studied photoelectric effect

904 Jules Poincaré (1854-1912),
rench mathematician,
scussed the relative nature of
me and motion

c. 1908 "Space-time" derived by Herman Minkowski (1864-1909), Russian-born German mathematician, from the special theory

904 Lorentz discussed how
lectric and magnetic forces are
fected by motion

1911 Rutherford proposed nuclear theory of the atom

1913 Niels Bohr (1885-1962), Danish physicist, proposed that an atom consists of a nucleus plus electrons and that radiation is absorbed, or emitted (as a quantum), every time an electron jumps outward from, or inward to, the nucleus

1917 Willem de Sitter (1872-1934), Dutch astronomer, showed that general relativity can lead to an expanding universe

1919 Arthur Eddington (1882-1944), British astrophysicist, found observational support for general relativity at time of solar eclipse

1919 Rutherford announced first artificial disintegration of an atom

1923 Nobel Prize won by Robert Millikan (1868-1953), US physicist, partly for work confirming Einstein's theories which he had sought to disprove

1924 Satyandra Nath Bose (1894-1974), Indian physicist, suggested method of applying statistics to subatomic particles

1927 Uncertainty principle relating to observation of subatomic particles, enunciated by Werner Heisenberg (1901-1976), German mathematical physicist

1929 Edwin Hubble (1889-1953), US astrophysicist, presented observations showing that the Universe is expanding

1934 Artificial radio isotopes first produced by Irène (1897-1956) and Frédéric (1900-1958) Joliot-Curie

1936 In *German Physics*, Lenard defended "true" science, attacked relativity theory as a Jewish conspiracy

1938 Otto Hahn (1879-1968), German chemist, and Austrian physicist Lise Meitner (1878-1968) and her nephew Otto Frisch (1904-1979) found that nuclei of uranium can be broken down by fission

1942 Nobel laureate Enrico Fermi (1901-1954), Italian physicist, produced first sustained nuclear chain reaction, working in United States

905 1910 1915 1920 1925

1948 Dennis Gabor (1900–1979), Hungarian-born British physicist, described method of holography, but it required use of lasers to work

1960 Theoretical proposals by Charles H. Townes (1915–), US physicist, Nikolai Basov (1922–) and Alexander Prokhorov (1916–), Soviet physicists, led to the development of the laser

1965 Arno Penzias (1933–), US astrophysicist, and Robert Wilson (1936–), US radio astronomer, discovered cosmic microwave background radiation, supporting "Big Bang" theory of origin of Universe

1974 Stephen Hawking (1942–), British astrophysicist, developed the theory of black holes

The ether and light

Like Newton, Einstein arrived early at one of the central problems he pursued throughout his life. At the age of 16 he wrote a letter to an uncle concerning the connection between electromagnetic waves and the ether. In so doing, he was touching on a matter immediately involved with his later work on relativity.

Until the latter half of the 19th century, electricity and magnetism were seen as separate things, though Michael Faraday (1791-1867) and others had shown (◆ page 139) how they could be converted into each other. However, James Clerk Maxwell (1831-1879), following up Faraday's ideas, showed theoretically that electromagnetic waves could exist in which the electrical and magnetic effects were inextricably linked together. He also showed that light represented one type of electromagnetic wave (◆ page 146). It seemed evident in the 19th century that, to have waves, there must be something present that could transmit them. How could you have waves at sea if you removed

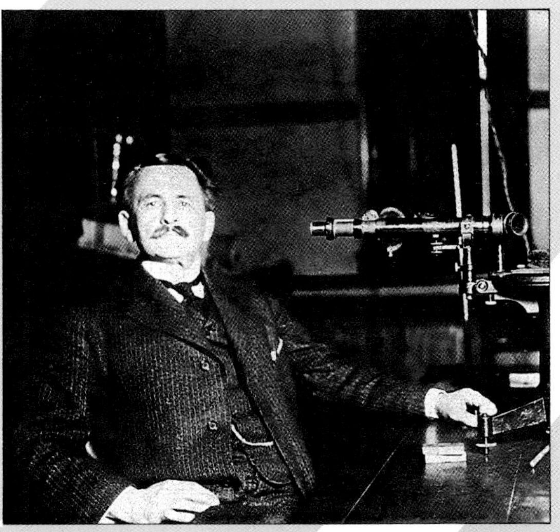

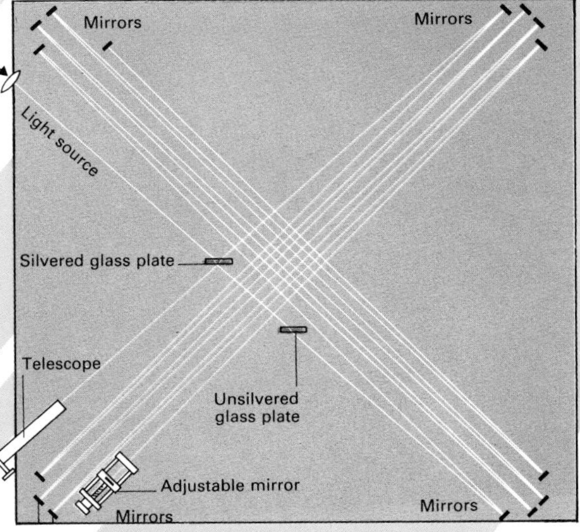

◀▲▶ To "dispel the ether" was not the intention of US scientists Albert Michelson (left) (1852-1931) and Edward Morley (1838-1931). But they did dispel it, in a celebrated experiment. In the tradition of Aristotle and Newton, 19th-century scientists believed in the existence of a transparent "ether" that permeated the Universe and transmitted electromagnetic waves including light. The experiment of 1887 set the interferometer devised by Michelson on a rotating table. A light beam was split into two parts moving at right angles to each other and then brought together again. Because of the Earth's motion in space, the "drag" of the ether was expected to produce a displacement of the interference fringes in the recombined beam. The experiment "failed", but this zero result began speculation that culminated in Einstein's 1905 special theory of relativity.

◀ Interference patterns occur when two (or more) coherent beams of light interact. Bright bands are seen where the amplitudes of the two waves reinforce each other and dark bands are seen where the waves cancel each other out. Because white light is made up of many colors of differing wavelengths (the spectrum), colored fringes are seen surrounding the bright bands. A slight change in phase between the two waves would result in a measurable shift of these colored fringes.

the sea? Maxwell had shown that light consisted of electromagnetic waves; but, since we can see the stars, light must be able to traverse the apparently empty space between the stars and us. In the 17th century, Newton had argued that some kind of material must exist throughout all space to transmit gravitational pull from one object to another. His argument was along the lines – how can you have a tug-of-war without a rope connecting the two teams to transmit the pull? This "ether", as it was called, seemed in the 19th century what was needed to transmit electromagnetic waves.

Newton recognized that most measurement was relative. We measure the motion of a ship relative to the Earth, although we know that the Earth itself is in motion – both spinning on its axis, and moving round the Sun. We could allow for these motions, but we would then need to allow for the fact that the Sun is moving, and so on. Newton wondered if it was possible to make any measurement that was not a series of relative measurements. The ether seemed to provide an answer. If it is thought of as a kind of very tenuous jelly through which the objects move, then, instead of measuring the movement of one object relative to another, we can measure all motions relative to the ether. These would represent *absolute* measurements.

Agreeing with Newton, 19th-century physicists turned the argument round – measurements of motion should make it possible to detect the ether. However, to the physicists' surprise, however much they refined their equipment, the experiments they carried out failed to reveal the ether. These experiments depended on observing the properties of light (electromagnetic waves), so Einstein's early thoughts concerned one of the most important and puzzling problems facing physics in the early 20th century.

Einstein realized that one of the key questions concerned the nature of light itself. He asked himself, what would happen if an observer could travel as fast as a ray of light? Evidently, if the observer and a particular wave started out together, they would remain together. The wave would therefore seem to be standing still, simply oscillating up and down.

However, work by James Clerk Maxwell had indicated that a stationary wave could not correspond to light: so an observer traveling with the speed of light should see something different from our observer traveling at less than the speed of light. But why should traveling at one particular speed make the whole world look different?

The innovations of 1905 – special theory of relativity

While he was at the Swiss Patent office, Einstein found the answer to the questions he was asking about the nature of light, which he published in 1905. The speed of light, and of electromagnetic waves in general, is unique because it is the maximum permissible speed in the Universe. No ordinary body can ever reach that speed, so no observer can ever keep up with a light wave. To arrive at this conclusion Einstein had to assume that however observers moved relative to the source of the light they always found the light moving relative to themselves at the same constant speed. This assumption was at variance not only with Newton, but also with common sense. If a stone is thrown from a passing vehicle, its speed, as seen by an observer standing by, will depend both on the speed with which it is thrown and the speed of the vehicle. Why should this not be true of a ray of light?

Einstein's assertion led to even stranger results. Speed is measured in terms of distance covered per unit time, in for example, miles per hour. Hence, if speed is relative, so is distance. One consequence, Einstein deduced, was that the apparent dimensions of an object depend on the relative motion of the observer. Equally, time is relative. So the length of time something takes to happen can also depend on relative motion of the observer.

It might be expected that Einstein's contemporaries would find these striking conclusions hard to swallow. In fact, some of them had been working along other similar lines. For example, Hendrik Lorentz (1853-1928) in the Netherlands and George Fitzgerald (1851-1901) in Ireland had already suggested that a moving body might appear to contract relative to a stationary observer. This is still sometimes referred to as "Lorentz-Fitzgerald" contraction. Einstein provided a logical explanation of why this should happen. His approach is nowadays called "the special theory of relativity." The word "relativity" reflects the fact that all motion is measured relative to some observer: Einstein denied the existence of both the ether and the possibility of absolute motion. The word "special" refers to a limitation of Einstein's theory. He and his contemporaries were only talking about bodies moving at constant speed.

For physicists, the fundamental properties of a body are typically related to its size and to the amount of material it contains (that is, its mass). Shortly after publishing his ideas on the measurement of length in relativity, Einstein extended his work to include mass. One of his results represents probably the most famous equation in the world today; it says that energy is equal to mass multiplied by the square of the speed of light. This is often written: $E = mc^2$. Because the speed of light is very high, destroying a very small amount of mass can lead to the release of a very large quantity of energy. At the time, this discovery was of academic interest, but we now know it as the source of the energy of the atomic bomb.

▲ Hendrik Lorentz suggested that dimensions are relative.

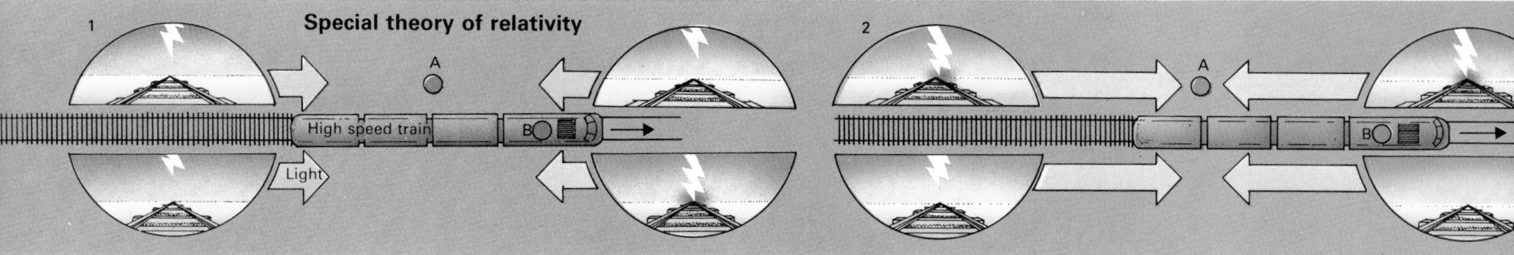

▲ Part of Einstein's special theory of relativity abolishes the Newtonian idea of time as an absolute. In Einstein's own "thought experiment" two bolts of lightning strike opposite ends of a railroad track and are seen by an observer "A" on the embankment and a passenger "B" on a high speed train. In situation (1), moving toward the light waves "B" sees the lightning strike the track only in front of him. In (2) "A" sees both lightning bolts striking the track simultaneously as the light waves arrive together. However "B" has yet to see the second lightning bolt strike. The light has still to arrive from that source. Which of these views is then correct? Neither. Einstein said that measurements of time depend on the frame of reference – whether the observer is moving or not.

► Einstein won the Nobel prize not for his work on relativity, but for explaining the long-standing puzzle of the photoelectric effect in 1905. If a metal surface is illuminated, electrons are given off. When the light is strengthened, more electrons are produced, as might be expected; but their energy levels are identical to those of electrons given off in dim light. The energy of the electrons depends not on the intensity of the light hitting the surface, but on its wavelength: the higher frequencies such as those from the blue end of the spectrum give higher energies than lower frequencies from the red end. Adapting Planck's idea, Einstein explained this by suggesting that radiation consists of "packets" or "quanta" of energy, each of which can be absorbed by an electron and dislodge it.

▲ *Max Planck's idea that energy is emitted in discrete "bundles" or "quanta" was taken up by Einstein and applied in his photoelectric theory. Niels Bohr used it in his model of the atom.*

Atoms and the quantum

The special theory of relativity would have been enough by itself to bring Einstein fame, but again in 1905 he published other work which was equally influential. Early in the 19th century, a Scottish botanist, Robert Brown (1773-1858), had noted that pollen grains suspended in water appeared to be in constant, but irregular motion when viewed through a microscope. Various attempts were made to explain this motion, without success. Now Einstein provided a satisfactory explanation in terms of the molecules which made up the water. These molecules were always in motion because of the heat of the water, and were therefore continually striking the pollen grains from all sides. However, sometimes more water molecules would strike one side of a pollen grain than the other, so causing it to move.

Some eminent scientists of the day still did not believe that matter is composed of atoms and molecules. Einstein's explanation of molecular effects was generally accepted as clear evidence that matter is formed from tiny particles too small to be seen by the human eye.

But the work specifically cited in the eventual (1922) award to Einstein of the Nobel Prize for physics appeared in yet another paper in 1905. A major interest for physicists, in the latter part of the 19th century was the way in which a hot body radiates energy into space. For example, a piece of metal heated to white heat radiates in a different way from one heated to red heat. It proved surprisingly difficult to provide a theoretical explanation of this difference. Finally, in 1900 the German physicist Max Planck (1858-1947) pointed out a possible solution. In order to fit the experimental data he found it necessary to suppose that radiation was emitted in bursts, rather than continuously. This seemed so contrary to common sense that most people saw it as just a device for the purposes of calculation, rather than as being physically meaningful. Einstein proposed that Planck's idea should be taken seriously: it actually did represent the true nature of radiation. For some purposes, light should not be thought of as waves, but as a collection of "particles", each produced by a "quantum" (i.e. discrete) change within an atom. Einstein went on to show that this approach could also explain the puzzling results of some experiments on the way in which light interacts with the surfaces of solids. His colleagues were less ready to accept this speculation than his theory of special relativity; yet, in retrospect, it represents a vital strand in the development of quantum mechanics, which along with relativity is one of the basic advances in 20th-century physics.

Photoelectric effect

High frequency light

High energy electrons

Metal plate

Low frequency light

Low energy electrons

Brownian motion

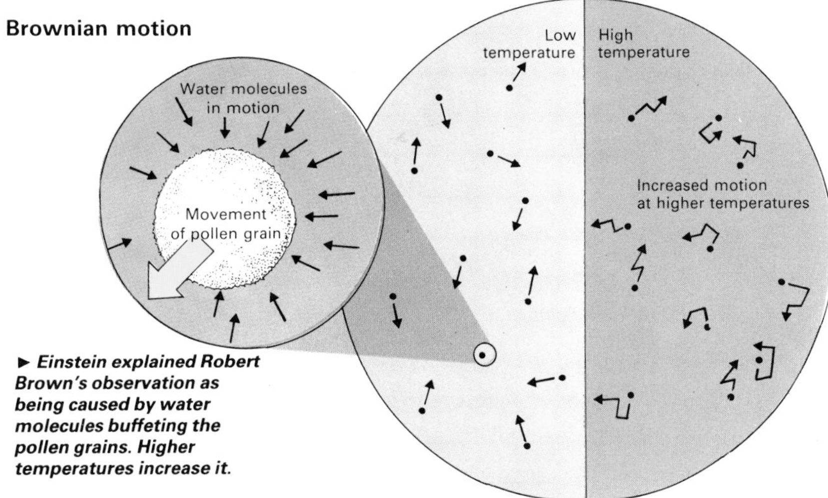

► *Einstein explained Robert Brown's observation as being caused by water molecules buffeting the pollen grains. Higher temperatures increase it.*

Ideology and Science

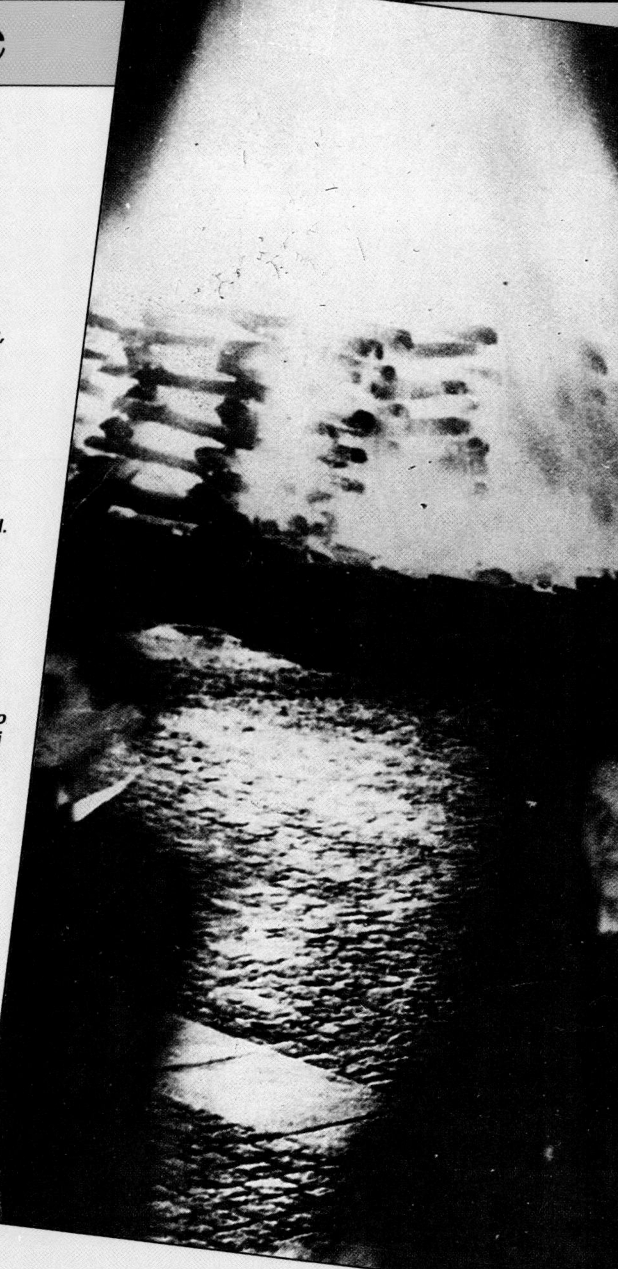

► 11 May 1933, Orpenplatz, Berlin: Nazis and students parade at the burning of "un-German" books. As Einstein finally abandoned Germany for the United States, scores of other academics with Jewish backgrounds either resigned or were dismissed. An April 7th law had decreed dismissal of Jews from government service and the universities.

In the face of post-World War I attacks on the "new physics" and Einstein's leading position in it, most German physicists rallied to his defense. But as the Nazi Party gained in strength so did the idea of a "Jewish science" which had to be eliminated in order to defend the superiority of the "Aryan race".

"Jewish science" in Nazi Germany

Einstein's immediate fame after World War I, coupled with his pacifism, fanned antisemitic opposition to his ideas. Antisemitism had existed for centuries. What was new was the attack on "Jewish science". Most of the people involved were not well known as scientists; but one, the German Philipp Lenard (1862-1947), was a Nobel Prize winner in physics. Indeed, Einstein had earlier been influenced by Lenard's experiments on the photoelectric effect. Lenard's original objection to Einstein's dismissal of the ether was shared with many older scientists who were not anti-semitic. But Lenard soon went far beyond this, and began to claim that relativity was part of a Jewish conspiracy to subvert "true" science.

Thus, new physics, especially relativity, had developed under the influence of Jewish scientists: its purpose was to pull German scientists away from the proper Aryan thinking found in classical physics. In his four-volume treatise "German Physics" (1936-37), Lenard explained: "German physics? one asks. I might rather have said Aryan Physics or the Physics of the Nordic Species of Man. The Physics of those who have fathomed the depths of Reality, seekers after Truth, the Physics of the very founders of Science. But I shall be answered, 'Science is and remains international.' It is false. Science, like every other human product, is racial, and conditioned by blood."

But so bizarre were Nazi leaders' ideas of science that attempts to establish a "German" science in its place achieved only very limited success. Some Nazis supported a curious pseudoscientific theory of the world, popular in Germany between the wars, called the "Cosmic Ice" theory. This supposed that the Earth had been subject to a variety of catastrophes which could explain everything from the ice ages to Noah's Flood.

Communism, "idealism" and Lysenkoism

The essence of the Nazi argument, that Jewish scientists had introduced a new and unacceptable way of looking at the world, was echoed in the Soviet Union from a different viewpoint. There, much of the new physics was thought to fall into the trap of "idealism": it failed to reflect everyday reality as required in Marxist-Leninist theory.

The most striking example of this occurred not in physics, but biology. In the latter part of the 1920s, Soviet newspapers began to mention an agricultural researcher, Trofim Lysenko (1898-1977), who was attacking currently accepted theories of heredity and evolution. Lysenko was concerned with the development of cold-resistant plants. He claimed that, by exposing plants to appropriate conditions, he could produce offspring which were more resistant to cold. The history of his ideas goes back to the end of the 18th century, when the French scientist Jean Lamarck (1744-1829) concluded that changes which occurred in a plant or animal during its lifetime could be passed on to subsequent generations. In the now classic example, he argued that all giraffes must stretch their necks upward to obtain their food, and that this stretching was passed on to the next generation in the form of slightly longer necks.

According to Charles Darwin (♦ page 152), however, it would take some time for organisms to change, because this required an alteration to the genetic material passed from parents to offspring. If Lamarck was right, the process of change could be speeded up by exposing organisms to the appropriate environment. Translated to the realm of human society, it meant that the efforts of a community could change the characteristics of the next generation. Communist politicians, who were calling for an heroic effort to build a new society, found Lysenko's claims for Lamarck's ideas much more helpful than traditional (Darwinian) science.

During the 1930s and 1940s, Lysenko's scientific opponents were banished or interned in labor camps. One, the distinguished biologist Nikolai Vavilov (1887-?1943), died in a Siberian labor camp. Lysenko's work had been praised by Stalin and in 1948 was officially accepted by the Soviet communist party. But the sheer lack of success of Lysenkoism gradually became evident, and, as the Cold War lessened a little in intensity, Lysenko's influence declined. Not until the end of the 1960s did mainstream ideas on genetics and evolution once again figure in Russian teaching and research.

► In the 1930s and 1940s, scientific opponents of Trofim Lysenko (right) were eliminated by dismissal, banishment or internment: the distinguished biologist Nikolai Vavilov (1887-?1943) died in a Siberian labor camp. Lysenkoism was endorsed by the Soviet Communist Party in 1948. But with sheer lack of practical results from the Russian plant improvement programs, Lysenko's influence declined.

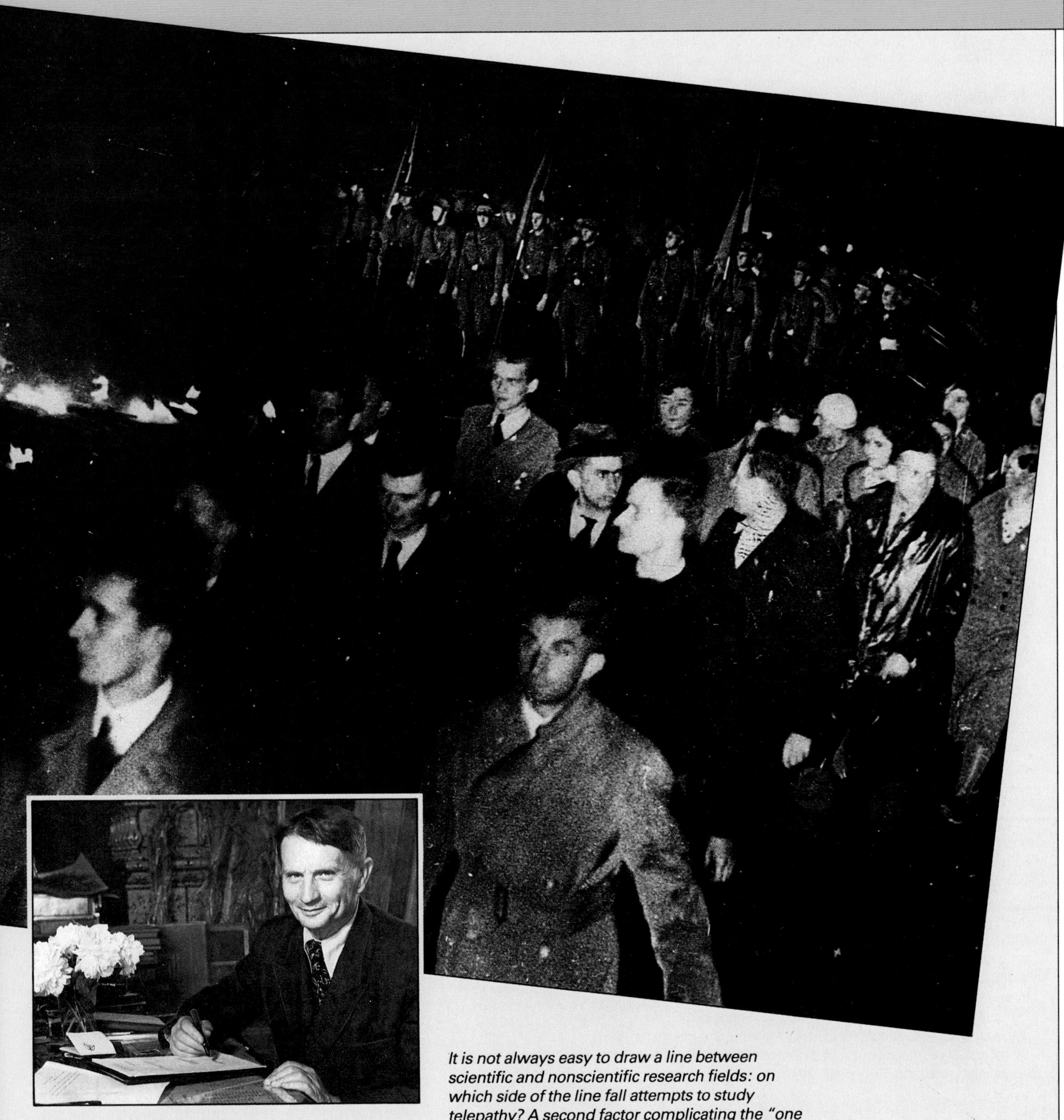

It is not always easy to draw a line between scientific and nonscientific research fields: on which side of the line fall attempts to study telepathy? A second factor complicating the "one science" picture is that, as most scientists would accept, external influences, such as social and economic, can determine which science develops, at what time and to what extent. For example, the rapid growth of nuclear physics since World War II can certainly be linked to the production of the atomic bomb during the war.

But virtually all scientists believe that the acid test of any theory is not how it is developed, but whether, once developed, it receives universal assent. A theory which is supported only by one national or political grouping over an extended period of time does not satisfy this requirement.

The uniqueness of science
Lysenkoism and Nazi attacks on "Jewish science" provide examples of what happens when politicians interfere with the process of scientific research. In both cases the interference ultimately proved to be sterile. Most scientists see this as strong support for their belief that only one science can exist the world over, regardless of the prejudices of the scientists themselves.

The word "science", however, has different meanings in different countries. Thus in some countries history is called a science, whereas in others it is thought to have little to do with science.

Einstein and Literature

As Canadian humorist Stephen Leacock wrote at the time, many people believed it was near-impossible to understand Einstein's work:
"But it was Einstein who made the real trouble. He announced in 1905 that there was no such thing as absolute rest. After that there never was...Einstein explains that there is no such place as here. *'But,' you answer, 'I'm here; here is where I am right now.' But you're moving, you're spinning round as the earth spins; and you and the earth are both spinning round the sun, and the sun is rushing through space towards a distant galaxy, and the galaxy itself is beating it away at 26,000 miles a second. Now where is that spot that is here!"*
In the 1700s Alexander Pope wrote of Newton:

"Nature and Nature's laws lay hid in night:
God said, 'Let Newton be!' and all was light."

Two centuries later, J.C. Squire (1884-1958) added another couplet:

"It did not last: the Devil howling 'Ho,
Let Einstein be,' restored the status quo."

Relativity has made the greatest impact on science fiction. The main problem for this genre is the limitations on travel imposed by the finite speed of light. Writers who want their characters to venture beyond the solar system immediately face the difficulty of the long travel times involved. As Kingsley Amis has pointed out:

"The fact is that to reach any but the nearest stars would take several hundred years even if one travelled at the speed of light, in the course of doing which one would, if I understand Einstein's popularisers correctly, become infinite in mass and zero in volume, and this is felt to be undesirable. A few writers simply accept this difficulty and arrange for their travellers to put themselves into some sort of deep-freeze until just before planetfall, or allow them to breed in captivity for the requisite number of generations. But most commonly, the author will fabricate a way of getting around Einstein: a device known typically as the space-warp or the hyper-drive will make its appearance, though without any more ceremony than 'He applied the space-warp,' or 'He threw the ship into hyper-drive.'"

More generally, writers' attention has been drawn to the problems of understanding space and time. In consequence, 20th-century references to space and time often carry overtones which cannot be found in earlier periods. An obvious example comes from the English poet, William Empson:

"Alas, how hope for freedom, no bars bind;
Space is like earth, rounded, a padded cell;
Plumb the stars' depth, your lead bumps you behind;
Blind Satan's voice rattled the whole of Hell."

The reference here to curved space-time is obvious; but even apparently straightforward references to space and time are sometimes written in the knowledge of relativity. After Freud, of 20th-century scientists, Einstein has probably most influenced nonscientists' thinking.

► ***In stories of lasers and intergalactic travel, science fiction writing is inspired by Einstein's work***

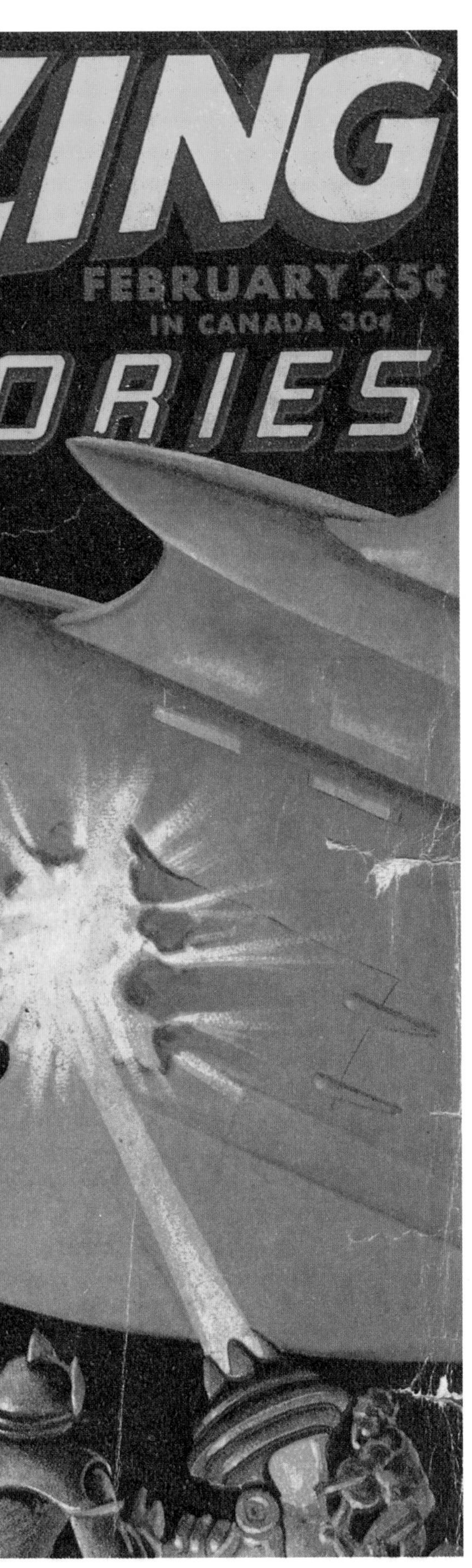

Einstein returns to Germany

Very rarely in history has a scientist produced such an array of fundamental ideas in one year as Einstein did in 1905. (Newton's *annus mirabilis* was comparable (◀ page 74), when in 1666 he laid the foundations of his future work on mathematics, gravitation and light.) Yet Einstein did not win immediate fame. His first application to teach in the University of Berne in 1907 was turned down; but he was accepted in the next year, and in 1909, as his reputation gradually grew, he was appointed to a post at the University of Zurich.

By this time his work was attracting particular attention in Germany. Hermann Minkowski (1864-1909), one of the leading German mathematicians, pointed out the way in which relativity mixed up space and time: measurements of distance involved time and *vice versa*. Minkowski therefore introduced the new concept of "space-time". "From now on," he declared, "space by itself and time by itself must sink into the shadows, while only a union of the two preserves independence."

In 1909 Einstein met Max Planck, author of the quantum theory, for the first time, and from then on came into increasingly close contact with German physicists. In 1911 Einstein moved to the German University in Prague, only to return to Zurich in the following year. Finally, in 1914 he moved to Berlin.

Einstein's appointment to one of the senior German academic posts, which he retained until 1933, reflected his now widely accepted eminence. He resumed his German nationality. However, though academically this position could hardly be bettered, Einstein was not totally happy.

His wife returned to Zurich with their two sons. This signaled the break up of his marriage, though the formal divorce did not come through until 1919. Moreover, he realized again how much he disliked the regimented atmosphere in Germany which had oppressed him as a boy. Almost at once this atmosphere became worse: at the outbreak of World War I, the vast majority of German scientists supported the war effort in every way they could. Einstein became increasingly at odds with the mood of his colleagues, and his attachment to pacifism increased.

General relativity

Hardly surprisingly, Einstein concentrated on his own research, examining again the basic limitation of his 1905 paper on relativity. How could this be extended to cover the case when bodies changed their speeds? People who are subject to an acceleration, for example in a car, can detect it by the presence of a force which pushes them back in their seats. As the car reaches a steady speed, this force disappears. Now the pull of gravity is also a force which we detect by the pressure it produces – on the soles of our feet when we are standing. Einstein argued that acceleration and gravity are interrelated; so, if he could generalize his theory of relativity to include acceleration, he would simultaneously have something new to say about gravity.

Einstein had been working on this extension of his theory since 1905, but his ideas came to fruition in the middle of World War I. In the paper he published in 1915 on his new "general theory of relativity" he showed that gravity, instead of being seen as a pull exerted by an object, can be viewed as a distortion of the space round the object. This distortion produces an acceleration in passing bodies, conventionally labeled "gravitational attraction".

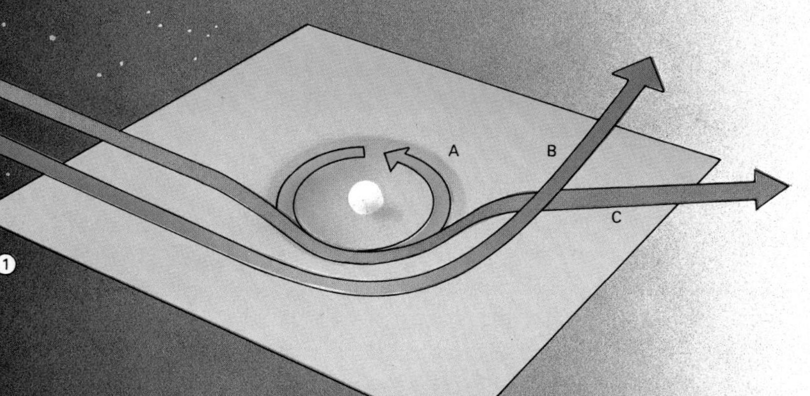

▲ ***The University of Berlin, where Einstein was a professor, and director of the Kaiser Wilhelm Institute for 30 years.***

◀ ***(1) Space can be visualized as like an elastic sheet on which a heavy mass has produced an indentation. On encountering this, bodies follow curved paths: a relatively slow-moving body (A) deep in the well follows a closed path (like a planet round the Sun), a faster body (B) follows an open curve, while a ray of light (C) is normally deflected only very slightly. (2) Until the 1919 eclipse the only important evidence for Einstein's general theory was that it accurately predicted the slow shift of Mercury's orbit round the Sun. (3) An object moving through distorted space seems to experience a "gravitational" pull. Einstein illustrated this equivalence in terms of a passenger isolated in an elevator, who cannot tell whether he is held on the floor by the lift's acceleration or the Earth's gravity.***

2
Mercury
Sun

3
Acceleration
9.8 meters per sec.
per sec.
Equivalent
"gravitational"
pull
Earth's gravity

Imagine a stretched elastic sheet with a weight placed on the center, depressing the sheet in its vicinity. A marble rolled across the sheet will be deflected by the depression in the sheet as it passes the weight. To an observer, it will appear that the weight has attracted the marble. So, Einstein would say, the space near the Sun is distorted by its presence in such a way that the much smaller planets are deflected into orbits round the Sun.

Einstein's description of gravity is obviously different from Newton's. Newton thought gravitational pull resided in bodies (◀ page 74), whereas Einstein saw it in the space around them. Yet this different viewpoint led initially to very few differences in the predictions the two theories made. They predicted identical results except for very high accelerations, or very massive bodies.

It was some years before the full impact of Einstein's 1915 paper on his general theory of relativity was fully comprehended by the scientific community. At first, the only important evidence in Einstein's favor came from a study of the path of the planet Mercury in its orbit round the Sun. Mercury is the closest planet to the relatively massive Sun, and moves more rapidly than any other planet. Hence, it is of all planets the one most likely to show deviations from Newton's predictions. The position of Mercury's elongated orbit round the Sun changes slowly with time. The reasons for this had already been partly understood by applying Newton's theory, but there was a small amount of shift in the orbit which could not be explained. Einstein was able to demonstrate that this small extra increment could be accounted for by his own theory.

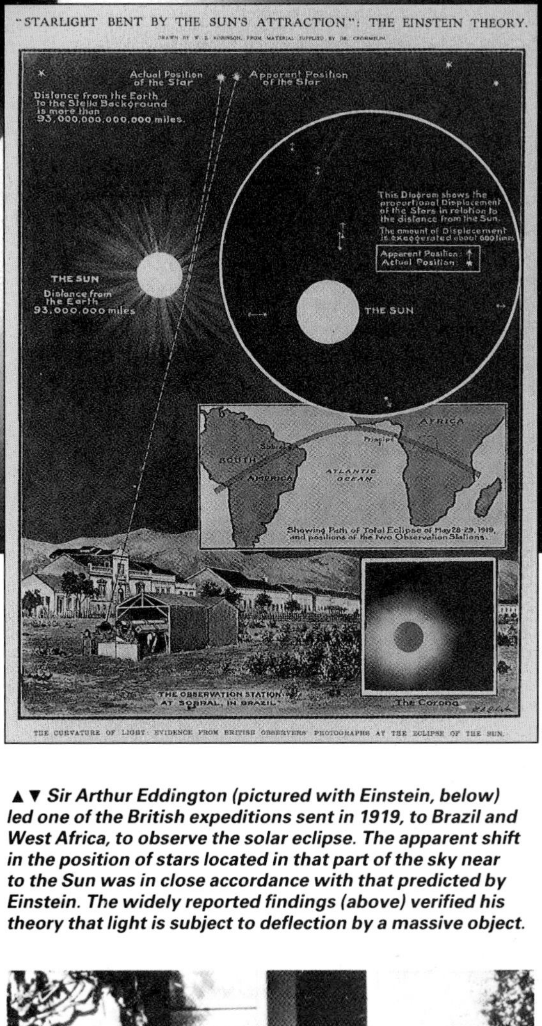

"STARLIGHT BENT BY THE SUN'S ATTRACTION": THE EINSTEIN THEORY.

DRAWN BY W. B. ROBINSON, FROM MATERIAL SUPPLIED BY DR. CROMMELIN.

THE CURVATURE OF LIGHT: EVIDENCE FROM BRITISH OBSERVERS' PHOTOGRAPHS AT THE ECLIPSE OF THE SUN.

▲▼ *Sir Arthur Eddington (pictured with Einstein, below) led one of the British expeditions sent in 1919, to Brazil and West Africa, to observe the solar eclipse. The apparent shift in the position of stars located in that part of the sky near to the Sun was in close accordance with that predicted by Einstein. The widely reported findings (above) verified his theory that light is subject to deflection by a massive object.*

Confirming general relativity

Though interesting, the provision of an explanation for a small change in Mercury's orbit was hardly striking enough for Einstein's contemporaries to abandon Newton's longstanding theory. For many, the clinching evidence in favor of Einstein came from another of his predictions, which was spectacularly confirmed after the war. Newton's theory indicated that a ray of light passing near a massive object like the Sun should be deflected slightly. Einstein's theory indicated a similar effect but predicted twice as much deflection as Newton's theory. Measurement of the deflection had never been attempted: a ray of light passing near the Sun could simply not be seen in the glare of the sunlight. The way forward was explored by a British astronomer, Arthur Eddington (1882-1944), who managed, despite the war, to obtain a copy of Einstein's paper. In 1919 the Sun would be eclipsed in a part of the sky where there were a number of bright stars. Consequently, once the Moon had totally eclipsed the Sun's disk, the light from these stars would be seen from the darkened Earth close to the edge of the Sun. The light rays would be slightly deflected by the Sun, giving the impression that the stars had altered their positions in the sky. A measurement of the apparent shift in position of the stars would test Einstein's prediction. Two British expeditions were sent out to observe the eclipse: one, led by Eddington, went to the island of Príncipe off West Africa, and the other to Sobral in northeast Brazil. The findings from their observations of the eclipse on 29 May proved to be in excellent agreement with the value derived from the general theory of relativity.

Exploration of the Atom

Today it may seem strange that Einstein received his Nobel Prize for work on the nature of radiation, rather than for work on relativity. But his development of quantum mechanics (how atoms and subatomic particles move and interact) proved to be of fundamental importance. He was the first to point out that light can be thought of, for some purposes, as a cloud of particles – each of these being one of Max Planck's quanta. Information obtained from the study, popular in 19th-century physics, of how the atoms in a cloud of gas move could now be applied to radiation. Einstein found that this approach explained why some metals became electrically charged when exposed to light – the "photoelectric effect" employed today, for example, in camera exposure meters.

Bohr's model of the atom

In 1905 hardly anyone accepted these ideas of Einstein: it seemed ridiculous that radiation could sometimes act like a wave and sometimes like particles. Robert Millikan (1868-1953) in the United States began work on the photoelectric effect in order to show that Einstein was wrong. When he eventually won a Nobel Prize (1923), it was in part because his experiments during World War I had shown Einstein to be right. Meanwhile, just before the war, the Danish physicist Neils Bohr (1885-1962) had suggested a picture of the atom which explained why radiation appears as quanta. In his model, the atom consists of a central nucleus (with a positive electrical charge) round which electrons circle (each with a negative electrical charge, so that the atom as a whole is electrically neutral). Bohr's atom therefore looks like a tiny solar system, with the difference that planets can move at any distance from the Sun, whereas Bohr's electrons can only orbit the nucleus at certain fixed distances. If they change orbit, they do so by jumping from one "acceptable" distance to another. If they jump toward the nucleus, radiation is emitted: if they jump away from the nucleus, radiation is absorbed. For Bohr the important point was this: the electrons only changed their distances at intervals, so the radiation was not produced continuously, but in bursts – each burst of radiation corresponding to one quantum.

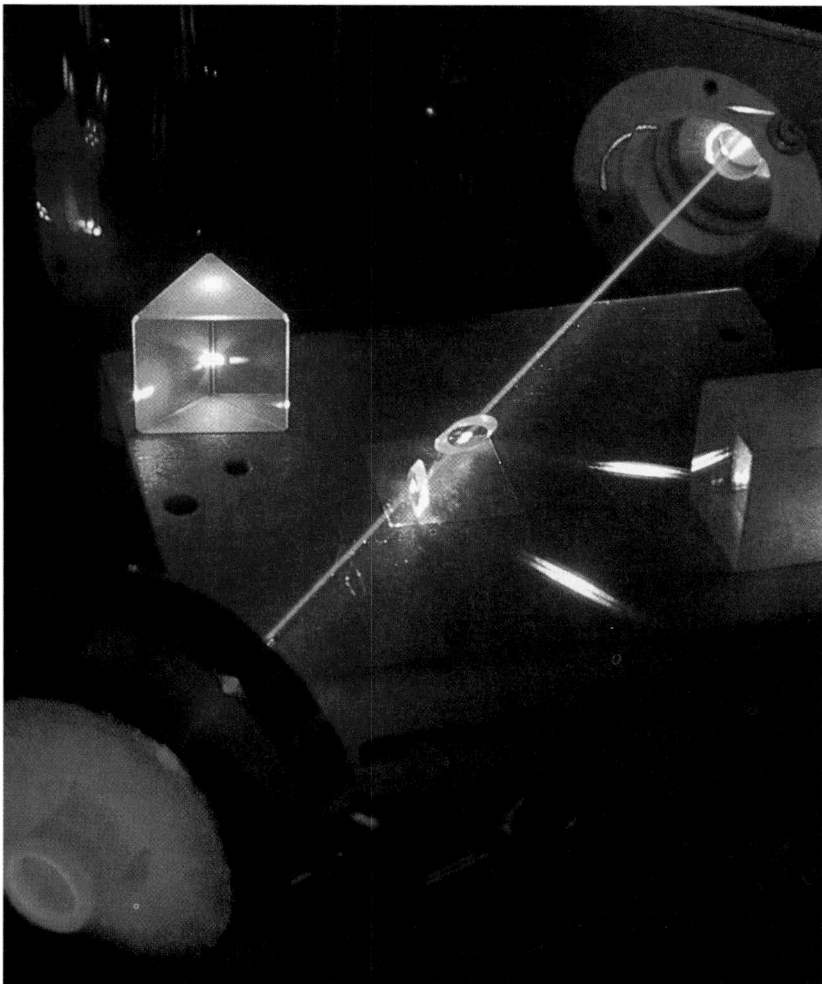

▼ The Danish physicist Niels Bohr deep in discussion with Einstein. Bohr proposed his revolutionary theory of the atom in 1913 while he was working in Manchester with the New Zealand-born Ernest Rutherford, discoverer of alpha and beta particles.

▲ A laser being used in quality control of quartz "windows". Einstein's idea of stimulated emission of quanta prompted the laser's development. Lasers may be used in communications, holography, welding and cutting (e.g. surgery), and "star wars" weaponry.

► 16 July 1945 – "Trinity", the world's first atomic bomb blast, in New Mexico. Einstein urged President Roosevelt to develop the bomb (his "one great mistake" he later called it). The range of application of physicists' work led to the war being named the "physicists' war".

Rutherford and the atomic nucleus

Bohr managed to bring together the idea of quanta, which interested Planck and Einstein, with the picture of an atom containing a nucleus and electrons, which was then of particular interest to physicists in Britain.

At the center of this work on the structure of the atom was Ernest Rutherford (1871-1937), a New Zealander who spent much of his life in England. Bohr worked with him in Manchester. After World War I, Rutherford moved to Cambridge, and there he and his colleagues concentrated mainly on exploring the nature of the atomic nucleus. They showed that it is made up of more than one type of particle (♦ page 200) and can, under some circumstances, be made unstable. During the 1930s this work was extended in a number of laboratories, and a variety of new, and often unstable, nuclei were created.

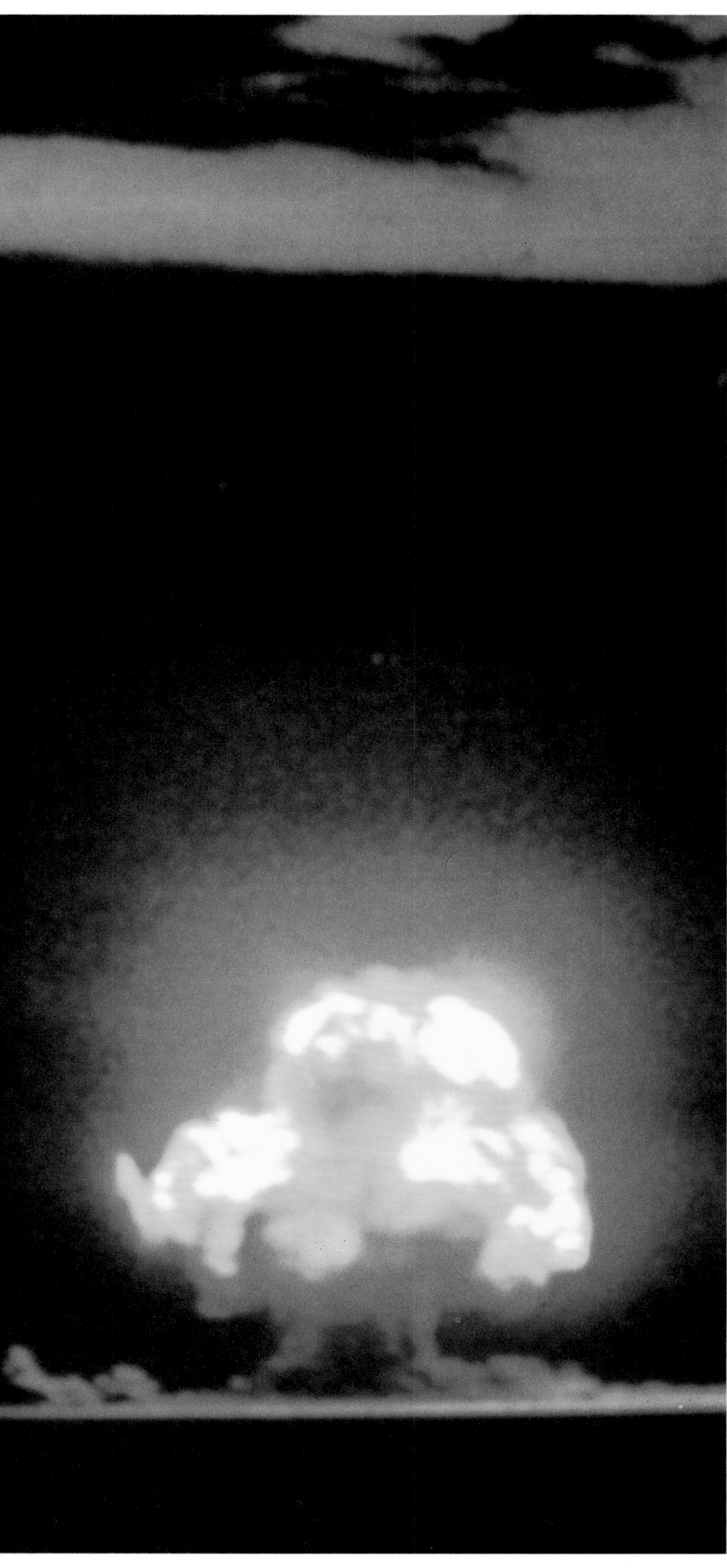

The atomic bomb

In 1938 an old friend of Einstein's, Otto Hahn (1879-1968), who was still working in Germany, found that the nuclei of the element uranium can be broken down by a newly discovered process, fission, in which the nucleus breaks down into two nearly equal fragments. It was rapidly realized by a number of physicists that fission might provide a source of tremendous energy $E = mc^2$*, since there was an appreciable loss of mass during fission. With Germany obviously preparing for war, physicists in both the United States and Britain feared that German scientists might rapidly capitalize on this discovery to build an entirely new type of bomb, much more destructive than any conventional weapon. Einstein gave his very influential support to the case for pressing ahead urgently with the "Manhattan Project" and played a significant part in ensuring its start.*

Einstein later regretted this: "I made one great mistake in my life – when I signed the letter to President Roosevelt recommending that atom bombs can be made." But, he added: "there was some justification – the danger that the Germans would make them." Yet the atomic bomb is the supreme example of the way in which the academic researches of Einstein and his generation of physicists became, within a few decades, applicable in ways they could never have foreseen. In fact, the range of applications of their work has sometimes led to World War II being called "the physicists' war".

Lasers and holograms

Einstein returned to the question of quanta in 1917, while he was still working on the implications of general relativity. In Bohr's theory of the atom, an electron jumping outward absorbs a quantum, whilst an electron jump inward emits a quantum. The process of absorption requires the atom to be bathed in radiation, but emission occurs spontaneously: it does not depend on what radiation is passing by the atom at the time. Einstein pointed out that if an atom is surrounded by appropriate radiation, it can be stimulated into emitting a quantum even when it would not normally do so.

This idea of "stimulated emission" lies at the heart of the modern laser ("Light Amplification by Stimulated Emission of Radiation"). Lasers use stimulated emission to produce very intense beams of radiation. Such beams have found many uses in recent years, from carrying out eye operations to projected attempts to blast enemy satellites in the United States' "Star Wars" program. Perhaps their greatest importance, however, will prove to be in transmitting messages.

Lasers have a second property of importance along with high intensity. All the waves in a laser beam are in step with each other – "coherent" radiation. It is such coherent light that is required in the new technique of holography. The laser beam is recorded as it is reflected from an object: the resultant image can be used to reconstruct a three-dimensional picture of the object. These pictures-in-the-round or "holograms" are becoming a common feature of our life.

▼ *In Brussels at the 1911 "Solvay Congress", sponsored by a Belgian industrialist, the young Einstein (second right) joins Planck, Lorentz, Curie, Rutherford and other top physicists to discuss the impact of quantum theory on the sciences.*

▲ *In 1921 Einstein and his second wife Elsa paid their first visit together to the United States accompanied by the Zionist leader Chaim Weizmann. The following year they visited Japan. On both occasions Einstein was showered with honors.*

Fame and travel

Up until 1919, Einstein's name had been familiar to physicists, but not to a wider audience. The results from the observations of the eclipse that confirmed Einstein's predictions, and the acceptance that they marked the overthrow of Newton's theory, catapulted Einstein to world fame. One of those present at the meeting in London where the eclipse expeditions reported their work later recorded: "The whole atmosphere of tense interest was exactly that of the Greek drama: we were the chorus commenting on the decree of destiny as disclosed in the development of a supreme incident."

In the year of the eclipse, Einstein married again. His new bride was his cousin Elsa, whom he had known since his boyhood. She was now a widow with two daughters and they had come together again a short time before when she had nursed Einstein through a serious illness. She proved an excellent manager and protector of Einstein at home and on his travels. These soon began in earnest: in 1921 he visited both the United States and Britain for the first time, and was also awarded the Nobel Prize for physics for his work on radiation.

Einstein went to the United States primarily to raise money for the Hebrew University then being built in Jerusalem. During the preceding decade Einstein had become increasingly conscious of his Jewish background; more particularly, he had become convinced of the need for a Jewish national home. The growing antisemitism in Germany certainly played a significant role in this new awareness.

Einstein established himself as an important supporter of the Zionist cause. After World War II his old acquaintance, the Zionist leader and chemist Chaim Weizmann, became President of the new state of Israel. On Weizmann's death in 1952 Einstein was invited to succeed him. He declined on the grounds of lack of aptitude and experience, but the offer gives some indication of his standing in the Jewish world by this time.

Einstein in the United States – the latter years

Early in 1933 Hitler came to power in Germany. Einstein was in the United States at the time, and soon decided to make his home there; he had already been approached to join the new Institute for Advanced Study at Princeton, and now decided to do so. He was deprived of his German citizenship, but did not become a United States citizen until 1940. At the time of his settling in the United States, Einstein was an outspoken pacifist, but subsequent events in Germany forced him to modify his stance (◀ page 243). Einstein was by now very isolated in the world of science. The last work that involved him fully with the scientific community had stemmed from his theory of general relativity. Since this was a theory of how gravitation acted, it obviously had implications for cosmology – the study of the Universe as a whole. Einstein's initial assumption that the Universe was stationary was challenged by the observations of the American astronomer Edwin Hubble (1889-1935), who showed that the Universe is expanding.

▼ Einstein lecturing in the United States at the California Institute of Technology, Pasadena, in 1930-1931. He undertook three such tours, the last being in 1933, when the Nazis came to power in Germany. Seven years later, having made the United States his home, Einstein was sworn in as an American citizen (below) together with his secretary Dukas and stepdaughter Margot.

Rise of a Scientific Power

During the 20th century, and particularly in the years leading up to World War II, the immigration of scientists into the United States played a significant role in developing scientific research there. This is a recent striking example of the perennial importance of population movements in affecting the growth of ideas.

Early practical science in the United States

In the 18th and 19th centuries, scientific interest in the United States concentrated on practical or environmental concerns. Benjamin Franklin (1706-1790), for example, studied electricity partly in order to answer questions about the nature of lightning and the design of lightning conductors. George Washington (1732-1799), like a number of fellow-Americans in his rapidly expanding country, was a trained surveyor. Since surveying involved astronomical measurements, astronomy soon became a topic for study. In fact, most of the sciences cultivated during the 19th century – astronomy, geology, meteorology and biology – reflected the needs of a country where exploration and understanding of the environment were at a premium.

The experimental tradition

In 19th-century America, individual scientists who became eminent in such subjects as physics were typically practical experimenters rather than theoreticians. One such was Albert Michelson (1852-1931), the first American to win a Nobel Prize for physics. Michelson was an immigrant to the United States: born in Poland, of Jewish descent, his family came to the country when he was about three years old. The results of the experiment he conducted with Edward Morley (1838-1923), to test the effects of the ether on the speed of light, were subsequently realized to be strong evidence in favor of Einstein's theory of special relativity (♦ page 233).

By the end of the 19th century the old connection between astronomy and surveying had essentially ceased: astronomers increasingly concentrated on the physical properties of distant, and therefore faint, objects. Such work required large telescopes which could collect large amounts of light. American scientists pioneered the building of these telescopes; they had the advantage of skilled mechanics, good observing sites and wealthy patrons who provided the money. By the early 20th century, American observers had become world leaders in astronomy. When Einstein's first model of the universe based on general relativity was finally discarded, it was on the basis of observations made at Mount Wilson in California.

The position in physics was different. In terms of Nobel Prizes, only three awards were made to American physicists up to the mid-1930s, even though, by this time, universities had created physics departments which could provide good research training. (In the 19th century, American scientists had gone to Germany to develop their research skills, ♦ page 121). However, the emphasis remained on experiment – all the Nobel Prizes in physics awarded to Americans up to World War II were for experimental work.

The new influx of Europeans

This situation changed as the political conditions in Europe worsened during the 1930s. In addition to Einstein arriving in Princeton, an array of other European physicists came to the United States. Not all were Germans. Perhaps the most eminent after Einstein was the Italian, Enrico Fermi (1901-1954), who came to the United States in 1938 directly from the ceremony in Sweden where he received the Nobel Prize for physics. Fermi was an outstanding experimental physicist – his group at Chicago produced the first nuclear energy generator in 1942. But he was also a highly competent theoretician who emphasized the importance of this type of work for the United States. Hans Bethe

▲ **Benjamin Franklin demonstrated the electrical nature of lightning, and invented the lightning conductor, Franklin stove and bifocal lenses.**

▼ **Prominent among the many physicists who came to the United States from troubled Europe in the 1930s was Italian Enrico Fermi.**

▼ **Germany was a world leader in scientific and technological research before World War II. It remained ahead in rocket development until the end of the war. The Americans and Russians then captured German researchers and used them to develop rocketry in their countries. Wernher von Braun and others helped launch American space exploration – shown here by an astronaut "space-walking".**

(born 1906), who had briefly worked with Fermi in Rome, was even more influential in this respect. Ejected from his post in Germany because his mother was Jewish, he established a research group in theoretical physics in the United States, ultimately receiving a Nobel Prize for his work. Bethe published some very influential teaching material on theoretical physics and, like Fermi, was closely involved in the atomic bomb project *(♦ page 243).*

The impact on American physics

The first, and more obvious consequence of this emigration of European physicists to the United States was the stimulation and leadership it gave to work on the atomic bomb. American preeminence in nuclear physics after the war would have stemmed from this alone. But for the physics community an equally important factor was the creation of theoretical groups who could work alongside, and on equal terms with, the experimental groups. A glance at the list of American postwar Nobel Prize winners quickly reveals how many were immigrants; but it also shows that there is now a fairly equal balance between experimenters and theoreticians. Immigration has not simply helped world leadership in physics to shift from Germany to the United States: it has also led to a more balanced and cohesive physics research program.

▲ A baggy sweater and the famous equation $E = mc^2$ (equivalence of mass and energy), recurrent motifs in stamps celebrating a giant of the 20th century.

◄ "I will a little tink." For ten years after his retirement from the Institute for Advanced Study at Princeton, New Jersey, Einstein continued his quest for a "unified field theory" bringing together quantum mechanics and gravitation theory.

▼ Einstein featured on an Israeli banknote, and being visited by Prime Minister Ben-Gurion in 1951. A year later Einstein declined a request to become President of Israel.

The Dutch astronomer Willem de Sitter (1872-1934), who had been responsible for informing Eddington of Einstein's work, had already shown that general relativity, when applied to the Universe, favored the idea of an expanding universe. Edwin Hubble's work confirmed this view. In 1930 Einstein declared his agreement with de Sitter in a paper that can be regarded as one of the foundation documents of modern cosmology.

From this point on, Einstein's concern was with something even wider and more fundamental. His research had dealt with two of the main areas of 20th-century theoretical physics, gravitation and quantum mechanics, but they had been treated quite separately. Einstein spent the last decades of his life trying to bring together into a unified theory these two basic aspects of modern physics, one dealing with very large amounts of matter and the other with very small amounts. Nowadays there is great interest amongst physicists in what are known as GUTs (Grand Unified Theories), but Einstein's contemporaries considered his attempts to be premature. In consequence, he came to have less and less contact with the world of physics. He had already accepted, some time before his death on 18 April 1955, that he was not going to reach his hoped-for goal: "As to my work, it no longer amounts to much. I don't get many results any more and have to be satisfied with playing the Elder Statesman and the Jewish Saint."

Credits

Key to abbreviations: AKG Archiv für Kunst und Geschichte, W. Berlin; AM Anthony Michaelis; ARPL Ann Ronan Picture Library; BAL Bridgeman Art Library; BHPL BBC Hulton Picture Library; BL British Library; BLO Bodleian Library, Oxford; BM British Museum; BN Bibliothèque Nationale, Paris; JLC Jean- Loup Charmet; MC Mansell Collection; MEPL Mary Evans Picture Library; MH Michael Holford; MHSO Museum of History of Science, Oxford; NPG National Portrait Gallery, London; RCPL Royal College of Physicians, London; RHPL Robert Harding Picture Library; RI Royal Institution, London; RS Royal Society London; SPL Science Photo Library. b bottom; bl bottom left; br bottom right; c center; cl center left; cr center right; t top; tl top left; tr top right; l left; r right.

1 Institut Pasteur, Paris **2-3** JLC **4-5** ARPL **7** US Army/SPL **8** The Metropolitan Museum of Art, purchase Mr & Mrs Charles Wrightman gift **9b** BM **12t** BM/MH **12b** BN **13t** BM/Ekdotike Athenon **13bl** Epidauros Museum **13br** G. Speake **14t** Metropolitan Museum of Art **14b** MC **15** SCALA **16t** Werner Forman Archive **16b** SCALA **17t** MC **17bl** Ekdotike Athenon **17br** BM **18bl** Hirmer Archive **18-19t** D. Harissiadis **19t** NHPA/G.J. Cambridge **18-19b** Leonard von Matt **20l** Östereichische Nationalbibliothek **20-21** ARPL **22-23b** BN **23l** Ashmolean Museum, Oxford **23r** SCALA **24b** Norbert Schimmel Collection, NY **25t** Bruce Coleman Ltd **26t** BPCC/Aldus Archive **26bl** MC **26-27** SCALA **27** Biblioteca Nazionale Marciana, Venice **28** ARPL **29,30** SCALA **32** Stanza della Segnatura, Vatican Palace, Rome **32-33t** SCALA **33c** AM **33b** ARPL **34** SCALA **35tl** Biblioteca dell'Accademia Nationale dei Lincei e Corsiniana **35tr** Istituto e Museo di Storia della Scienza **35b** Edimedia **36** MHSO **36-37** Elsevier, Amsterdam **37tl** MEPL **37tr** Biblioteca Nazionale, Firenze **38tl** MC **38b** Science Museum, London **38tr,39tl** ARPL **39b** SCALA **39insets** Istituto e Museo di Storia della Scienza **40bl** Biblioteca Nazionale, Firenze **40-41t** SPL **40-41b** Istituto e Museo di Storia della Scienza **42t** ARPL **42b** MC **43** SCALA **44-45** BAL **45t** ARPL **46l** JLC **46t** MC **46-47t** ARPL **48t** Biblioteca Nazionale, Firenze **48b** Archivo Alinari **49** RCPL **49b** MC **52bl** ARPL **52-53t** RCPL **52br** ARPL **53br** RCPL **54-55** BL **54c** RCPL **54b** BHPL **55b** ARPL **56t** BPCC/Aldus Archive **56-57b** RCPL **57t** BL **57b** RCPL **58tl** ARPL **58tr** BHPL **58c** ARPL **58b** BHPL **58-59t** SCALA **59b** BHPL **60l** RCPL **60-61** CNRI/SPL **61t,61bl** RCPL **61br** BHPL **62-63** SCALA **64-65,65t** BL **65b** Zentralbibliothek, Zürich **66** RCPL **67t** Rijksmuseum, Amsterdam **67c** SPL **67b** ARPL **68c** MC **68b** CNRI/SPL **69l** RS **69r** NPG **72t** RS **72b** National Maritime Museum, Greenwich **73tl** MC **72-73b** By permission of the Master & Fellows, Magdalene College, Cambridge **73tr** ARPL **75,76t** RS **76c** BLO **76b** ARPL **77tl** Frank Spooner Pictures **77bl,tr,br,78** ARPL **79l** MH **79r** RS **80-81t** SPL **80b** BLO **81c** RS **81b** ARPL **82l** Derby Museum & Art Gallery **82-83** BAL **84-85t** David Parker/SPL **84c,84bl** BL **84br** Cambridge University Library **85t** ARPL **86t** JLC **86b** NPG **86-87t** BN **87b** Yale University Art Gallery, New Haven **88tl** ARPL **88tr** AM **88bl** RS **89t** JLC **89b** Eidgenössische Technische Hochschule Bibliothek **92t** ARPL **92b,92-93** JLC **93b** BHPL **94t** Giraudon **94b** JLC **95** Roger Viollet **96t** Metropolitan Museum of Art **96b** Giraudon **97t** MHSO **97b** Science Museum, London **98t** Derby Museums & Art Gallery **98b** ARPL **99t** MHSO **99b** Archiv Gerstenberg **100t** JLC **100-101b** BPCC/Aldus Archive **101tl** Cavendish Laboratory, University of Cambridge **101tr** BLO **101cl** BAL **102tl** Roger-Viollet **102bl** ARPL **102-103,104t** JLC **104-105** MC **105tl** BN/Sonia Halliday Photographs **105tr,106** JLC **107t** BLO **107c** MC **107b** Royal Swedish Academy of Sciences, Berzelius Museum, Stockholm **108t** ARPL **108b** Roger-Viollet **109** RHPL **110** NPG **112t** AKG **112cl,112bl** BM (Natural History) **112br** Universitetsbiblioteket, Uppsala **113t** Bergakademie Freiberg **113c** National Galleries of Scotland **113b** ARPL **114** T. Morrison/South American Pictures **115tl** Holt Studios **115tr,115c** Royal Botanical Gardens, Kew **115b** Landsat/Eros Data Image **118tl** BM (Natural History)/G. Rainbird/RHPL **118-119t** Bruce Coleman Ltd **118b** RHPL **119b** BAL **120b** JLC **120r** Dr Ernst Berninger **121t** Gesellschaft Liebig-Museum **121c,121b** Bildarchiv Preussischer Kulturbesitz **122-123** AKG **123r** RHPL **124** Illustrated London News **125** AM **126l** NASA/SPL **126r** BM (Natural History) **127t** Trans Antarctic Expedition **127b** ARPL **128t** Museum für Kunst und Gewerbe **128b** RHPL **129l** BLO **129r** NPG **130** ARPL **132t** NPG **132b** ARPL **133t** MH/RI **134t** MH/Science Museum, London **134c,134b** ARPL **135t** BPCC/Aldus Archive **135b** MEPL **136-137, 137t** ARPL **137b** MEPL **138t** RI **138-139** MH/RI **139t** MC **140l** Smithsonian Institution/National Portrait Gallery, Washington **140r** MH/RI **141** ARPL **142-143** Giraudon **142bl** BHPL **142br** MC **144-145t** ARPL **144b** BHPL **145tr** ARPL **145c,145b** The Electricity Council **146t** Colchester & Essex Museum **146b** ARPL **147t,147bl** RI **147br** BAL **148t** AM **148c** ARPL **148b** SPL/Don Thomson **149** SCALA **152t** RHPL/Royal College of Surgeons **152b** BHPL **154-155** RHPL **154t** Royal Naval College/Rainbird/RHPL **155t,158tr,158c** G. Rainbird/RHPL **158b** RHPL/Witty **160bl** RHPL **160br** Darwin Museum, Down House **160-161t** Biofotos/H. Angel **161bl** Rainbird/RHPL **161bc** BHPL **161br** Royal Botanic Gardens, Kew **162tl** MEPL **162bl,162-163t** BHPL **162-163b** BAL **163l** MEPL **164b** Kobal Collection **165b** Weidenfeld & Nicolson **166-167** BAL **166bl** Punch **167** NHPA **168** BHPL **169,172t** Institut Pasteur, Paris **172b,173t** BHPL **172-173b** Paul Brierley **174tl** CNRI/SPL **174cl** John Walsh/SPL **174bl** MC **174br** BHPL **175t** ARPL **175b** SCALA **176-177t** JLC **176c** ARPL **176b** Biofotos **177tr** ARPL **177br** National Dairy Council **178tl** Institut Pasteur, Paris **178-179t** Dr R. King/SPL **178-179b** MC **179tl** ARPL **179tr** Musée Pasteur, Paris **180tr** John Durham/SPL **180c** AKG **180bl** Paul Brierley **180br** Dr J. Burgess/SPL **181** L. Migdale/SPL **182-183** ARPL **183t** BHPL **183b** MEPL **184t** BHPL **184b,185b** Musée Pasteur, Paris **185t** Tektoff-RM/CNRI/SPL **186** BHPL **186-187t** A. McClenaghan/SPL **186-187b** BHPL **187tr** Hoechst **188t** Musée Pasteur, Paris **188b** ARPL **189** MEPL **192t** Association Frédéric et Irène Joliot-Curie **192c** ARPL **192b** BHPL **192-193** JLC **193t** SPL **194b** Paul Brierley **194t** BHPL **195t** MEPL **195b** Bildarchiv Preussischer Kulturbesitz **196,197** Novosti Press Agency **198bl** ARPL **198tr** Musée du Laboratoire Curie **198-199b,199tl** ARPL **199r** MEPL **200tl** JLC **200tr** ARPL **200cr,200-201** Cavendish Laboratory, University of Cambridge **201tl** L.A. Herald Examiner **201tr** Marie-Curie Museum, Warsaw **202** MEPL **203** R. Adler Creutz/Tiofoto **204t** ARPL **204bl** MHSO **204br** Marie-Curie Museum, Warsaw **205t** Ohio Nuclear Corp/SPL **205b** Zefa Picture Library **206bl** AKG **206-207t** BHPL **206-207b,207tr** Marie-Curie Museum, Warsaw **207cl** ARPL **208t** Popperfoto **208b** United Kingdom Atomic Energy Authority **209l** MEPL/Sigmund Freud Copyrights **209r** AM **212l** MEPL/Sigmund Freud Copyrights **212r** MEPL **212-213,213b** JLC **215t** BN/Sonia Halliday Photographs **215b** MEPL **216** ARPL **216-217t** CNRI/SPL **216-217b** Bildarchiv des Instituts für Geschichte der Medizin der Universität Wien **218-219** Nellens Collection, Knokke **218b** MEPL/Sigmund Freud Copyrights **219b** Rijksmuseum, Amsterdam **220t** SCALA **220-221** R. Reid/SPL **221tl** World Health Organization **221tr** MEPL/Sigmund Freud Copyrights **222-223** Giraudon **222b** BHPL **223c** ARPL **224t** MEPL/Sigmund Freud Copyrights **224b** Walter Verlag **225t** Guttmann-Maclay Collection, Bethlem Royal Hospital **225b** BPCC/Aldus Archive **226t** Tate Gallery, London **226br** NPG **227** Graphische Sammlung Albertina, Vienna **228t** MEPL/Sigmund Freud Copyrights **228b** Bildarchiv Preussischer Kulturbesitz **229,230** With permission of the Hebrew University of Jerusalem, Israel **232t** The Bettmann Archive Inc **233** Mount Wilson & Las Campanas Observatories **234** AKG **235** Mrs Otto Frisch **236-237** The Photosource **237b** Novosti Press Agency **238-239** MEPL **240t** Bildarchiv Preussischer Kulturbesitz **241t** Royal Observatory, Greenwich **241c** Illustrated London News **242t** SPL/Photo Researchers Inc **242b** AIP Niels Bohr Library **243** SPL/US Army **244t** AIP Niels Bohr Library **244-245** Institut Internationale de Physique Solvay **245t** Brown Brothers **245b** Werner Wolff/Time Magazine **246t** Philadelphia Museum of Art: The Mr & Mrs Wharton Sinkler Collection **246-247** Daily Telegraph Colour Library **246b** Brown Brothers **248t** Magnum/Haas **248c** AM **248b** AIP Niels Bohr Library

Artists Trevor Boyer; Simon S.S. Driver; Mick Loates; Location Map Services; Kevin Maddison; Denys Ovenden; David Smith
Cartographic editor Nicholas Harris
Editorial assistant Monica Byles
Art assistant Frankie Macmillan
Indexer John Baines
Media conversion and typesetting Peter MacDonald, Ron Barrow

Further Reading

Barber, Lynn *The Heyday of Natural History* (Jonathan Cape)
Bernal, John D. *Science in History* (C.A. Watts)
Bowler, Peter J. *Evolution: The History of an Idea* (University of California Press)
Brent, Peter *Charles Darwin* (Heinemann)
Brock, William H. *From Protyle to Proton* (Hilger)
Clark, Ronald W. *Einstein: The Life and Times* (Hodder & Stoughton)
Clark, Ronald W. *Freud: The Man and the Cause* (Jonathan Cape)
Clark, Ronald W. *The Survival of Charles Darwin: A Biography of a Man and an Idea* (Weidenfeld & Nicolson)
Cohen, I. Bernard *The Birth of a New Physics* (Penguin)
Cohen, I. Bernard *Album of Science: From Leonardo to Lavoisier 1450-1800* (Scribner)
Cohen, I. Bernard *Revolution in Science* (Belknap, Harvard University Press)
Fancher, Raymond E. *Pioneers of Psychology* (W.W. Norton)
Farley, John *The Spontaneous Generation Controversy from Descartes to Oparin* (Johns Hopkins University Press)
Feldman, Anthony and Ford, Peter *Scientists and Inventors* (Aldus)
Gillispie, Charles C. (ed.) *The Dictionary of Scientific Biography* (16 volumes) (Scribner)
Gjertsen, Derek *The Classics of Science: A Study of Twelve Enduring Scientific Works* (Lilian Barber Press)
Guerlac, Henry *Antoine-Laurent Lavoisier: Chemist and Revolutionary* (Scribner)
Hall, Marie Boas *All Scientists Now: The Royal Society in the Nineteenth Century* (Cambridge University Press)
Harré, Rom *Great Scientific Experiments* (Phaidon Press)
Ihde, Aaron J. *The Development of Modern Chemistry* (Harper & Row)
Keller, Alex *The Infancy of Atomic Physics* (Clarendon Press, Oxford)
Kevles, Daniel J. *The Physicists: The History of a Scientific Community in Modern America* (Vintage)
Knight, David *The Age of Science: The Scientific World View in the Nineteenth Century* (Basil Blackwell)
Koestler, Arthur *The Sleepwalkers: A History of Man's Changing Vision of the Universe* (Hutchinson, Penguin)
Lloyd, Geoffrey E.R. *Early Greek Science: Thales to Aristotle* (Chatto & Windus)
Lloyd, Geoffrey E.R. *Greek Science after Aristotle* (Chatto & Windus)
Lyons, Albert S. and Petrucelli II, R. Joseph *Medicine: An Illustrated History* (Harry N. Abrams)
Marks, John *Science and the Making of the Modern World* (Heinemann)
Newby, Eric *The World Atlas of Exploration* (Mitchell Beazley)
Pyenson, Lewis *The Young Einstein: The Advent of Relativity* (Hilger)
Reid, Robert *Marie Curie* (Collins)
Ronan, Colin A. *The Cambridge Illustrated History of the World's Science* (Cambridge University Press, Newnes Books)
Russell, Colin A. *Science and Social Change 1700-1900* (Macmillan)
Westfall, Richard S. *Never at Rest: A Biography of Isaac Newton* (Cambridge University Press)
Whitteridge, Gweneth *William Harvey and the Circulation of the Blood* (Macdonald, London and Elsevier, New York)
Williams, L. Pearce *Michael Faraday: A Biography* (Chapman & Hall)
Williams, L. Pearce *Album of Science: The Nineteenth Century* (Scribner)
Yule, John-David (ed.) *Concise Encyclopaedia of the Sciences* (Phaidon Press, Facts on File)
Ziman, John *The Force of Knowledge: The Scientific Dimensions of Society* (Cambridge University Press)

Glossary

Academy
The word derives from the garden in Athens where Plato taught; medieval academies were schools; new academies of the 15th to 18th centuries were associations promoting science, literature and technology.
Achromatic microscope
Microscope incorporating compound lenses to correct CHROMATIC ABERRATION.
Alchemy
Blend of philosophy, mysticism, chemical technology, aiming to transmute base metals to gold, and to prolong life.
Alienists
Specialized medical practitioners of the 19th and early 20th century who treated mentally ill people; forerunners of today's psychiatrists.
Allopathy
Disease treatment by adding (e.g. drugs) or removing (e.g. blood, germs) something in order to restore equilibrium to the body's system.
Altruism
Behavior which benefits another individual at the cost of the first.
Amplitude
The maximum deviation from its mean value of a physical property which is subject to modulation.
"Animalcule"
A minute, usually microscopic, organism – e.g. a BACTERIUM or PROTOZOON.
"Animal electricity"
Electricity generated in the body of animals (18th century).
Anode
The positive electrode of a BATTERY, cell or electron tube; the electrode at which electrons leave the system to an external circuit.
Antibiotic
Chemical produced by a MICROORGANISM and used as a drug to kill or inhibit the growth of other microorganisms.
Apothecary
Preparer or dealer in medicinal drugs.
Acquired characteristics
It was the view of Jean Lamarck (1744-1829) that changes which took place in a plant or animal in its lifetime could be passed on to later generations.
Astrolabe
Instrument used to measure altitude of celestial bodies, comprising a vertical disk and a pointer.
Astrology
The science-art of divining the future from study of the heavens.
Atlantis
A mythical island west of Gibraltar that sank beneath the sea.
Atom
According to Leucippus (Greece, 8th century BC) atoms were indivisible units differing only in size and shape; 20th-century ideas of the atom stem from the models of E. Rutherford and N. Bohr.
Bacteria (singular **bacterium**)
A large and varied group of MICROORGANISMS, classified by their shape and staining ability. They live in many environments; only a few are PATHOGENS.
Battery
A device for converting internally-stored chemical energy into direct-current electricity.
Big bang
The hot, dense explosive event which is widely believed to have been the origin of the Universe.
Biogeography
Science, established largely through the work of A. von Humboldt, of determining the distribution of living things over the Earth's surface.
Black Hole
A region of space surrounding an old, collapsed star, from which not even light can escape.
Calculus
Mathematics dealing with continually changing quantities.
"Caloric"
Heat seen as a fluid; term used by (among others) A. Lavoisier in the 18th century.
"Calx"
Oxide (18th century).
Capillaries
The smallest blood vessels forming a fine network in all tissues.
Catastrophist geology
The interpretation of fossils as signs of the violent extinction of life in past epochs, through floods, earthquakes and other cataclysms (cf. UNIFORMITARIAN GEOLOGY).
Cathartic method
J. Breuer's term for a treatment of symptoms of HYSTERIA using recollection of past emotional crises under HYPNOSIS.
Cathode
The negative electrode of a BATTERY, cell or electron tube. The electrode at which electrons enter the system from an external circuit.
Cathode ray tube
An evacuated glass tube containing at one end a cathode and anode; electrons from the cathode are accelerated through the anode to form a beam that hits a screen. It forms the principal component of a television set.
Celestial sphere
An imaginary sphere surrounding the Earth, whose center is coincident with the center of the Earth; a part of pre-Copernican COSMOLOGY.
Chromatic aberration
In early microscopes and telescopes, blurring of the image by a surrounding rainbow of colors.
Chromosome
A thread of genetic material contained in the cell nucleus and duplicated when the cell divides.
Conditioned reflex
In physiology, the triggering of automatic reflexes by substitute stimuli.
Corpuscular (or **emission**) **theory of light**
A theory that light comprises small particles that are shot out from a source much like shot from a gun.
Cosmology
Study of the structure and evolution of the Universe.
Creationism (**creationist science**)
The reemergence in the USA in recent years of FUNDAMENTALISM, claiming equal time with evolutionary biology in school curricula.
Daguerrotype
The first practical photographic process, invented by L.J.M. Daguerre in 1837.
"Dephlogisticated air"
Oxygen (term used by J. Priestley in 18th century).
Dextro-rotatory
Describes a type of crystal or molecule that rotates to the right the direction of polarization of a beam of POLARIZED LIGHT passing through it. (Cf. LAEVO-ROTATORY.)
Distillation
Process in which substances are vaporized, then condensed by cooling.
Electromagnetic radiation
RADIATION consisting of an electric and a magnetic disturbance which travels in a vacuum at a constant speed known as the speed of light (about 300,000km/s). Visible light and radio waves are examples.
Electron micrograph
Highly magnified image obtained by using a beam of electrons rather than light.
Element
Defined in 1787 by A. Lavoisier as a substance that was chemically "simple", i.e. could not be decomposed; over 2,000 years before the Greek Empedocles defined four "roots" (called elements by Plato): earth, fire, air, water.
Enzyme
An organic catalyst, usually a protein.
Epigenesis
Succession of distinct stages through which an embryo passes.
Ether
To Aristotle, a rigid transparent substance filling the Universe; to R. Descartes, a transparent fluid; in the 19th century, existence of a tenuous ether was still accepted by most scientists.
False-color photography
Processing of special photographic or other emulsions to display information, using "unnatural" color.
Fermentation
The decomposition of carbohydrates by MICROORGANISMS in the absence of oxygen.
Fission
In nuclear physics, the changing of an ELEMENT into two or more elements of lower atomic weight, with the release of energy.
"Fixed air"
Term used in the 18th century for carbon dioxide.
Fluorescence
The emission of light from a substance being stimulated by light or other forms of electromagnetic radiation (e.g. X-rays) or by certain other means. If the luminescence continues after the stimulus ceases it is known as phosphorescence.
Fluoroscope
Device containing a FLUORESCENT screen which converts the X-ray into visible light.
Fluxions
Isaac Newton's term for CALCULUS.
Frequency
The rate at which a wave motion completes its cycle.
Fundamentalism
Religious movement developed in the USA after World War I; fundamentalists insist on the absolute truth of the Bible, including the Creation story.
Geocentric
Centered on the Earth.
Germ theory
Theory of L. Pasteur that diseases are spread by living germs (BACTERIA).
Gravimetric (**experiment**)
Refers to experiments in which accurate weighing is important.
Gravitation
The force of attraction existing between all matter.
GUT (**grand unified theory**)
Attempt to explain the forces of nature in terms of one underlying force.
Half-life
A measure, devised by E. Rutherford, taken from the time required by a substance to lose half its radioactivity.
Heliocentric
Sun-centered.
Hellenistic
Relating to ancient Greek civilization after Alexander the Great (died 323 BC), to 1st century BC.
Herbarium
Systematically arranged collection of dried and preserved plant specimens.
Heterogenesis
The appearance of living entities from a reconstitution of dead organic matter. (Cf. SPONTANEOUS GENERATION.)
Holography
The creation of three-dimensional images by photographing the subject when illuminated by a split LASER beam, and reproducing the image by recreating the beam.
Homeopathy
Treatment of medical symptoms by giving minute amounts of a drug which produces similar symptoms.
Homocentric
Man-centered.
Humors
The four body fluids whose relative quantities governed the working of the human body, according to a doctrine of ancient Greek medicine.
Hypnosis
Artificially induced mental state characterized by the subject's openness to suggestion.
Hysteria
In the 19th century, a loose term for physical symptoms and behavior (e.g. tics, blindness, numbness, lameness) which had no apparent physical cause.
Induction
The phenomenon in which an electric field is generated in an electric circuit when the number of magnetic field lines passing through it changes.

Inertia
The resistance of matter to change in its state of motion.
Integrated circuit
A single structure in which a large number of electronic components are assembled.
Interferometer
An instrument employing interference effects, used for measuring wavelength or for detecting the direction of a RADIO source.
Isochronism
Recurrence at regular intervals of time, as in a pendulum's swing.
Isomers
Chemical compounds of identical composition but differing in the arrangement of atoms within the molecule, and having different properties.
Isotherm
An imaginary line joining points on Earth experiencing an equal temperature.
Isotopes
Atoms of a chemical element with the same number of protons but different number of neutrons.
Kin selection
Selection in favor not necessarily of the individual but of relatives carrying the same genes; may be favored by behavior such as ALTRUISM.
Laevo-rotatory
The converse of DEXTRO-ROTATORY.
Laser
Light Amplification by Stimulated Emission of Radiation; a device producing a very intense beam of parallel light with a precisely defined wavelength.
Lesser circulation
Circulation of the blood between heart and lungs.
Mechanics
Study of the action of forces on bodies.
Mesmerism (or **"animal magnetism"**)
The system invented by Anton Mesmer in the late 18th century for influencing beneficially the flow of ether through the body – a forerunner of HYPNOSIS.
Metaphysics
Study of or speculation on the fundamentals of existence or reality.
Meteorology
Study of the atmosphere, weather and climate.
Miasma
A disease-causing vapor present in the air; the "miasmatic theory" of disease was predominant up to the mid-19th century.
Microbe
MICROORGANISM.
Microorganism
A living organism too small to be seen without a microscope.
Moderator
In a nuclear reactor, a substance used to slow the neutrons released by FISSION.
"Naturalism"
Doctrine, which gained acceptance in the 19th century, that the cause-and-effect laws of natural sciences such as physics and chemistry suffice to explain all phenomena.
Natural philosophy
Until the 19th century "natural philosophy" meant something similar to our word PHYSICS.
Natural selection
The mechanism of evolution discovered by Charles Darwin.
Naturopathy
A system of treating disease that emphasizes assistance to nature.
Neptunist geology
Theory, current c. 1860, that all rocks are sedimentary, deposited from water. (Cf. PLUTONIST GEOLOGY.)
Neurosis
Disorder of the nervous system without physical lesion.
Optics
Science of light and vision.
Paleomagnetism
Study of past changes in Earth's magnetic field by examining the magnetic properties of old rocks.
Pasteurization
Technique of heating milk, beer, wine etc. to destroy bacteria it may contain; named after L. Pasteur.
Pathogen
An organism that produces disease.
Periodic table
A table of elements arranged by atomic number (the number of protons in the nucleus) to show similarities and trends in physical and chemical properties.
Phlogiston
One of three varieties of "earth", a combustible material lost by the substance that burns (18th century).
Phosphorescence
See FLUORESCENCE.
Photoelectric effect
A phenomenon whereby some metals become electrically charged when exposed to light.
Photon
A quantum of electromagnetic energy, often thought of as the particle associated with light.
Phrenology
A 19th-century "science of the mind" in which the brain is mapped into regions that control different aspects of behavior.
Physics
Science of the interaction of matter and energy (not chemistry); originally, the study of natural things.
Piezoelectricity
The relationship between mechanical stress and electric charge exhibited by certain crystals.
Pile
A Voltaic BATTERY; a nuclear reactor.
Plutonist geology
Theory, current c. 1800, that changes in rocks are effected by heat. (Cf. NEPTUNIST GEOLOGY.)
Pneumatic chemists
Chemists of the 18th century who studied airs released from solids and fluids, including J. Black, J. Priestley, H. Cavendish, C. Scheele.
Polarized light
Light in which the orientation of wave vibrations displays a definite pattern, for example being in a single plane.
Protozoa
The single-celled animals.
Psychoanalysis
The free-association therapy developed by S. Freud.
Quantum theory
Theory of M. Planck that ELECTROMAGNETIC RADIATION exists in discrete bundles (quanta).
Racemic
Describes a compound that contains equal amounts of DEXTRO- and LAEVO-ROTATORY forms and is therefore optically inactive toward POLARIZED LIGHT.
Radiation
Any form of energy, particularly heat or light, that can be transmitted through a medium without having an effect on that medium.
Radio
Communication between distant points using radio waves, ELECTROMAGNETIC RADIATION of a certain frequency.
Radioactivity
The spontaneous disintegration of unstable nuclei, accompanied by the emission of RADIATION of alpha particles, beta particles or gamma rays.
Radioelement
A radioactive element.
Radiography
The use of X-rays for diagnostic purposes.
Radiotherapy
The use of X-rays for curative purposes.
Radon
Radium emanation – used in therapy.
Rationalism
Theory that reason alone, unaided by experience, can arrive at basic truth; the rationalism of R. Descartes, G. von Leibniz and B. Spinoza stands in opposition to the empiricism of J. Locke.
Relativity
Theory of the nature of space, time and matter, enunciated by A. Einstein.
Scientific naturalism
A term current in the 19th century for a belief that human reason, through application of scientific method, could understand everything.
Space-time
Concept introduced by H. Minkowski (1864-1909) to show how Einstein's ideas of RELATIVITY necessarily mix up space and time.
Spectroscopy
The production, measurement and analysis of SPECTRA, much used by astronomers, chemists and physicists.
Spectrum
The distribution of the intensity of ELECTROMAGNETIC RADIATION with wavelength. The visible spectrum encompasses those wavelengths to which the human eye responds, the different wavelengths corresponding to the different colors.
Spontaneous generation
A theory that living entities can arise either from inorganic elements (abiogenesis) or from a reconstitution of dead organic matter (HETEROGENESIS) without a living "parent".
Stereochemistry
The study of the spatial arrangement of atoms in molecules.
Stereoisomerism
Property of ISOMERS having the same molecular structure but differing in the spatial arrangement of their atoms.
Strata
The various layers in which a sedimentary rock is formed. The singular "stratum" is rarely used.
Stratigraphy
The study of the chronological sequence and the correlation of rock STRATA in different districts.
Teleology
The use of predetermined purpose to explain the Universe.
Thermodynamics
Science of the interconversion of heat, work and other forms of energy.
Transit
The passage of a celestial body across the observer's meridian or of Mercury and Venus when they pass across the Sun's disk as seen from the Earth.
Transmutation
Alchemists believed that one substance could be "transmuted" into another – hence their efforts to transmute into gold. The same term was used by Darwin to label the process of one species giving rise to another.
Triangulation
Finding a position by taking bearings from two points a known distance apart.
Trigonometry
Study of the ratios of the sides of right-angled triangles.
Uniformitarian geology
A "steady-state" view of geological history, whereby the present-day landscape developed through the long-term uniform action of forces including heat, cold, Sun, wind, rain, sea.
"Uranic rays"
A. Becquerel's name for the natural radioactivity in a uranium salt he discovered in 1896.
Vaccination, vaccine
Originally, the introduction of matter from cowpox pustules to lessen the danger of catching smallpox; by extension, vaccines are attenuated forms of disease used to confer immunity.
Virus
The smallest form of living organism, dependent on living cells for replication.
Vivisection
Strictly, dissection of living animals; usually, all experimentation by humans on other animals.
X-rays
ELECTROMAGNETIC RADIATION with a wavelength between that of ultraviolet radiation and gamma rays.
Zodiac
A belt stretching round the sky to 8 degrees on either side of the ecliptic, in which the Sun, Moon and bright planets are always to be found.

Index

Page numbers in *italics* refer to illustrations, captions or special features.